LOEB CLASSICAL LIBRARY

FOUNDED BY JAMES LOEB 1911

EDITED BY

JEFFREY HENDERSON

HIPPOCRATES

VI

LCL 473

HIPPOCRATES

VOLUME VI

WITH AN ENGLISH TRANSLATION BY

PAUL POTTER

HARVARD UNIVERSITY PRESS

CAMBRIDGE, MASSACHUSETTS

LONDON, ENGLAND

First published 1988

LOEB CLASSICAL LIBRARY® is a registered trademark
of the President and Fellows of Harvard College

ISBN 978-0-674-99522-2

*Composed in ZephGreek and ZephText by
Technologies 'N Typography, Merrimac, Massachusetts.
Printed on acid-free paper and bound by
The Maple-Vail Book Manufacturing Group*

CONTENTS

PREFACE TO VOLUMES V AND VI

In his preface to volume IV (1931), W. H. S. Jones writes: "This book completes the Loeb translation of Hippocrates," offering no explanation why the rest of the Collection is to be ignored, unless it is implied in his next sentence: "The work of preparing the volume has taken all my leisure for over five years . . . "

Whatever Jones' reasons for stopping may have been, the lack of a complete English translation has been noted and regretted by classicists and historians of medicine alike. A plan to continue the Loeb *Hippocrates* has now existed in America for several decades, and it is chiefly due to the untiring efforts of Dr. Saul Jarcho and Mr. Richard J. Wolfe that volume V sees the light of day.

The cost of preparing and publishing volumes V and VI has been met by NIH Grant LM 02813 from the National Library of Medicine, and the examination of Hippocratic manuscripts in Florence, Paris, Rome, Venice and Vienna made possible by grants generously provided by the Jason A. Hannah Institute for the History of Medicine.

PREFACE

Work on volumes V and VI was greatly facilitated by the use of computer texts and indexes kindly furnished by Prof. Gilles Maloney and his team at the *Laboratoire de recherches hippocratiques* in Quebec.

Finally, it is my pleasant duty to thank Prof. G. P. Goold, Associate Editor of the series, Prof. Dr. Fridolf Kudlien, Prof. Wesley D. Smith, William B. Spaulding M.D., F.R.C.P.(C.), and Lynn Wilson Ph.D., all of whom read the volumes in various stages of their preparation, for their manifold helpful comments.

Rome, November 1983   Paul Potter

INTRODUCTION TO VOLS V
AND VI[1]

These volumes contain the most important Hippocratic works on the pathology of internal diseases. Presumably in consequence of their common purpose, these six
treatises tend to share the same general structure: independent chapters of constant form each devoted to one
specific nosological entity.[2]

About the treatises' interdependencies, authors, and
relative dates of composition, nothing can be said with any
degree of certainty. There is neither any evidence that

---

[1] This introduction deals only with the treatises in volumes
V and VI; for an orientation to Hippocrates and the Hippocratic
Collection in general, the reader is referred to W. H. S. Jones'
"General Introduction" (Loeb *Hippocrates* I. ix-lxix) and "Intro-
ductory Essays" (Loeb *Hippocrates* II. ix-lxvi). Useful guides to
Hippocratic scholarship since Jones are Ludwig Edelstein's arti-
cle "Hippokrates" in *Paulys Real-Encyclopädie der classischen
Altertumswissenschaft*, Supplement VI, Stuttgart, 1935, cols.
1290–1345, H. Flashar (ed.), *Antike Medizin*, Darmstadt, 1971,
Robert Joly's article "Hippocrates of Cos" in the *Dictionary
of Scientific Biography*, vol. VI, New York, 1972, 418–31, and
G. Maloney and R. Savoie, *Cinq cents ans de bibliographie
hippocratique*, Quebec, 1982.

[2] The individual works are analysed in more detail in their particular introductions.

Fig. 1

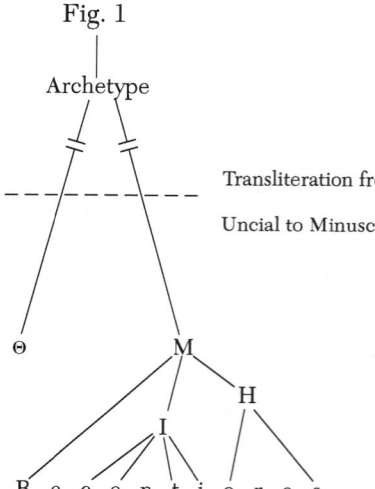

Archetype

Transliteration from

Uncial to Minuscule

10th c.

12th c.

14th c.

16th c.

Θ

M

H

I

R e c e n t i o r e s

Θ[3] = Vindobonensis Medicus Graecus 4

M = Marcianus Venetus Graecus 269

H = Parisinus Graecus 2142

I = Parisinus Graecus 2140

Recentiores = approximately twenty manuscripts

[3] Littré (VI. 139) assigned the siglum θ to this manuscript, but several later editors and translators, to whose number I belong, prefer Θ in order to avoid possible confusion with a lost manuscript.

would confirm, nor any evidence that would call into doubt, their traditional time of origin about 400 B.C.

In the first century A.D. Erotian knew *Diseases I* and *III* and *Regimen in Acute Diseases (Appendix)*, and Galen (129–199) makes reference, in addition, to *Affections*, *Diseases II* and *Internal Affections*.

MANUSCRIPT TRADITION

Five of the six works in these volumes (*Affections*, *Diseases I–III* and *Internal Affections*) share a transmission that can be represented by the *stemma codicum* that appears as Fig. 1 (p. x).

The transmission of the sixth work, *Regimen in Acute Diseases (Appendix)*, is more complex both because of the existence of a commentary by Galen, which provides a fertile source of variant readings, and also because it was translated into Latin at an early date.[4] The *stemma codicum* that appears as Fig. 2 (p. xii) indicates the relationships among the Greek manuscripts upon which the critical editions, including this one, are based.

Furthermore a papyrus (Rylands Greek Papyrus 56)[5] of the first half of the second century A.D. containing two fragments[6] of the text of *Regimen in Acute Diseases (Ap-*

---

[4] See Hermann A. Diels, *Die Handschriften der antiken Ärzte*, Berlin, 1905–1907, pp. 8 f. and Supplement p. 25.

[5] Edited by A. S. Hunt in *Catalogue of the Greek Papyri in the John Rylands Library at Manchester*, vol. I, Manchester, 1911, 181 f.

[6] Chapter 24 φιλέει τῷ τοιῷδε — (25) ὑγρὰ διαχωρήσῃ καὶ and Chapter 26 τὸ ἕτερον παρὰ τὸ ἕτερον — (27) Τοὺς τοιούσδε.

Fig. 2

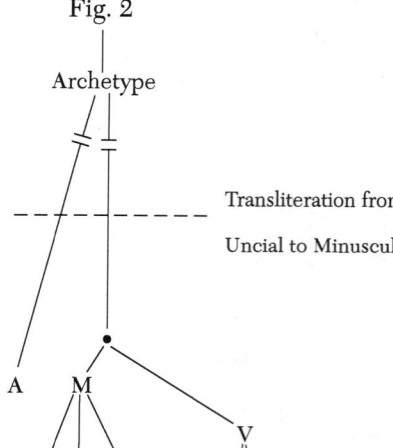

Archetype

Transliteration from
Uncial to Minuscule

10th c.

A   M

12th c.

V

14th c.

R e c e n t i o r e s

16th c.

A = Parisinus Graecus 2253
M = Marcianus Venetus Graecus 269
V = Vaticanus Graecus 276
Recentiores = approximately twenty manuscripts

pendix) makes a limited but significant contribution to the establishment of the text.

TEXT AND TRANSLATION

For *Diseases I* and *III* and *Regimen in Acute Diseases (Appendix)* I have generally relied on the collations given in the critical editions.[7] For the other three works, which lack critical editions since Littré, collations of Θ and M have been made from microfilms and supplemented by inspection of the actual manuscripts.

In establishing the Greek text and making the English translation, I have consulted many earlier texts, translations and commentaries, among which the most important are:

Hippocratis Coi ... opera ... per M. Fabium [Calvum] ... *Latinitate donata ...* , Basel, 1526. (= Calvus)

Hippocratis Coi ... libri omnes, ad vetustos Codices ... collati & restaurati [per Ianum Cornarium], Basel, 1538. (= Cornarius)

Hippocratis Coi ... opera ... omnia. Per Ianum Cornarium ... Latina lingua conscripta, Lyons, 1554. (= Cornarius/Latin)

Magni Hippocratis ... opera omnia ... latina interpretatione & Annotationibus illustrata Anutio Foesio ... Oeconomia Hippocratis alphabeti serie distincta, Geneva, 1657–1662. (= Foes)

---

[7] See vol. V p. 87, and vol. VI pp. 5 and 227.

Magni Hippocratis Coi Opera omnia edita . . . industria & diligentia Joan. A. Vander Linden, Leiden, 1665. (= Vander Linden)

Hippokrates Werke aus dem Griechischen . . . von J. F. C. Grimm. Revidiert . . . von L. Lilienhain, Glogau, 1837–1838. (= Grimm)

E. Littré, *Oeuvres complètes d'Hippocrate*, Paris, 1839–1861. (= Littré)

F. Z. Ermerins, *Hippocratis . . . reliquiae*, Utrecht, 1859–1864. (= Ermerins)

Car. H. Th. Reinhold, Ἱπποκράτης / Ψευδωνύμως Ἱπποκράτεια, Athens, 1865–1867. (= Reinhold)

H. Kuehlewein, *Hippocratis Opera omnia*, Leipzig, 1894–1902. (= Kuehlewein)

R. Fuchs, *Hippokrates, sämmtliche Werke. Ins Deutsche übersetzt . . .* , Munich, 1895–1900. (= Fuchs)

The English translation attempts to be as close to the original as possible while still remaining readable. In matters of vocabulary, I have taken the *Shorter Oxford English Dictionary* and *Dorland's Illustrated Medical Dictionary* as a basis.

Bibliographical Note: Supplementary bibliographical information is to be found in the introductions to individual treatises and in notes to the Indexes, which are printed at the end of volume VI.

DISEASES III

INTRODUCTION

Diseases III was known to Erotian, Galen and Caelius Aurelianus,[1] although, as I argue elsewhere, probably under the title *Diseases II*.[2]

In its present state, the treatise consists of three parts: a two-line tag, attaching it to a preceding work on fevers; a nosological work (1–16); a collection of cooling agents (17).

The first sentence of *Diseases III* is identical to the last sentence of the Hippocratic *Sevens*, a work preserved, for the most part, only in Latin translation:

> *De febribus quidem omnibus <dixi>; de ceteris autem iam dicam.*[3]

However, for a number of reasons it seems impossible that the two treatises share a common origin[4]; rather, we must suppose that two independent works were joined at a later date and fitted with a suitable connecting link.

[1] See Potter pp. 46–50.
[2] Potter pp. 55 ff.
[3] Littré VIII. 673 = Littré IX. 466 = Roscher p. 80.
[4] See J. Mansfeld, *The Pseudo-Hippocratic Tract* ΠΕΡΙ ἙΒΔΟΜΑΔΩΝ *ch. 1–11 and Greek Philosophy*, Assen, 1971, pp. 11–15.

The second part of *Diseases III*, which makes up the bulk of the treatise, has sixteen chapters, each dealing with one particular nosological entity:

1: Swelling of the Brain
2: Intense Headache
3: The Stricken
4: Sphacelus of the Brain
5: Lethargy
6: Ardent Fever
7: Swelling of the Lung
8: Headache and Speechlessness
9: Phrenitis
10: Angina
11: Jaundice
12: Tetanus
13: Opisthotonus
14: Ileus
15: Pneumonia
16: Pleurisy

Each of these chapters is organized according to the following scheme: name or identifying feature; symptoms and course; prognosis; treatment. In most chapters, because of their simplicity, this scheme is immediately apparent, but even in longer chapters, which are complicated by many remarks about variant forms of the disease and about complications, it is strictly followed.

To judge from its clear arrangement and lack of speculative aetiology, it would seem that this part *of Diseases III* was a physician's handbook.

Chapter 17 is a collection of about twenty-five cooling agents, to be given "against every ardent and other fever".

This chapter is neither closely related to, nor obviously incompatible with the earlier part of the treatise. Perhaps it was added at the same time that *Sevens* and *Diseases III* were joined.

Diseases III is represented in all the standard collected Hippocratic editions and translations, and in the two renaissance works on Hippocratic diseases.[5]

In 1973 it was edited and translated into German in a Kiel dissertation subsequently published as:

Paul Potter, *Hippokrates Über die Krankheiten III*, CMG I 2,3, Berlin, 1980 (= Potter).

[5] See vol. V p. 86 f.

ΠΕΡΙ ΝΟΤΣΩΝ Γ

1. [Περὶ μέν νυν πυρετῶν ἁπάντων εἴρηταί μοι· ἀμφὶ δὲ τῶν λοιπῶν ἐρέω ἤδη.]

Ἐπὴν ὁ ἐγκέφαλος οἰδήσῃ ὑπὸ φλεγμασίης, ὀδύνη ἴσχει πᾶσαν τὴν κεφαλήν, μάλιστα δ' ὅπῃ ἵσταται ἡ φλεγμασίη· ἵσταται δ' εἰς τοὺς κροτάφους.[1] τὰ δ' οὔατα ἠχῆς πλέα γίνεται, καὶ ἀμβλὺ ἀκούει, καὶ αἱ φλέβες τέτανται καὶ σφύζουσι, πυρετός τε καὶ ῥῖγος ἐνίοτε ἐπιλαμβάνει· ἤ τ' ὀδύνη ἐκλείπει οὐδέποτε, ἀλλὰ ποτὲ μὲν ἀνίησι, ποτὲ δὲ πιέζει μᾶλλον· βοᾷ τε καὶ ἀναΐσσει ὑπὸ τῆς ὀδύνης, καὶ ὅταν ἀναστῇ, αὖτις σπεύδει ἐς τὴν κλίνην πεσεῖν καὶ ῥιπτάζει ἑωυτόν. οὗτός ἐστι μὲν θανατώδης, ὁπόσων δ' ἡμερέων ἀποθανεῖται, οὐκ ἔχει κρίσιν· ἄλλοι γὰρ ἄλλως ἀπόλλυνται· ὡς μέντοι ἐπὶ τὸ πολὺ ἐν τῇσιν ἑπτὰ ἡμέρῃσι τελευτῶσιν· εἴκοσι δὲ καὶ μίαν διαφυγόντες ὑγιέες γίνονται.

Χρὴ δ' ἐπὴν περιωδυνέῃ ψύχειν τὴν κεφαλήν— μάλιστα μὲν ξυρήσαντα—ἢ ἐς κύστιν ἢ[2] ἐς ἔντερα ἐγχέαντα τῶν ψυκτικῶν,[3] οἷον χυλὸν στρύχνου καὶ γῆν κεραμίτιδα, τὸ μὲν προστιθέναι, τὸ δ' ἀφαιρέειν

---

[1] Θ: ἐν τῷ κροτάφῳ Μ.

6

DISEASES III

1. [Having spoken about all the febrile diseases, I come now to speak about the rest.]

When the brain swells up as the result of phlegmasia, pain occupies the whole head, but particularly the region where the phlegmasia is located, the temples. The patient's ears are filled with ringing, he hears unclearly, and the vessels in his head are stretched, and throb; sometimes fever and chills occur as well. The pain never ceases completely, but sometimes it becomes milder, while at other times it presses the patient more intensely; he cries out and starts up from the pain, but as soon as he has arisen, he is eager to fall back into bed, and he casts himself about. This patient will certainly die, although the number of days in which he will die cannot be judged; for different patients die in different lengths of time; in most cases, however, they succumb within seven days. If they survive for twenty-one days, they recover.

When the patient is suffering intense pain, you must cool his head—best after shaving it—by pouring cooling agents such as nightshade juice and potter's earth into a bladder or length of gut; alternately apply and remove this

---

2 ἤ . . . ἤ om. Θ.
3 M adds τι.

πρὶν χλιαρὸν εἶναι· καὶ τοῦ αἵματος ἀφαιρέειν[4] καὶ
τὴν κεφαλὴν καθαίρειν σελίνου χυλοῖσι μιγνύντα
εὐώδεα. ἀοινεῖν δὲ τὸ πάμπαν· ῥυφεῖν δὲ πτισάνης
χυλὸν ψυχρόν, καὶ τὴν κάτω κοιλίην λύειν.

2. Ὅταν δὲ περιωδυνέῃ ἡ κεφαλὴ ὑπὸ πληρώσιος
τοῦ ἐγκεφάλου, ἀκαθαρσίην σημαίνει· καὶ τὴν κεφαλὴν ὅλην περιωδυνίαι ἴσχουσι,[5] καὶ παραφρονέει καὶ
120 ἀποθνῄ|σκει ἑβδομαῖος, καὶ οὐκ ἂν ἐκφύγοι, εἰ μὴ
ῥαγείη τὸ πύον[6] κατὰ τὰ οὔατα· οὕτω δὲ ἡ ὀδύνη
παύεται, καὶ ἔμφρων γίνεται· ῥεῖ δὲ πολλὸν καὶ ἄν-
οσμον.

Τοῦτον μάλιστα μὲν μὴ θεραπεύειν, πρὶν ἂν ἐρ-
ρωγὸς εἰδῇς τὸ πύον. ἢν δὲ βούλῃ καθῆραι τὴν ἄνω
καὶ τὴν κάτω κοιλίην, λῦσον χωρὶς ἑκατέρην.[7] ἔπειτα
πυριᾶν τὴν κεφαλὴν ὡς μάλιστα ὅλην τε καὶ διὰ τῶν
ὤτων καὶ διὰ τῶν ῥινῶν. ῥυφεῖν δὲ πτισάνης χυλὸν
καὶ ἀοινεῖν τὸ πάμπαν. ἐπὴν δὲ ῥαγῇ τὸ πύον, ἐπί-
σχειν, ἔστ᾽ ἂν ἡ πολλὴ τῆς ῥύσιος παύσηται. ἔπειτα
κλύζειν τὰ ὦτα οἴνῳ γλυκεῖ ἢ γάλακτι γυναικείῳ ἢ
ἐλαίῳ παλαιῷ, χλιαροῖσι δὲ κλύζειν· καὶ πυριᾶν θαμινὰ τὴν κεφαλὴν μαλθακῇσι πυρίῃσι καὶ εὐόδμοισιν,
ἵνα θᾶσσον καθαίρηται ὁ ἐγκέφαλος. οὗτος τὰ μὲν
πρῶτα οὐδὲν ἀκούει·[8] προϊόντος δὲ τοῦ χρόνου ἥ τε
ῥύσις ἥσσων ἐστί, καὶ ἡ ἀκοὴ ἅμα τῇ ῥύσει παντελῶς
παυομένη παραγίνεται· ὁμοιοῦταί τε μάλιστα ἑωυτῷ.
φυλάσσεσθαι δὲ χρὴ ἡλίους, ἀνέμους, πῦρ, καπνόν,

---

[4] καὶ . . . ἀφαιρέειν om. Θ.

before it becomes warm. Also, draw off blood, and clean out the head with fragrant substances mixed in celery juice. Let the patient abstain totally from wine, give him cold barley-water as gruel, and empty his lower cavity.

2. When the head suffers violent pain due to a filling of the brain, this indicates uncleanness. Pains are present through the entire head, the patient becomes deranged, and on the seventh day he dies. He has no chance of recovering unless pus breaks out through his ears; if this happens, the pain stops and he recovers his wits; such a flux is copious and odourless.

It is best not to treat this patient until you know that the pus has broken out. If, however, you wish to clean out his upper and lower cavities, evacuate each separately. Then treat the head—as far as possible the whole of it—with vapour-baths both through the ears and through the nostrils. Have the patient drink barley-water, and abstain totally from wine. When the pus has broken out, stop treatment until the major part of the flow ceases. Then wash the ears with sweet wine, woman's milk, or old olive oil—use these fluids warm. Treat the head frequently with mild and fragrant vapour-baths in order that the brain will be cleaned out more quickly. At first this patient hears nothing, but as time passes the flux diminishes, his hearing, which had stopped completely during the flux, resumes, and he becomes for all intents and purposes normal. He must avoid sun, wind, fire, smoke, and sharp substances

---

5 Θ adds καὶ παρακοπαί.

6 πύον Θ: ὑγρὸν ἑβδομαίῳ Μ.

7 λῦσον χωρὶς ἑκατέρην Μ: χωρὶς ἑκατέρη Θ.

8 Θ: οὐκ ἐσακούει Μ.

ΠΕΡΙ ΝΟΥΣΩΝ Γ

δριμέων ὀδμὰς καὶ αὐτά· καὶ ἡσυχάζειν διαίτῃ μαλθακῇ χρώμενον·[9] ‹καὶ›[10] ὑπὸ κενεαγγειῶν ἡ κοιλίη εὔλυ-
τος ἔστω ἡ κάτω.

3. Οἱ δὲ βλητοὶ λεγόμενοι· ὅταν ὁ ἐγκέφαλος πλησθῇ πολλῆς ἀκαθαρσίης, ὀδύνην παρέχει τῷ πρόσθεν
τῆς κεφαλῆς, καὶ ἀναβλέπειν οὐ δύνανται, οἱ μὲν
ἀμφοτέροις τοῖς ὀφθαλμοῖς, οἱ δὲ θατέρῳ· καὶ κῶμά
μιν ἔχει, καὶ ἄφρονές εἰσι, καὶ οἱ κρόταφοι πηδῶσι,
καὶ πυρετὸς λεπτὸς ἔχει καὶ τοῦ σώματος ἀκρασίη.
οὗτος ἀποθνήσκει τριταῖος ἢ πεμπταῖος· ἐς δὲ τὰς
ἑπτὰ οὐκ ἀφικνεῖται· ἢν δ᾽ ἄρα ἀφίκηται, ἐξάντης
γίνεται.

122   Τοῦτον ἢν βούλῃ θεραπεύειν, | πυριᾶν τὴν κεφαλὴν
καὶ ταμὼν ἀνάπνευσιν ποιέειν· ἢν δ᾽ ἡ ὀδύνη ἐστη-
ριγμένη ᾖ,[11] καὶ πταρμοὺς ἐμποιέειν, καὶ τὴν κεφαλὴν
καθαίρειν κούφοισι καὶ εὐώδεσι, καὶ τὴν κάτω κοιλίην
καθαίρειν. ἀοινεῖν δὲ τὸ πάμπαν, πτισάνης δὲ χυλῷ
χρῆσθαι.

4. Ἢν δὲ σφακελίσῃ ὁ ἐγκέφαλος, ὀδύνη ἔχει τὴν
κεφαλὴν καὶ διὰ τοῦ τραχήλου φοιτᾷ ἐς τὴν ῥάχιν·
καὶ ἐπιλαμβάνει αὐτὸν ἀνηκουστίη, καὶ ψῦχος ἐπέρχεται ἐπὶ τὴν κεφαλήν, καὶ ἰδίει[12] ὅλος, καὶ ἐξαίφνης
ἄφωνος γίνεται, καὶ ἐκ τῶν ῥινῶν αἷμα ῥεῖ, καὶ πε-
λιδνὸς γίνεται. τοῦτον ἢν μὲν ἡ νοῦσος χαλαρῶς
λάβῃ, τοῦ αἵματος ἀπελθόντος, ῥαΐζει· ἢν δὲ σφόδρα
ἀπειλημμένος ᾖ, ἀποθνήσκει τάχα.

9 Cornarius: -ος ΘΜ.

10

and their vapours. Have him rest and employ a mild diet; let the lower cavity be thoroughly opened by evacuants.

3. The "stricken", as they are called: when the brain becomes filled with great uncleanness, this produces pain in the front of the head, and patients cannot look up—some with both eyes, others with one. Drowsiness befalls this patient, he is senseless, his temples throb, mild fever is present, and his body is powerless. He dies on the third or fifth day, and generally does not reach the seventh; if he does, he recovers.

If you wish to attend this patient, treat his head with vapour-baths, and fashion a connection to the external air by incising.[1] If the pain remains fixed, induce sneezing, clean out the head with mild and fragrant substances, and also clean out the lower cavity. Have the patient abstain totally from wine, and drink barley-water.

4. If the brain becomes sphacelous, pain occupies the head, and moves through the neck to the spine. Deafness befalls the patient, coldness comes over his head, and he sweats all over; he suddenly becomes speechless, blood flows from his nostrils, and he becomes pale. If he has been attacked by a mild form of the disease, he will become better after the blood has flowed out, but if he is severely affected, he soon dies.

---

[1] Presumably the same kind of procedure as that described in *Diseases II* 15, 18, and 25.

---

[10] Added by I (= Par. Graec. 2140, 14th cent.).
[11] Littré: ἐστηρημένη ἢ Θ: ἐστήκη M.
[12] Θ: οἰδέει M.

Τούτῳ πταρμούς τε ἐμποιέειν διὰ τῶν εὐωδέων καὶ
τὰς κοιλίας ἄμφω καθαίρειν ἑκατέρην ἐν μέρει· κού-
φας δὲ ὀδμὰς τῇσι ῥισὶν ἀείρειν, καὶ ῥύφημα λεπτὸν
χλιαρόν· ἀοινεῖν δὲ τὸ πάμπαν.

5. Οἱ δὲ λήθαργοι· στάσις μὲν ἡ αὐτὴ τοῦ κακοῦ τῇ
περιπλευμονίῃ, χαλεπωτέρη[13] δὲ καὶ οὐ πάμπαν ἀπήλ-
λακται ὑγρῆς περιπλευμονίης· βαρυτέρη δὲ ἡ νοῦσος
πολλόν. πάσχει δὲ τάδε· βὴξ καὶ κῶμά μιν[14] ἔχει, καὶ
τὸ σίαλον ὑγρὸν καὶ πολὺ ἀνάγει, καὶ ἀδυνατέει
σφόδρα, καὶ ὅταν μέλλῃ ἀποθανεῖσθαι, κάτω ὑποχωρέει ἐπὶ πολὺ καὶ ὑγρόν. τούτῳ ἐλπὶς μὲν πάνυ
βραχέη περιγενέσθαι· ὅμως δὲ πτύειν τε ποιέειν ὡς
πλεῖστον καὶ θερμαίνειν καὶ ἀοινεῖν· ἢν δὲ ἐκφύγῃ,
ἔμπυος γίνεται.

6. Ἡ δὲ καυσώδης λεγομένη· δίψα τε ἔχει πολλή,
καὶ ἡ γλῶσσα πέφρικε· τὸ δὲ χρῶμα αὐτῆς τὸν μὲν
πρῶτον χρόνον | οἷόν περ εἴωθε, ξηρὴ δὲ σφόδρα·
προϊόντος δὲ τοῦ χρόνου σκληρύνεται καὶ τρηχύνεται
καὶ παχύνεται· ἔπειτα μελαίνεται. ἢν μὲν οὖν ἐν ἀρχῇ
ταῦτα πάθῃ, θάσσους αἱ κρίσιες γίνονται· ἢν δ' ὕστερον, χρονιώτεραι. τῆς δ' ἀφέσιος πάντα ταὐτὰ ἡ
γλῶσσα σημαίνει ἅπερ ἐν τῇ περιπλευμονίῃ. καὶ τὰ
οὖρα, χλωρὰ[15] μὲν ἢ αἱματώδεα ἐόντα, ἐπίπονα· ξανθὰ
δέ, ἀπονώτερα. καὶ τὸ πτύσμα ὑπὸ θερμασίης καὶ
ξηρασίης ξυγκεκαυμένον παχύ ἐστι. πολλάκις δὲ καὶ
ἐς τὴν περιπλευμονίην μεθίσταται· καὶ ἢν μεταστῇ,
τάχα ἀποθνήσκει.

124

12

In this patient induce sneezing by means of fragrant substances, and clean out the two cavities, one at a time. Hold mild fragrances up to his nostrils, and have him drink thin warm gruel; let him abstain totally from wine.

5. Lethargy: the same state of evil as pneumonia, but more severe, and not altogether removed from moist pneumonia[2]: this disease is much more violent. The person suffers the following: he is subject to coughing and drowsiness, he expectorates plentiful moist sputum, and he is quite powerless. When he is about to die, he passes copious watery stools. His hope of survival is very slight; nevertheless, induce him to cough as much as possible, warm him, and have him abstain from wine. If he survives, he suppurates internally.

6. Ardent fever, as it is called: there is great thirst, and the tongue becomes rough; at first, it stays its normal colour, but is very dry; with time, it becomes hard, rough and thick, then dark. Now if the tongue suffers these things at the beginning, the crises occur sooner; if it suffers them later, the crises take longer. Concerning recovery, the tongue gives all the same indications as in pneumonia. Urines that are green or bloody are painful, but if yellow, they are less so. The sputum, being burnt up by the heat and dryness, is thick. There is often a change to pneumonia, too, and if this occurs, the patient soon dies.

[2] For the particular severity of moist pneumonia see chapter 15 below.

---

[13] M: χαλαρωτέρη Θ.
[14] καὶ κῶμά μιν om. Θ.
[15] Θ: χολώδεα M.

Τοῦτον ὧδε χρὴ θεραπεύειν· λούειν θερμῷ δὶς ἢ
τρὶς ἡμέρης ἑκάστης πλὴν τῆς κεφαλῆς, καὶ ἐν τῇσι
κρίσεσιν οὐ χρὴ λούειν· καὶ τὰς μὲν πρώτας τῶν
ἡμερέων ὑποκαθαίρειν καὶ ὑδροποτέειν· καὶ γὰρ ἔμετον ἄγει τὸ ὕδωρ ὡς ἐπὶ τὸ πολύ· τὰς δ' ὑστέρας μετὰ
τὴν κάθαρσιν ὑγραίνειν, καὶ ῥυφήμασι χρῆσθαι καὶ
οἴνοισι γλυκέσιν. ἢν δὲ μὴ ἐξ ἀρχῆς παραλάβῃς, ἀλλ'
ἤδη τῶν ἐν τῇ γλώσσῃ σημείων ἐόντων, ἐὰν χρή, ἕως
ἂν αἱ κρίσιες παρέλθωσι καὶ τὰ τῆς γλώσσης σημεῖα
ἠπιώτερα γένηται, καὶ μήτε φάρμακον δῷς μήτε κλύ-
σῃς ἐς κάθαρσιν, πρὶν αἱ κρίσιες παρέλθωσιν.

7. Ὅταν δ' ὁ πλεύμων πρησθῇ[16] ὑπὸ φλεγμασίης[17]
καὶ οἰδήσῃ, βὴξ ἔχει ἰσχυρὴ καὶ σκληρὴ καὶ ὀρ-
θοπνοίη· καὶ ἀναπνεῖ ἀθρόον, καὶ πυκνὸν ἀσθμαίνει,
καὶ ἰδίει καὶ τοὺς μυκτῆρας ἀναπετάννυσιν ὡς ἵππος
δραμών, καὶ τὴν γλῶσσαν θαμινὰ ἐκβάλλει· καὶ τὰ
στήθεα αὐτῷ ἀείδειν δοκέει καὶ βάρος ἐνεῖναι, διὸ
χωρέειν οὐ δύναται τὰ στήθεα, ἀλλὰ διαρρήγνυται
καὶ ἀδυνατέει. ἡ δ' ὀδύνη ὀξέη ἴσχει τὸν νῶτον καὶ τὰ
126 στήθεα, | καὶ τὰς πλευρὰς ὡς βελόναι κεντέουσι, καὶ
καίεται ταῦτα ὡς πρὸς πυρὶ καθιζόμενος· καὶ ἐρυθήματα ἐκφλύει ἐς τὸ στῆθος καὶ τὸν νῶτον ὡς φλογο-
ειδέα.[18] καὶ δηγμὸς ἰσχυρὸς ἐμπίπτει καὶ ἀπορίη,
ὥστε οὔτε κατακεῖσθαι οὔθ' ἵστασθαι οὔτε καθίζεσθαι
οἷός τ' ἐστίν, ἀλλ' ἀπορέει ἀλύων ῥιπτάζει τε ἑωυτόν,
καὶ δοκέει ἤδη ἀποθανεῖσθαι. ἀποθνήσκει δὲ μάλιστα
τεταρταῖος ἢ ἑβδομαῖος· ἢν δὲ ταύτας ἐκφύγῃ, οὐ
μάλα ἀποθνήσκει.

14

You must treat this patient as follows: wash him with hot water two or three times every day, except for his head—during the crises, you must not wash. The first days clean him downwards, and give him water to drink, for water usually provokes vomiting. The days after the cleaning, moisten him and employ gruels and sweet wines. If you have not attended the patient from the beginning, but only after the signs in his tongue were already present, you must withhold treatment until his crises arrive and the signs become fainter; before the crises, for the purpose of cleaning give neither a medication nor an enema.

7. When the lung is distended with phlegmasia and swells up, a violent harsh cough and orthopnoea set in. The patient respires rapidly, gasps frequently for breath, sweats, dilates his nostrils like a running horse, and continually protrudes his tongue. His chest seems to sing and to contain a heaviness that prevents it from moving; in fact, it feels torn, and is powerless. Sharp pain is present in the patient's back and chest, needles, as it were, prick his sides, and he burns in these areas as though he were sitting next to a fire; red patches like flames erupt on his chest and back. A violent gnawing pain attacks the patient, and he is in such straits that he can neither lie down, nor stand up, nor sit; he is distraught and casts himself about, and seems already on the point of death. He usually dies on the fourth or seventh day; if he survives that many, death is rare.

---

16 Jouanna (p. 376): πρισθῇ Θ: πλησθῇ M.

17 Θ: θερμασίης M.

18 M: φολιδοειδές Θ.

Τοῦτον ἢν θεραπεύῃς, τὴν κάτω κοιλίην ὡς τάχιστα καθῆραι κλύσματι εὖ, καὶ ἀπὸ τῶν ἀγκώνων καὶ
τῆς ῥινὸς καὶ τῆς γλώσσης καὶ πάντοθεν αἷμα ἀφι-
έναι· καὶ πώματα διδόναι ψυκτικὰ καὶ ῥυφήματα τὰ
αὐτὰ δυνάμενα, καὶ τῶν οὐρητικῶν, μὴ θερμαινόντων
δέ, πολλάκις διδόναι. καὶ πρὸς μὲν τὰς ὀδύνας αὐτάς,
ὅταν καταιγίζωσι, χλιάσματα κοῦφα καὶ ὑγρὰ χρὴ
προσφέροντα χλιαίνειν καὶ ὑγραίνειν τὸν τόπον, οὗ ἂν
ᾖ ὀδύνη·[19] πρὸς δὲ τὰ ἄλλα ψυκτήρια προσίσχειν, τὸ
μὲν ἀφαιρέοντα, τὸ δὲ προστιθέντα, καὶ ἢν[20] κατακαί-
ηται, ψῦχος ποιέειν· ἀοινεῖν δὲ τὸ πάμπαν.

8. Ὅταν δ' ἀπὸ τῆς κεφαλῆς ἀρξαμένη ὀδύνη ὀξέη
ἄφωνον ποιήσῃ παραχρῆμα—ἄλλως τε καὶ ἐκ μέθης—οὗτος ἀποθνήσκει ἑβδομαῖος, ἧσσον δὲ τοῖσιν
ἐκ τῆς μέθης θανάσιμα· ἢν γὰρ ῥήξωσι φωνὴν αὐθη-
μερὸν ἢ τριταῖοι,[21] ὑγιέες εἰσί· ποιέουσι δ' ἐκ τῆς
μέθης ἔνιοι τοῦτο, οἱ δ' ἕτεροι[22] ἀπόλλυνται.

Τούτοισι πταρμούς τε ἐμποιέειν ἰσχυροὺς καὶ ὑπο-
κλύσαι, ὅ τι χολὴν ἄξει σφόδρα· καὶ | ἢν ἐπαίσθηται,
ὀπὸν θαψίης δοῦναι ἐν πολλῷ τῷ ὑγρῷ καὶ θερμῷ, ἵνα
ὡς τάχιστα ἀπεμέσῃ· ἔπειτα ἀπισχναίνειν καὶ ἀοινεῖν
ἑπτὰ ἡμέρας· ἀφαιρέειν δὲ καὶ ἀπὸ τῆς γλώσσης
αἷμα, ἢν δύνῃ λαβεῖν φλέβα.

9. Φρενῖτις δὲ γίνεται καὶ ἐξ ἑτέρης νούσου·
πάσχουσι δὲ τάδε· τὰς φρένας ἀλγέουσιν, ὥστε μὴ

128

---

[19] προσφέροντα . . . ὀδύνη Μ: πρὸς ἴσχειν Θ.
[20] ἢν Θ: ὅκου ἂν Μ.

16

If you treat this patient, clean out his lower cavity thoroughly as quickly as possible with an enema, and draw blood from the bends of his arms, his nose, his tongue, and in fact all over. Give him cooling drinks, gruels that have the same effect, and also frequent diuretics that are not warming. Against the pains themselves, when they are pressing, you must apply light moist fomentations to warm and moisten the place where the pain happens to be. Against the rest apply cooling agents; apply and remove these alternately. If the patient is consumed with heat, cool him. Have him abstain totally from wine.

8. When a sharp pain beginning from the head suddenly makes a person speechless—both in consequence of drunkenness and otherwise—the person dies on the seventh day. This condition is less often fatal in cases where it arises from drunkenness; for if patients' voices break through on the same day, or on the third day, they have recovered; some who have the condition from drunkenness do this, the rest die.

In these patients induce energetic sneezing, and give an enema that will draw bile effectively. If a patient recovers his senses, give him thapsia juice in adequate warm fluid, in order that he will vomit as soon as possible. Then make him lean, and have him abstain from wine for seven days. Also draw blood from his tongue, if you can catch hold of the vessel.

9. Phrenitis can also develop out of another disease. Patients suffer as follows: they experience such pain in

---

²¹ τριταῖοι Θ: τῇ ὑστεραίῃ ἢ τῇ M.
²² οἱ δ' ἕτεροι Θ: οὐδέτεροι M.

ἐᾶσαι ἂν ἅψασθαι, καὶ πῦρ ἔχει, καὶ ἔκφρονές εἰσι,
καὶ ἀτενὲς βλέπουσι, καὶ τἆλλα παραπλήσια ποιέουσι τοῖσιν ἐν τῇσι περιπλευμονίῃσιν, ὅταν [οἱ ἐν τῇ
περιπλευμονίῃ]²³ ἔκφρονες ἔωσι.

Τοῦτον χλιαίνειν χλιάσμασιν ὑγροῖσι καὶ πώμασι
πλὴν οἴνου· καὶ ἢν μὲν οἷός τε ᾖ, ἀποκαθαίρειν ἄνω,
βηχί τε καὶ πτύσει ἀνάγειν χρὴ ὥσπερ ἐν τῇ περι-
πλευμονίῃ· εἰ δὲ μή, τὴν κάτω κοιλίην παρασκευάζειν
ὅπως ὑποχωρέῃ· ὑγραίνειν δὲ πώματι· ἀγαθὸν γάρ. ἡ
δὲ νοῦσος θανατώδης· ἀποθνήσκουσι δὲ τριταῖοι ἢ
πεμπταῖοι ἢ ἑβδομαῖοι· ἢν δὲ ἠπίως ληφθῇ, κρίνει ὡς
περιπλευμονίη.

10. Ὑπὸ δὲ τῆς κυνάγχης λεγομένης πνίγεταί τε
ὥνθρωπος καὶ ἐν τῇ φάρυγγι ὡς μῆλον²⁴ δοκέει ἐνέχεσθαι καὶ κατασπᾷ οὔτε τὸ σίαλον οὔτ' ἄλλ' οὐδέν·
καὶ οἱ ὀφθαλμοὶ πονέουσί τε καὶ ἐξέχουσιν ὡς ἀγχο-
μένοισι, καὶ ἐκβλέπει αὐτοῖσιν ἀτενὲς καὶ στρέφειν
σφέας οὐχ οἷός τε· καὶ ‹ἀ›λύζει²⁵ καὶ ἀναΐσσει θαμι-
νά· καὶ τὸ πρόσωπον καὶ ἡ φάρυγξ πίμπραται, ἀτὰρ
καὶ ὁ τράχηλος· ὑπὸ δὲ τοῖσιν οὔασιν²⁶ οὐδὲν κακὸν
ἔχειν δοκέει· καὶ ὁρᾷ καὶ ἀκούει ἀμβλύτερον, καὶ ὑπὸ
130 τοῦ πνιγμοῦ οὐκ ἔννοός ἐστιν, οὔτ' ἤν τι λέγῃ | οὔτ' ἤν
τι ἀκούῃ ἢ ποιέῃ· ἀλλὰ κεχηνὼς κεῖται σιαλοχ‹ο›έων.
τοιάδε ποιέων οὗτος ἀποθνήσκει πεμπταῖος ἢ ἑβδομαῖος ἢ ἐναταῖος.²⁷ ὅταν δὲ τούτων τι ἀπῇ τῶν ση-

²³ Del. Potter.
²⁴ ὡς μῆλον Θ: οἱ μᾶλλον Μ.

18

the diaphragm that they will not allow themselves to be touched, there is fever, they are deranged, they stare fixedly, and for the rest they resemble patients with pneumonia that are deranged.

Warm this patient with warm moist fomentations and with drinks other than wine. If he can stand it, clean him upwards; he must bring up material by coughing and expectoration just as in pneumonia; if he fails to do this, prepare the lower cavity in order to evacuate it; moisten the patient with drink, for that helps. The disease is usually mortal, and patients die on the third, fifth or seventh day. If the case is a mild one, it has its crisis the way pneumonia does.

10. With angina, as it is called, the person chokes and seems to have something like an apple caught in his throat, so that he is unable to swallow either his saliva or anything else. His eyes hurt and protrude as in those that are being strangled, they stare fixedly, and he cannot turn them. The patient is distraught, and casts himself about incessantly. His face, throat and neck are distended,[3] but below his ears he appears normal. He sees and hears less keenly than before and, because of the strangulation, is not aware of what he says or hears or does; he just lies there with his mouth open, drooling. In this case the patient dies on the fifth, seventh or ninth day. When any of the signs men-

---

[3] The verb πιμπράναι can also mean "burn", and may here indicate the presence of inflammation.

---

25 Cornarius/Latin: λύζει ΘΜ.
26 ὑπὸ . . . οὔασιν Θ: τοῖσι δὲ ὀρέωσι Μ.
27 ἢ ἐναταῖος om. Θ.

μείων, χαλαρωτέρην δηλοῖ τὴν νοῦσον, καὶ καλέουσι
παρακυνάγχην.

Τούτων φλεβοτομέειν χρὴ μάλιστα μὲν ὑπὸ τὸν
τιτθόν· συνακολουθεῖ γὰρ ταύτῃ ἐκ τοῦ πλεύμονος
θερμὸν πνεῦμα· χρὴ δὲ καὶ τὰ κάτω καθαίρειν φαρμάκῳ ἢ κλύσματι καὶ τοὺς αὐλίσκους παρῶσαι ἐς τὴν
φάρυγγα κατὰ τὰς γνάθους, ὡς ἕλκηται τὸ πνεῦμα ἐς
τὸν πλεύμονα· καὶ ποιέειν ὡς τάχιστα πτύσαι καὶ
ἰσχναίνειν τὸν πλεύμονα· καὶ ὑποθυμιᾶν ὕσσωπον
Κιλίκιον καὶ θεῖον καὶ ἄσφαλτον, καὶ ἕλκειν διὰ τῶν
αὐλίσκων καὶ διὰ τῶν ῥινῶν, ὡς ἐξίῃ φλέγμα· καὶ τὴν
φάρυγγα καὶ τὴν γλῶσσαν ἀνατρίβειν τοῖσι τὸ
φλέγμα ἄγουσι· καὶ τὰς φλέβας τὰς ὑπὸ τῇ γλώσσῃ
τάμνειν, ἀφιέναι δὲ καὶ ἐκ τῶν ἀγκώνων, ἢν ἰσχύῃ·
ἀοινεῖν δὲ καὶ ῥυφεῖν πτισάνης χυλὸν λεπτόν·[28] ἐπειδὰν δὲ ἀνῇ ἡ νοῦσος καὶ σιτίων γεύηται, ἐλατηρίῳ νέῳ
περικαθῆραι, ἵνα μὴ ἑτέρῳ κακῷ περιπέσῃ.

11. Ἴκτερος δὲ τοιόσδ᾽ ἐστὶν ὀξὺς καὶ διὰ τάχεος
ἀποκτείνων· ἡ χροιὴ ὅλη σιδιοειδὴς σφόδρα ἐστί,
χλωροτέρη ἢ οἱ σαῦροι οἱ χλωροί· παρόμοιος δὲ καὶ ὁ
χρώς. καὶ τῷ οὔρῳ ὑφίσταται οἷον ὀρόβιον πυρρόν·
καὶ πυρετὸς καὶ φρίκη βληχρὴ ἴσχει· ἐνίοτε δὲ καὶ τὸ
ἱμάτιον οὐκ ἀνέχεται ἔχων, ἀλλὰ δάκνεται καὶ ξύεται·
καὶ ἄσιτος ἐὼν τὰ ἑωθινὰ τὰ σπλάγχνα ἀμύσσεται ὡς
ἐπὶ τὸ πολύ, καὶ ὅταν ἀνιστῇ τις αὐτὸν ἢ προσδιαλέ-
γηται, οὐκ ἀνέχεται. οὗτος ὡς ἐπὶ τὸ πολὺ ἀποθνήσκει
ἐντὸς τεσσερεσκαίδεκα ἡμερέων· ταύτας δὲ διαφυγὼν
ὑγιής.

20

tioned is absent, it shows that the disease is milder, and people call it "parangina".

You must phlebotomize this patient, best of all under the nipple; for there warm breath follows, out of the lung. You must also clean out the lower cavity with a medication or enema, and insert tubes into the throat behind the jaws, in order that air may be drawn into the lung. Induce expectoration as soon as possible, and dry up the lung: burn Cilician hyssop, sulphur and asphalt, and have the patient draw the vapours through the tubes into his nostrils, in order to discharge the phlegm; also anoint the throat and tongue with agents that draw phlegm. If the patient is strong, incise the vessels under his tongue, and draw blood from the bends of his arms; let him abstain from wine, and drink thin barley-water gruel. When the disease goes away, and the patient has tasted food, clean him out thoroughly with fresh squirting-cucumber juice, in order that he does not fall into some new evil.

11. The acute and rapidly fatal jaundice is as follows: the whole skin is very much the colour of pomegranate-peel, greener than green lizards, and the body the same. In the urine a reddish sediment like vetch-meal precipitates; fever and mild shivering are present. Sometimes the patient will not even tolerate having his blanket on, but it scratches and irritates him; in the morning, before he has eaten, his inward parts usually suffer tearing pains, and when anyone wakes him up or talks to him, he will not tolerate it. The patient generally dies within fourteen days; if he survives that many, he recovers.

---

²⁸ λεπτόν om. Θ.

Χρὴ δὲ θερμολουτέειν τε καὶ πίνειν μελίκρη|τον
σὺν καρύων Θασίων λεπισθέντων²⁹ καὶ ἀψινθίου κό-
μης ἴσῳ, ἀννίσου σεσησμένου ἡμίσει·³⁰ πίνειν ὁλκῆς
τριώβολον νῆστις καὶ πάλιν ἐς κοίτην τὸ μελίκρητον
τοῦτο³¹ καὶ οἶνον λεπτὸν παλαιὸν καὶ ῥυφήματα· ἀσι-
τέειν δὲ μή.

12. Οἱ δὲ τέτανοι ἢν ἐπιλάβωσιν, αἱ γέννες πεπή-
γασιν ὡς ξύλιναι, καὶ τὸ στόμα διαίρειν οὐ δύνανται,
καὶ οἱ ὀφθαλμοὶ δακρύουσί τε καὶ ἰλλαίνονται· καὶ τὸ
μετάφρενον πέπηγε, καὶ τὰ σκέλεα οὐ δύνανται συνάγειν, ὁμοίως³² οὐδὲ τὰς χεῖρας· καὶ τὸ πρόσωπον
ἐρεύθει, καὶ³³ σφόδρα ὀδυνᾶται, καὶ ὅταν ἀποθνήσκειν
μέλλῃ, ἀνεμέει διὰ τῶν ῥινῶν τὸ πῶμα καὶ τὸ ῥύφημα
καὶ τὸ φλέγμα. οὗτος τριταῖος ἢ πεμπταῖος ἢ ἑβδομαῖος ἢ τεσσερεσκαιδεκαταῖος ἀπόλλυται· ταύτας δὲ
διαφυγὼν ὑγιὴς γίνεται.

Τούτῳ διδόναι κατάποτα πέπερι καὶ ἐλλέβορον
μέλανα, καὶ ζωμὸν ὀρνίθειον πίονα θερμόν· καὶ πταρμοὺς ἰσχυροὺς καὶ πολλοὺς ἐμποιέειν καὶ πυριᾶν·
ὅταν δὲ μὴ πυριᾷς, τὰ χλιάσματα προστιθέναι ὑγρὰ
καὶ λιπαρὰ ἐν κύστεσι καὶ ἀσκίοισι πανταχόθεν,
μάλιστα δὲ πρὸς τὰ ὀδυνώμενα, καὶ ἀλείφειν θερμῷ
καὶ πολλῷ πολλάκις.

13. Ὁ δὲ ὀπισθότονος τὰ μὲν ἄλλα ὡς ἐπὶ τὸ πολὺ
ὡσαύτως, σπᾶται δ᾽ εἰς τοὔπισθε· καὶ βοᾷ ἐνίοτε, καὶ
ὀδύναι ἴσχουσιν ἰσχυραί, καὶ συνάγειν ἐνίοτε οὐκ ἐᾷ
τὰ σκέλεα οὐδὲ τὰς χεῖρας ἐκτεῖναι· ξυγκεκαμμένοι
γὰρ οἱ ἀγκῶνες γίνονται, καὶ τοὺς δακτύλους πὺξ

The patient must employ warm baths, and drink melicrat with equal amounts of shelled Thasian nuts[4] and wormwood leaves, and half as much sifted anise; he must drink three obols' weight in the morning before eating, and on retiring this melicrat again together with light aged wine, and gruels. Let him not go without eating.

12. When tetanus occurs, the jaws become as hard as wood, and patients cannot open their mouths. Their eyes shed tears and look awry, their backs become rigid, and they cannot adduct their legs; similarly, not their arms either. The patient's face becomes red, he suffers great pain and, when he is on the point of death, he vomits drink, gruel and phlegm through his nostrils. This patient generally dies on the third, fifth, seventh or fourteenth day; if he survives for that many, he recovers.

Give him pills of pepper and black hellebore, and warm fat bird soup. Induce frequent energetic sneezing, and treat with vapour-baths; when you do not employ vapour-baths, apply moist rich fomentations in bladders and small leather skins to all parts of the body, especially the painful ones, and anoint often with plentiful warm oil.

13. Opisthotonus is mainly the same, except that the patient is drawn backwards. He sometimes cries out, his pains are violent, and sometimes the disease does not allow him to adduct his legs or to extend his arms; for the

[4] Almonds.

---

29 I²: λεπτισθέντων M.
30 ἴσῳ . . . ἡμίσει Potter: ἴσον . . . ἥμισυ M.
31 σὺν καρύων Θασίων . . . τοῦτο om. Θ.
32 ὁμοίως om. Θ.   33 ἐρεύθει, καὶ om. Θ.

ἔχει, καὶ τὸν μέγαν δάκτυλον τοῖσιν ἄλλοισι κατέχει
ὡς ἐπὶ τὸ πολύ· καὶ βοᾷ καὶ φλυηρέει ἐνίοτε, καὶ οὐ
δύναται ἑωυτὸν κατέχειν, ἀλλ' ἀναΐσσει ἐνίοτε, ὅταν ἡ
ὀδύνη ἔχῃ· ὅτε δὲ ἀνίησιν ἡ ὀδύνη, ἡσυχίην ἔχει·
134 ἐνίοτε δὲ καὶ ἄφωνοι γίνονται ἅμα ἁλισκόμενοι | ἢ
μανικοί τι ἢ[34] μελαγχολικοί· οὗτοι τριταῖοι ἀποθνήσκουσι τῆς φωνῆς λυθείσης· καὶ ἀνεμέουσι διὰ τῶν
ῥινῶν καὶ οὗτοι· τὰς δὲ τεσσερεσκαίδεκα διαφυγὼν
ὑγιής.

Θεραπεύειν δὲ ὡς τὸν ἄνω· ἢν δὲ βούλῃ, καὶ ὧδε
ποιέειν· ὕδωρ ὡς πλεῖστον ψυχρὸν καταχέας ἔπειτα
ἱμάτια θερμὰ καὶ καθαρὰ καὶ πολλὰ καὶ λεπτὰ ἐπι-
βάλλειν, πῦρ δὲ τότε[35] μὴ προσφέρειν. ὧδε καὶ τοὺς
τετανικοὺς καὶ τοὺς ὀπισθοτονικοὺς[36] ποιέειν.

14. Εἰλεοὶ δὲ γίνονται τῆς ἄνω κοιλίης θερμαι-
νομένης καὶ τῆς κάτω ψυχομένης· συναναίνεται γὰρ
τὸ ἔντερον[37] ὥστε μήτε τὸ πνεῦμα μήτε τὰς τροφὰς
διεξιέναι, ἀλλὰ τὴν γαστέρα ξηρὴν εἶναι, καὶ ἐμέει[38]
ἐνίοτε, πρῶτον μὲν φλεγματώδεα, ἔπειτα χολώδεα,
τελευτῶν δὲ κόπρον. καὶ δίψα ἔχει, καὶ ὀδύνη ἔχει,
μάλιστα μὲν πρὸς τὰ ὑποχόνδρια· ἀλγέει δὲ καὶ ὅλην
τὴν γαστέρα καὶ πεφύσηται καὶ λύζει,[39] καὶ πυρετοὶ
ἐπιλαμβάνουσι. γίνεται δὲ μάλιστα μετοπώρου· ἀποθνήσκει δὲ μάλιστα[40] ἑβδομαῖος.

---

34 Potter: τε ἢ Θ: τε καὶ Μ.
35 τότε om. Θ.
36 Θ: τετάνους . . . ὀπισθοτόνους Μ.

elbows become flexed, and he holds his fingers in a fist, usually enclosing the thumb inside the other digits. The patient cries out and sometimes talks nonsense; when the pain is present, he is unable to restrain himself, casting himself about, but when it remits, he is still. Sometimes they may also become speechless during an attack, at the same time being seized by some sort of rage or melancholy; such patients generally die on the third day after becoming speechless. These patients, too, vomit through their nostrils. If one survives for fourteen days, he recovers.

Treat as the patient above; if you wish, do the following as well: pour a very large amount of cold water over the patient, then cover him with large clean warm light blankets, and during this time do not bring fire close to him. Do this both for tetanus and for opisthotonus.

14. Ileus occurs when the upper cavity is heated, and the lower one is cooled; for the intestine dries up, so that neither air nor food can pass through it, and the belly becomes costive. The patient sometimes vomits first material that is like phlegm, then like bile, and finally faeces. There are thirst and pain, especially in the hypochondrium; the patient also has pain throughout his whole belly with distention, and hiccups; fever, too, comes on. This disease occurs in fall, in most cases, and the patient generally dies on the seventh day.

---

37 M adds καὶ συμπιλέεται ὑπὸ τῆς φλεγμασίης.
38 Ermerins: ἐμέειν ΘΜ.
39 καὶ πεφύσηται καὶ λύζει om. Θ.
40 μετοπώρου . . . μάλιστα om. Μ.

Τοῦτον ὧδε θεραπεύειν· καθῆραι τὴν ἄνω κοιλίην
ὡς τάχιστα, καὶ αἷμα ἀφαιρέειν ἀπὸ τῆς κεφαλῆς καὶ
τῶν ἀγκώνων, ἵνα παύσηται ἡ ἄνω κοιλίη θερμαινο-
μένη,[41] καὶ ψύχειν τὰ ἄνω τῶν φρενῶν πλὴν τῆς
καρδίης· τὰ δὲ κάτω θερμαίνειν ἐν σκάφῃ ὕδατος
θερμοῦ καθίζων· καὶ ἀλείφειν ἀεί, καὶ χλιάσματα
ὑγρὰ προστιθέναι· καὶ βάλανον μέλιτος μόνου ποιέων
ὡς δέκα δακτύλων καὶ ἄκρῳ[42] χολὴν ταύρου ἐς[43] τὸ
πρόσθεν προσπλάσσων πρόσθες καὶ[44] δὶς καὶ τρίς, ὡς
πάντα τὰ συγκεκαυμένα περὶ τὸν ἀρχὸν ἐξαγάγῃς τῆς
κόπρου· καὶ ἢν μὲν οὕτως ὑπακούῃ, κλύζειν ἐπὶ |
136 τούτοισιν· εἰ δὲ μή, φῦσαν λαβὼν χαλκευτικὴν ἐσ-
ιέναι καὶ φυσᾶν ἐς τὴν κοιλίην, ἵνα διαστήσῃς τήν τε
κοιλίην καὶ τὴν τοῦ ἐντέρου συστολήν·[45] εἶτα πάλιν
ἐξελὼν τὴν φῦσαν κλύσαι εὐθύς· ἕτοιμον δ᾽ ἔστω τὸ
κλύσμα, μὴ πολὺ τῶν θερμαντικῶν, ἀλλὰ διαλυόντων
τὰς κόπρους καὶ τηκόντων· εἶτα βύσας τὴν ἕδρην
σπόγγῳ, καθήσθω ἐν ὕδατι θερμῷ κατέχων τὸ κλύσμα, καὶ ἢν δέξηται τὸ κλύσμα καὶ πάλιν μεθῇ, ὑγιής.
ἐν δὲ τῷ πρόσθεν χρόνῳ μέλι τε ὡς κάλλιστον λειχέ-
τω, καὶ οἶνον αὐτίτην[46] πινέτω εὔζωρον. ἢν δὲ τοῦ
εἰλεοῦ ἀφέντος πυρετὸς ἐπιλάβῃ, ἀνέλπιστος· ἴσως
γὰρ ἡ κάτω κοιλίη λυθεῖσα συναποκτείνειεν ἄν.

15. Ἡ δὲ περιπλευμονίη τοιάδε ποιέει· πυρετός τε
ἰσχυρὸς[47] ἴσχει, καὶ πνεῦμα πυκινὸν καὶ θερμὸν ἀνα-

[41] Μ: -νουσα Θ.
[42] καὶ ἄκρῳ Potter: ἐξ ἄκρου ΘΜ.

26

Treat this patient as follows: clean out his upper cavity as quickly as possible, draw blood from his head and the bends of his arms, in order to remove the overheating of the upper cavity, and cool the region above the diaphragm, except for the heart. Warm the lower cavity by sitting the patient in a basin of hot water; anoint him often and apply moist fomentations. Form a suppository of pure honey ten fingers long and, smearing the anterior tip with bull's gall, introduce it two or three times, in order to remove all the fecal material that has been burnt dry about the anus. If this succeeds, follow it with an enema; if not, take a bronzesmith's bellows and, introducing this, blow air into the cavity in order to open up the cavity and the intestinal contraction. Then remove the bellows and immediately administer an enema; let the enema be prepared beforehand and not too warming, but capable of dissolving and melting the faeces; then stop the anus with a sponge, and let the patient sit in hot water while retaining the enema; if he takes the enema fluid in and then sends it forth again, he recovers. Before this let the patient take some of the best honey you have, and drink new wine unmixed with water. If, after the obstruction has been resolved, fever comes on, the case is hopeless, because the lower cavity, being relaxed, will in all likelihood combine with the fever to kill the patient.

15. In pneumonia the following happens: there is violent fever, and the patient's breathing is rapid and hot; he is

---

43 M: ὡς Θ.   44 M: ὡς Θ.
45 Potter: ξυνσταλειν Θ: σύσταλσιν M.
46 Vander Linden: ναυτιτην Θ: αὐγιτην M.
47 Θ: ὀξὺς M.

πνεῖ, καὶ ἀπορίη καὶ ἀδυναμίη καὶ ῥιπτασμὸς καὶ
ὀδύνη ὑπὸ τὴν ὠμοπλάτην καὶ ἐς τὴν κληῖδα καὶ ἐς
τὸν τιτθόν, καὶ βάρος ἐν τοῖσι στήθεσι[48] καὶ παραφροσύνη. ἔστι δὲ οἷσιν ἀνώδυνός ἐστιν, ἕως ἄρξωνται
βήσσειν· πολυχρονιωτέρη δὲ καὶ χαλεπωτέρη κείνης.

Τὸ δὲ σίαλον λευκὸν καὶ ἀφρῶδες πτύει πρῶτον,
καὶ ἡ γλῶσσα ξανθή· προϊόντος δὲ τοῦ χρόνου μελαί-
νεται· ἢν μὲν οὖν ἐν ἀρχῇ μελαίνηται, θάσσους αἱ
ἀπαλλάξιες· ἢν δ' ὕστερον, σχολαίτεραι· τελευτῶσι δὲ
καὶ ῥήγνυται ἡ γλῶσσα, καὶ ἢν προσθῇς τὸν δάκτυλον, θράσσεται.[49] τὴν δ' ἀπάλλαξιν τῆς νούσου σημαίνει ἡ γλῶσσα, ὥσπερ καὶ ἐν τῇ πλευρίτιδι ὁμοίως.

Ταῦτα δὲ πάσχει ἡμέρας τεσσερεσκαίδεκα τοὐλά-
χιστον, τὸ πλεῖστον δὲ εἴκοσι καὶ μίαν· καὶ βήσσει
τοῦτον τὸν χρόνον σφόδρα, καὶ καθαίρεται ἅμα τῇ
138 βηχὶ τὸ μὲν πρῶτον πολὺ | καὶ ἀφρῶδες σίαλον,
ἑβδόμῃ δὲ καὶ ὀγδόῃ—ὅταν ὁ πυρετὸς ἐνακμάζῃ—ἢν
ὑγρὴ ἡ περιπλευμονίη ᾖ, παχύτερον· εἰ δὲ μή, οὔ·
ἐνάτῃ δὲ καὶ δεκάτῃ ὑπόχλωρον καὶ ὕφαιμον, δωδε-
κάτῃ δὲ μέχρι τεσσερεσκαιδεκάτης, πολὺ καὶ πυῶδες.

Ὧν ὑγραί εἰσιν αἵ τε φύσιες καὶ αἱ διαθέσιες τοῦ
σώματος, ἀτὰρ καὶ ἡ νοῦσος ἰσχυρή· ὧν δ' ἥ τε φύσις
καὶ ἡ στάσις τῆς νούσου ξηρή, ἧσσον οὗτοι.

Ἢν μὲν οὖν πέμπτη καὶ ἕκτη ἐπὶ δέκα[50] ξηρανθῇ
καὶ μηκέτι ἀποβήσσῃ πυῶδες, ὑγιής ἐστιν· εἰ δὲ μή,

[48] M adds ἐνίοτε δέ.   [49] Θ: ἔχεται Μ.
[50] πέμπτη . . . δέκα Θ: τετάρτη καὶ δεκάτη Μ.

28

distraught, weak and restless, and beneath his shoulder-blade he suffers pain that radiates toward his collar-bone and nipple; he has a heaviness in his chest, and he is deranged. In some patients, there is no pain until they begin to cough; this pneumonia lasts longer and is severer than the one with pain from the beginning.

The patient first expectorates white frothy sputum, and his tongue is yellow; as time passes, the tongue becomes dark. Now if it becomes dark at the beginning, recovery is more rapid, but if it becomes dark later, recovery is slower; in the end, the tongue also develops fissures, and if you touch it with your finger, the patient is irritated. The tongue gives an indication of recovery in this disease just as in pleurisy.

The patient suffers these things for at least fourteen days, at most twenty-one; he coughs hard during this time, clearing with his cough first copious frothy sputum, and then on the seventh or eighth day—whenever the fever reaches its high point—a thicker sputum, if the pneumonia happens to be a moist one; not, however, if the pneumonia is not moist. On the ninth and tenth days, the sputum is somewhat yellow-green and charged with blood; on the twelfth to the fourteenth days, it is copious and purulent.

In patients whose natures and bodily propensities are moist, the disease is severe; in those whose nature and state of disease is dry, less so.

Now if on the fifteenth and sixteenth days the patient becomes dry and no longer coughs up purulent sputum, he

πρόσεχε πρὸς τὰς εἴκοσι δυοῖν δεούσας καὶ τὰς εἴκοσι
καὶ μίαν τὸν νόον, καὶ ἢν μὲν ἐνταῦθα παύσηται τοῦ
πτύσματος, ἐκφεύγει·[51] εἰ δὲ μή, ἔρου αὐτὸν εἰ γλυκύ-
τερον τὸ σίαλον, καὶ ἢν μὲν φῇ, ὁ πλεύμων αὐτῷ
ἔμπυός ἐστι, καὶ ἡ νοῦσος καθίσταται ἐνιαυσίη, ἢν μὴ
ἐν τῇσι τεσσεράκονθ᾽ ἡμέρῃσι σπεύδων ἀναγάγῃ τὸ
πύον· ἢν δὲ φῇ ἀηδὲς εἶναι τὸ σίαλον, θανατώδης ἡ
στάσις τῆς νούσου. ἀλλ᾽ ἐν τῇσι πρώτῃσιν κβ΄[52]
ἡμέρῃσι μάλιστα διαδηλοῖ· ἢν γὰρ ἐκπτύσῃ τὸ σαπὲν
καὶ τὸ πυωθὲν ἐν ἡμέρῃσι δύο καὶ εἴκοσι καὶ μὴ
ἑλκωθῇ, ἐκφεύγει· εἰ δὲ μή, οὔ.

Αὕτη ἡ περιπλευμονίη οὐδὲν ἀπολείπει τῶν ἐκ
περιπλευμονίης[53] κακῶν· ἢν οὖν τι τούτων ἀπῇ τῶν
κακῶν, εἰδέναι χαλαρωτέρην ἐοῦσαν[54] τῷ κάμνοντι
ἔχειν καὶ τῷ ἰητρῷ μεταχειρίζεσθαι· ἢν δὲ ὀλίγα ἔχῃ
140 τούτων | τῶν σημείων, μὴ ἐξαπατάτω ὡς οὐ περιπλευ-
μονίη ἐστίν· ἔστι γὰρ μαλθακή.

Θεραπεύειν δ᾽ ὧδε τὴν περιπλευμονίην—οὐ μέντοι
ἐξαμαρτήσῃ καὶ πλευρῖτιν καὶ φρενῖτιν οὕτω μεταχει-
ριζόμενος· τὴν κεφαλὴν ἄρχου κουφίζων, ἵνα μηδὲν
ἐπιρρέῃ ἐς τὸ στῆθος· τὰ δὲ ῥυφήματα τὰς μὲν πρώτας ἡμέρας γλυκύτερα· οὕτω γὰρ ἂν μάλιστα τὸ
συγκαθήμενον καὶ τὸ συνεστηκὸς ἀποπλύνοις καὶ
κινοίης· τεταρταίοισι δὲ καὶ πεμπταίοισι καὶ ἑκταίοισι
μηκέτι γλυκύτερα, ἀλλὰ λιπαρά· ἐς γὰρ τὴν ἄνω
πτύσιν ὑποχρέμπτεσθαι ξυμφέρει· ἢν δὲ μὴ δύνηται
κατὰ λόγον πτύειν, τῶν ἀναγόντων φαρμάκων διδό-

has recovered. If not, turn your attention to the eighteenth and twenty-first days, and if he stops expectorating then, he survives. If not, ask him if his sputum is sweetish, and if he says it is, his lung is suppurating, and the disease will last for a year, unless he exerts himself to bring up the pus in forty days. If he says the sputum has a foul taste, the state of his disease is mortal. The sputum generally gives an indication in the first twenty-two days, for, if putrefied purulent material is coughed up in twenty-two days, and if there is no ulceration, the patient escapes; otherwise, he does not.

This pneumonia lacks none of the evils of pneumonia; now if any of these evils should be absent, know that the disease will be easier for the patient to bear and for the physician to treat. However, if the patient has only a few of the signs, do not be deceived into thinking that it is not really pneumonia; for it is, only a mild one.

Treat pneumonia as follows (indeed, it would not be a mistake to handle pleurisy and phrenitis in the same way): begin by lightening the head, in order that no flux to the chest will occur. On the first days, gruels should be sweet-ish, for with these you will best wash away and remove what has been deposited and congealed in the chest; on the fourth, fifth and sixth days, change from sweet to rich ones, for this helps the patient to cough up sputum gently; if he is unable to expectorate as he should, give expecto-

---

51 τοῦ πτύσματος, ἐκφεύγει Potter: ἐ. τ. π. ΘΜ.
52 κβ′ Potter: ἡ δευτέρησιν Θ: om. Μ.
53 ἐκ π. Θ: ἐν τῷ πλεύμονι ἐόντων Μ.
54 χ. ἐ. Potter: χρὴ δεουσαν Θ: χρὴ ὅσα τε Μ.

ναι. τὰς δὲ κοιλίας ἐν μὲν τῇσι πρώτῃσιν ἡμέρῃσι
τέσσερσιν ἢ πέντε ὑποχωρέειν χρὴ καὶ ὀλίγῳ μᾶλλον,
ἵνα οἵ τε πυρετοὶ ἀμβλύτεροι ἔωσι καὶ τἀλγήματα
κουφότερα· ὅταν δὲ κεκενωμένος ᾖ καὶ ἀσθενὴς τὸ
σῶμα, τὴν κάτω κοιλίην διὰ τρίτης ὑποκινέειν, ἵνα τό
τε σῶμα μὴ ἀδύνατον ᾖ καὶ τὰ ἄνω⁵⁵ χωρία ἔννγρα·⁵⁶
ἢν γὰρ κάτω τὸ ὑγρὸν πολλὸν ὑποχωρέῃ ἀπὸ τῆς
πέμπτης ἡμέρης, θάνατον ποιέει· κάτω γὰρ τοῦ ὑγροῦ
ὑποχωρέοντος, τὰ ἄνω ξηραίνεται, καὶ ἡ κάθαρσις ἡ
τοῦ πτύσματος οὐ χωρέει ἄνω. δεῖ οὖν καὶ τὴν κάτω⁵⁷
κοιλίην μήθ᾽ ἑστάναι, ἵνα μὴ ὀξέες ἔωσιν οἱ πυρετοί,
μήτε λίην ὑποχωρέειν, ἵνα τὸ σίαλον ἀνιέναι δύνηται
καὶ ἰσχύῃ ὁ κάμνων. φάρμακα δὲ τῆς ἀναγωγῆς
ἑκταίοισι καὶ ὀγδοαίοισι⁵⁸ καὶ ἔτι περαιτέρω ἐοῦσι τῆς
νούσου δίδου· τὸ δὲ φάρμακον ἔστω ἐλλέβορος λευ-
κός, θαψίη, ἐλατήριον νέον, ἴσον ἑκάστου.

Ἢν δὲ τὸ σίελον μὴ καθαίρηται εὖ καὶ τὸ πνεῦμα
πυκνὸν ᾖ καὶ τῆς καθάρσιος μὴ ἐπικρατέῃ, προειπεῖν
ὅτι ἀνέλπιστός ἐστι ζῆν, ἢν μὴ τῇ καθάρσει δύνηται
ὑπουργέειν. ποιέειν δὲ καὶ τὰ ἐν τῇ περιπλευμονίῃ, ἢν
142 *σοι τὰ τῆς κοιλίης τῆς κάτω καλῶς ὑπουρ|γέῃ.*

Ποιέειν δὲ καὶ ἄλλως ἀπὸ τῆς πρώτης ἡμέρης
ἀρξάμενος· δίδου ἄρου τοῦ μεγάλου [κόγχην]⁵⁹ χηρα-
μύδα καὶ δαύκου καὶ ἀκαλήφης μίαν καὶ νάπυος καὶ
πηγάνου ὅσον τοῖσι τρισὶ δακτύλοισι λαβεῖν καὶ
ὀπὸν σιλφίου ὅσον κύαμον· ταῦτα ἐν ὄξει γλυκεῖ καὶ

---

⁵⁵ Μ: κάτω Θ.   ⁵⁶ Θ: ἄνικμα Μ.   ⁵⁷ Μ: ἄνω Θ.

rant medications. In the first four or five days, you must
evacuate the cavities, and quite well, in order that the
fevers will be blunted and the pains lightened. However,
when the body has been emptied and is weak, move the
lower cavity down gently only every other day, in order that
the body will retain some strength, and that the upper regions will remain adequately moist; for if too much moisture passes off below from the fifth day onwards, it leads to
death; for as the moisture passes off below, the upper regions become dry, and cleaning by expectoration does not
take place. In short, the lower cavity can neither be al-
lowed to remain inactive—to prevent the fevers from being too sharp—nor be too thoroughly evacuated—in order
that the sputum will be able to be expectorated, and the
patient will remain strong. Give expectorant medications
on the sixth and eighth days, and even later in the course of
the disease; let these be equal amounts of white hellebore,
thapsia, and fresh squirting-cucumber juice.

If the sputum is not being cleaned out effectively, if res-
piration is rapid, and if expectoration is failing, announce
that there is no hope of survival unless the patient can help
with the cleaning. But still treat as is appropriate for pneumonia, if the lower cavity cooperates with you.

Alternatively, do the following, beginning on the first
day: give a cheramys each of cuckoo-pint, dauke and sting-
ing nettle, good pinches of mustard and rue, and silphium
juice to the amount of a bean[5]; mix these in sweetened

---

[5] Presumably a reference to the size of drop.

---

[58] ὀγδοαίοισι Θ: ἑβδομαίοισι καὶ ἐναταίοισι M.
[59] Del. Ermerins.

ὕδατι κεράσας καὶ διηθήσας νήστι δίδου χλιαρόν.
ἐπειδὰν δὲ ἄρχηται καθαρὸν ἐκπτύειν, ἄρου χηραμύδα
καὶ σήσαμον καὶ ἀμυγδάλας καθαρὰς ἐν ὄξει γλυκεῖ
κεκρημένῳ πίνειν· ἢν δὲ μᾶλλον βούλῃ ἄγειν, καππάριος τῆς ῥίζης φλοιὸν μιγνύναι τούτοισιν.

16. Ὅταν δὲ πλευρῖτις λάβῃ, τάδε πάσχει· ὀδύνη
τὴν πλευρὴν καὶ πυρετὸς καὶ φρίκη ἴσχει, καὶ ἀναπνεῖ
πυκινόν, καὶ ὀρθοπνοίη ἴσχει, καὶ ἀποβήσσει ὑπόχο-
λον οἷον ἀπὸ σιδίου, ἢν μὴ ῥήγματα ἔχῃ· ἢν δ' ἔχῃ,
καὶ αἷμα ἀπὸ τῶν ῥηγμάτων·[60] ἐν δὲ τῇ αἱματώδει
ὕφαιμον. ἔστι δὲ ἡ μὲν χολώδης ἠπιωτέρη, ἢν μὴ
ῥήγματα ἔχῃ ὁ κάμνων· εἰ δὲ μή, ἐπιπονωτέρη μέν,
θανατωδεστέρη δὲ οὔ. ἡ δὲ αἱματώδης ἰσχυρὴ[61] καὶ
ἐπίπονος καὶ θανατώδης.

Ὅταν οὖν προσῇ καὶ λύγξ ἅμα καὶ αἵματος θρόμβους ἀποβήσσῃ ἅμα τῷ σιάλῳ μέλανας, οὗτος ἀποθνήσκει ἑβδομαῖος· δέκα δ' ἡμέρας διαφυγὼν τὴν μὲν
πλευρῖτιν ὑγιὴς γίνεται, εἰκοστῇ δ' ἐμπυΐσκεται, καὶ
ἀποβήσσει πύον, τελευτῶν δὲ καὶ ἀπεμέει, καὶ οὐ
πάνυ εὐθεράπευτος γίνεται.

Εἰσὶ δὲ καὶ ξηραὶ πλευρίτιδες ἄπτυστοι, χαλεπαὶ
<δ'>[62] αὗται· αἱ δὲ κρίσιες ὅμοιαι τῇσιν ἄλλῃσιν·
ὑγρασίης <δὲ>[63] πλέονος δέονται τῶν ἄλλων ἐν τῷ
ποτῷ.

144   Αἱ δὲ χολώδεες καὶ αἱματώδεες κρίνουσιν ἐνα|ταῖαι
ἢ δεκαταῖαι,[64] καὶ οὗτοι ὑγιέες μᾶλλον γίνονται. ἢν δὲ

[60] M adds ἔστι δὲ καὶ αἱματώδης.

34

vinegar and water, sieve, and give warm to the fasting patient. When he begins to cough up material that is clean,
have him drink a cheramys of cuckoo-pint, sesame, and
shelled almonds in sweetened vinegar mixed with water; if
you want to promote expectoration even more, mix root
bark of the caper-plant in with these.

16. When pleurisy arises, a person suffers the following: he has pain in his side, fever and shivering, he respires
rapidly, and he has orthopnoea. He coughs up somewhat
bilious material the colour of pomegranate-peel, unless
he has tears; if he has tears, then he coughs up blood,
too, from the tears; in sanguinous pleurisy, the sputum is
diffused with blood. The bilious variety of pleurisy is rela-
tively mild, unless the patient has tears; if he has tears, it is
more painful but not more mortal. The sanguinous variety
of pleurisy is severe, painful and mortal.

Now when in addition hiccups are present, and the
patient coughs up dark clots of blood in his sputum, he
succumbs on the seventh day. If he survives for ten days,
he recovers from the pleurisy, but on the twelfth day suppurates internally, coughs up pus, and finally vomits as
well; this patient is not especially easy to treat.

There are also dry pleurisies without expectoration,
and these are severe. The crises are the same as in the
other varieties, but these patients require more moisture
in their drinks than do the others.

The bilious and sanguinous varieties of pleurisy have
their crises on the ninth or tenth day, and such patients

---

61 ἰσχυρὴ om. Θ.
62 Added by I.   63 Foes (n. 36).
64 ἢ δεκ. Θ: καὶ ἑνδεκαταῖαι M.

κατ' ἀρχὰς μὲν μαλθακαὶ ἔχωσιν ὀδύναι, ἀπὸ τῆς
πέμπτης δὲ καὶ ἕκτης ὀξέαι, αὗται τελευτῶσι μέχρι
δωδεκάτης, καὶ οὐ μάλα ἀποθνήσκουσι·⁶⁵ κίνδυνος δὲ
μάλιστα μὲν μέχρι ἑβδόμης, ἀτὰρ καὶ ἐς τὴν δωδε-
κάτην· μετὰ δὲ ταύτας ὑγιαίνονται. αἱ δ' ἐξ ἀρχῆς μὲν
μαλθακαί, ἀπὸ δὲ τῆς ἑβδόμης καὶ ὀγδόης ὀξέαι, πρὸς
τὰς τεσσερεσκαίδεκα κρίνουσί τε καὶ ὑγιαίνονται.

Ἡ δ' ἐς τὸν νῶτον πλευρῖτις τοσόνδε διαφέρει τῶν
ἄλλων· τὸν νῶτον ὀδυνᾶται ὡς ἐκ πληγῆς καὶ στένει
καὶ ἀναπνεῖ ἀθρόον· εὐθὺς δὲ πτύει ὀλίγα, καὶ κοπιᾷ
τὸ σῶμα· τρίτῃ δ' ἢ τετάρτῃ οὐρέει ἰχῶρα ὕφαιμον·
ἀποθνήσκει δὲ μάλιστα πεμπταῖος· εἰ δὲ μή, ἑβδο-
μαῖος· ταύτας δὲ διαφυγὼν ζώει, καὶ ἡ νοῦσος ἠπίη
καὶ ἧσσον θανατώδης· φυλάσσειν δὲ μέχρι τεσσερεσ-
καίδεκα· μετὰ δὲ ταύτας ὑγιής.

Ἐνίαις⁶⁶ δὲ τῶν πλευριτίδων τὸ μὲν σίαλον καθαρόν, ἡ δ' οὔρησις αἱματώδης, οἷον ἀπὸ κρεῶν ὀπτῶν
ἰχῶρες, ὀδύναι τε ὀξέαι διὰ τῆς ῥάχιος ἐς τὸ στῆθος
καὶ ἐς τὸν βουβῶνα τείνουσιν· οὗτος τὴν ἑβδόμην
διαφυγὼν ὑγιής.

Ὅταν δὲ τούτων τῶν πλευριτίδων τινὶ προσγένηται
τὸν νῶτον ἐρυθριᾶν καὶ τοὺς ὤμους θερμαίνεσθαι καὶ
ἀνακαθίζοντα βαρύνεσθαι καὶ ἡ γαστὴρ ἐκταράσ-
σηται χλωρῷ καὶ δυσώδει σφόδρα, οὗτος διὰ τὴν
ὑποχώρησιν εἰκοστῇ καὶ μιῇ ἀποθνήσκει· ταύτας δὲ
διαφυγὼν ὑγιής.

⁶⁵ Θ: ἐκφεύγουσι Μ.

usually recover. If their pains are mild at the beginning, but sharp from the fifth or sixth day on, the disease ends by the twelfth day and is not very mortal; danger is greatest up to the seventh day, although some danger persists until the twelfth day; after that patients recover. If the pains are mild at the beginning, but sharp from the seventh or eighth day on, patients have their crises and recover about the fourteenth day.

Pleurisy in the back differs from the other varieties in the following: the patient suffers pain in his back as if from a blow, he groans, and he respires rapidly; he immediately coughs up small amounts of sputum, and his body is weary. On the third or fourth day, he passes serous urine charged with blood; he usually dies on the fifth day; if not on the fifth day, then on the seventh; if he escapes for that many, he lives, and the disease is mild and less mortal. Protect him until the fourteenth day; after that he has recovered.

In some pleurisies the sputum is clean, but the urine bloody, resembling the fluid that runs out of roasted meat; sharp pains extend along the spine to the chest and groin. If this patient survives until the seventh day, he recovers.

When, in one of these pleurisies, in addition the back becomes red, the shoulders are warm, the patient feels a heaviness on sitting up, and his belly is set in violent motion by yellow-green foul-smelling stools, he dies on the twenty-first day as a result of the evacuation. If he survives after that, he recovers.

---

66 Potter: ἐνιαι Θ: ἐνίη M.

Οἶσι δ' αἱ πτύσιες εὐθὺς παντοδαπαί εἰσι καὶ τὰ ἀλγήματα πάνυ ὀξέα, οὗτοι τριταῖοι ἀποθνήσκουσι·[67] ταύτας δὲ διαφυγόντες ὑγιέες· μὴ γενόμενος δὲ ὑγιὴς τῇ ἑβδόμῃ ἢ ἐνάτῃ ἢ δεκάτῃ ἄρχεται ἐμπυΐσκεσθαι· κρέσσον δ' ἐμ|πυῆσαι· ἧσσον γὰρ θανατῶδες, ἐπίπονον δέ.

146

Πρὸς δὲ τοῖσι σημείοισι τοῖσιν εἰρημένοισιν ἐν ἑκάστῃ τῶν πλευριτίδων καὶ τάδε χρὴ σκοπεῖσθαι τὴν γλῶσσαν·[68] εἰ μὲν ἐν ἀρχῇ γίνοιτο τρηχείη, χαλεπωτέρη ἡ ἀπάλλαξις τῆς νούσου, καὶ ἀνάγκη αἷμα ἀποβῆξαι ἐν τῇσιν ἡμέρῃσιν, ἐν ᾗσι δεῖ· εἰ δὲ προκεχωρηκυίης τῆς νούσου γίνοιτο, αἱ μὲν κρίσιες ἐς τὴν τετάρτην καὶ δεκάτην ἡμέρην, ἀνάγκη δὲ πτύσαι αἷμα.

Ἔχει δὲ ὧδε περὶ τῆς ἀπαλλάξιος· εἰ μὲν τριταίῳ ἄρχοιτο πεπαίνεσθαι καὶ πτύεσθαι, θάσσους αἱ ἀπαλλάξιες· εἰ δ' ὕστερον πεπαίνοιτο, ὕστερον καὶ αἱ κρίσιες γίνονται, ὡς ἐν τοῖσι τῆς κεφαλῆς σημείοισι. τὰ δ' ἀλγήματα τὰ ἐν ἁπάσῃσι τῇσι πλευρίτισιν ὡς ἐπὶ τὸ πολὺ κουφίζει μεθ' ἡμέρην μᾶλλον ἢ νύκτωρ.

Θεραπεύειν δὲ χρὴ τὰς πλευρίτιδας ὧδε· τὰ μὲν πολλὰ ὡς τὴν φρενῖτιν[69] καὶ τὴν περιπλευμονίην, πλὴν λουτροῖσί τε χρῆσθαι θερμοῖσι καὶ οἴνοισι γλυκέσιν. ἢν μὲν οὖν τῇ πρώτῃ ἢ τῇ ἐπιούσῃ λάβῃς τῆς λήψιος, ἢν μὲν ὑπεληλύθῃ ἡ κόπρος καθαρὴ ἢ ἀτρέμα χολώδης καὶ ὀλίγη, ὑποκλύσαι θαψίῃ· ἢν δὲ

---

[67] Potter: φεύγουσι Θ θνήσκουσι Μ.

Patients whose expectorations are manifold from the start, and whose pains are very sharp, die on the third day; if they survive for that many, they recover. If one does not recover, he begins to suppurate internally on the seventh, ninth or tenth day; in fact, it is better to suppurate, for this is less mortal, even though painful.

Besides the signs already mentioned in each of the pleurisies, you must also observe the following signs of the tongue. If the tongue becomes rough at the beginning, recovery from the disease is difficult, and it is imperative for the patient to cough up blood on the days when he should. If this sign appears when the disease is already advanced, the crises will be toward the fourteenth day, and the patient inevitably expectorates blood.

The manner of recovery is as follows: if on the third day the sputum begins to reach maturity and to be coughed up, recovery is faster; but if the sputum matures later, then the crises too occur later, just as with the signs in the head. The pains in all pleurisies are generally lighter during the day than at night.

You must treat pleurisies as follows: for the most part, just as phrenitis and pneumonia, except that you must also administer warm baths and sweet wines. Now when you have taken the case on the first day of the disease's onset, or on the day after that, if the stools pass clean or slightly bilious, and scanty, administer an enema of thapsia; if the

---

68 M adds πομφόλυγος γὰρ ὑποπελίδνου γινομένης ἐπὶ τῆς γλώσσης, οἷα σιδηρίου βαφέντος εἰς ἔλαιον. Cf. *Coan Prenotions* 378.

69 M: πλευρῖτιν Θ.

κινηθεῖσα ἢ λυθεῖσα⁷⁰ ἡ κοιλίη τὴν μὲν νύκτα χαλά-
σῃ, τῇ δ' ὑστεραίῃ ὀδύνη καὶ στρόφος ἔχῃ, πάλιν
ὑποκλύσαι.

Ἢν δ' ὁ κάμνων χολώδης ᾖ τῇ φύσει καὶ ληφθῇ τῇ
νούσῳ ἀκάθαρτος ἐών, πρὶν ἀναπτύεσθαι τὸ σίαλον
χολῶδες, καὶ τῷ φαρμάκῳ καθῆραι τὴν χολὴν εὖ·
πτύοντι δὲ ἤδη χολώδεα, μὴ δίδου τὸ φάρμακον· ἢν
γὰρ δῷς, τὸ πτύσμα οὐ δυνήσεται ἄνω ἀνιέναι, ἀλλ'
ἑβδομαῖος ἢ ἐναταῖος ἀποπνιγήσεται. ἢν δὲ πρὸς τῇ
ἐν τῇσι πλευρῇσιν ὀδύνῃ καὶ τὰ ὑποχόνδρια ἀλγέῃ,
148   ὑπο|κλύσαι τε καὶ πιεῖν νήστι δοῦναι ἀριστολοχίαν
καὶ ὕσσωπον καὶ κύμινον καὶ σίλφιον καὶ μήκωνα
λευκὴν καὶ ἄνθος χαλκοῦ καὶ μέλι καὶ ὄξος καὶ ὕδωρ.

Πρὸς μὲν τὰ φάρμακα οὕτω δεῖ ποιέεσθαι τὰς
θεραπείας τὰς πρώτας· τὰ δ' ἄλλα ὧδ' ἔχει· λούειν
πολλῷ θερμῷ πρὸς δύναμιν τὴν τοῦ κάμνοντος πλὴν
κεφαλῆς, καὶ ὅταν αἱ κρίσιες ὦσι, τὰ ὀδυνώμενα
χλιαίνειν ὑγρῇσι πυρίῃσιν ὑπαλείφων ἐλαίῳ. ὅταν δὲ
καταιγίζωσιν αἱ νοῦσοι, ἡσυχάζειν καὶ τὸν κάμνοντα
καὶ τὸν ἰητρὸν τῇσι θεραπείῃσιν, ὅπως μὴ ἐξεργάση-
ταί τι κακόν· πτισάνης δὲ χυλὸν κάθεφθον διδόναι
ὀλίγῳ παχύτερον μελιχρὸν ποιέων. μετὰ δὲ τὰ λουτρὰ
καὶ οἶνον γλυκὺν ὑδαρέα προπίνειν, μὴ ψυχρόν, ὀλίγον ἐκ βομβυλίου ⟨οὐκ⟩⁷¹ εὐρυστόμου. καὶ ὅταν βῆχες ἐπίωσιν, ἐπιπίνειν καὶ χρέμπτεσθαι ὡς μάλιστα,
καὶ τῷ ποτῷ ὑγραίνειν ἵνα ὁ πλεύμων ὑγρότερος ἐὼν
ῥᾷον καὶ θᾶσσον ἀποδιδῷ τὸ πτύσμα καὶ ἡ βὴξ
ἦσσον πονέῃ· καὶ ῥοιῆς δὲ γλυκείης ἢ οἰνώδεος χυλὸν

cavity, on being set in motion, evacuates during the night, but on the following day pain and colic are present, administer another enema.

If the person is bilious by nature, and has been taken by the disease when in an unclean state, before he expectorates bilious sputum clean out bile thoroughly with a medication; but to a patient already expectorating bilious material, do not give a medication, because, if you do, he will be unable to discharge his sputum upwards, and will choke to death on the seventh or ninth day. If, besides the pain in the side, the hypochondrium too is in pain, administer an enema, and give the patient in the fasting state aristolochia, hyssop, cummin, silphium, white poppy, flower of copper, honey, vinegar, and water to drink.

This is the first treatment you must apply, as far as medications are concerned; the other measures are the following: wash with plentiful water as hot as the patient can stand, except for his head; when his crises are occurring, warm the painful areas with moist vapour-baths and anoint them with olive oil. When diseases are pressing, let the patient rest and the physician suspend treatment lest he do any harm; give only boiled barley-water slightly thickened and sweetened with honey. After bathing, let the patient first drink sweet wine mixed with water, not too cold, a small amount from a narrow-necked bottle. When coughing is present, have him drink more of this, expectorate as much as possible, and moisten himself by drink, in order that his lung, becoming moister, will discharge its sputum more easily and quickly, and the cough will be less painful; give juice of the sweet or vinous pomegranate mixed with a

---

70 ἢ λυθεῖσα om. M.   71 Foes (n. 42).

γάλακτι αἰγείῳ ὀλίγῳ καὶ μέλιτι μιγνύς, κατὰ σμικρὸν πολλάκις δίδου νύκτωρ τε καὶ μεθ' ἡμέρην· καὶ
ὕπνον δ' ὡς μάλιστα διακωλύειν, ἵνα κάθαρσις γίνηται θάσσων τε καὶ πλείων.

Τὴν δὲ αἱματώδεα πλευρῖτιν θεραπεύειν οὕτω· μετὰ
τὰς κρίσιας ἀνακομίζειν σιτίοισι κούφοισι καὶ ἡσυχάζειν καὶ φυλάσσεσθαι περισσῶς ἀνέμους, ἡλίους,
πλησμονάς, ὀξέα, ἁλυκά, λιπαρά, καπνόν, φύσας τὰς
ἐν τῇ κοιλίῃ, πόνους, λαγνείας· ἢν γὰρ ὑποτροπιάσῃ,
ἀποθνῄσκει.

Ἐν δὲ τῇσι πτύσεσιν, ἢν ὀδύνη τ' ἔχῃ καὶ μὴ
δύναται ἀποπτύειν, νήστι δίδου ἄνθος χαλκοῦ ὅσον
150 κοτινάδα | καὶ ὁποῦ σιλφίου ἥμισυ καὶ τριφύλλου
καρπὸν ὀλίγον ἐν μέλιτι λείχειν· ἢ πεπέριος κόκκους
πέντε καὶ ὁποῦ σιλφίου ὅσον κύαμον καὶ μέλι καὶ
ὄξος καὶ ὕδωρ πίνειν χλιαρὸν νήστι δίδου· τοῦτο καὶ
τὰς ὀδύνας παύει.

Ἢν δὲ μὴ δύναται πτύειν κατὰ λόγον, ἀλλ' ἐνίσχη-
ται αὐτῷ καὶ ῥέγκῃ ἐν τοῖσι στήθεσιν, ἄρου τοῦ
μεγάλου ῥίζης χηραμύδα καὶ ἔλαιον ἐν μέλιτι λεῖξαι,
ἐπιρρυφεῖν δὲ ὄξος κεκρημένον. ἄλλο ἰσχυρόν· ἄνθος
χαλκοῦ ὅσον κύαμον καὶ λίτρον ὀπτὸν διπλάσιον καὶ
ὕσσωπον, ὅσον τοῖσι τρισὶ δακτύλοισι λαβεῖν, μέλιτι
μίξας, καὶ ὕδωρ καὶ ἔλαιον σμικρὸν ἐπιστάξας, χλιήνας ἐν χηραμύδι, ἐγχεῖν ἵνα μὴ ἀποπνίγῃ. καὶ ἐν
περιπλευμονίῃ, ἢν μὴ καθαίρηται, τοῦτο ἐγχεῖν.

Ἢν δὲ μήτε ῥέγκῃ μήτε πτύῃ ὡς δεῖ, καππάριος
καρποῦ ὅσον τοῖσι τρισὶ δακτύλοισι λαβεῖν, καὶ πέ-

little goat's milk and honey, administering it often in small amounts both at night and during the day. Prevent sleep as much as possible, in order that cleaning will be more rapid and complete.

Treat sanguinous pleurisy thus: after the crises, restore the patient with light foods, make him rest, and protect him strictly from wind, sun, repletion, foods that are acid, salty or rich, smoke, wind in the cavity, exertions, and venery; for if he has a relapse, he dies.

If pain is present during coughing and makes expectoration impossible, give the patient in the fasting state flower of copper in the amount of a wild olive, half as much silphium juice, and a little clover seed in honey to eat; or give him in the fasting state five corns of pepper, silphium juice to the amount of a bean, honey, vinegar and water to drink warm; this also stills pains.

If the patient is unable to expectorate as he should, but his sputum is caught fast and stertorous breathing is heard in his chest, have him take a cheramys of cuckoo-pint root and olive oil in honey, and afterwards drink a mixed vinegar potion. Another strong medication: mix flower of copper to the amount of a bean, twice as much burnt soda, and a pinch of hyssop in honey, sprinkle water and a little olive oil over it, warm in a mussel-shell, and infuse in order to prevent the patient from choking. Make this infusion in pneumonia, too, if the patient is not being cleaned.

If the patient does not breathe stertorously, but also does not expectorate as he should, mix a good pinch of

περι καὶ λίτρον ὀλίγον[72] καὶ μέλι καὶ ὄξος καὶ ὕδωρ
μίξας, τοῦτο χλιαρὸν ἐπιρρυφεῖν· τὴν δὲ ἄλλην ἡμέρην ὕσσωπον[73] ἐν ὄξει καὶ μέλιτι καὶ ὕδατι ἀναζέσας
ἐπιρρυφεῖν. τοῦτο καὶ τοῖσι ῥέγκουσι διδόναι καὶ μὴ
δυναμένοισι καθαίρεσθαι. εἰ δ' ἰσχυρότερον βούλοιο
ποιέειν, ὑσσώπου καὶ νάπυος καὶ καρδάμου [κόγ-
χην][74] χηραμύδα τρίψας ἐν μέλιτι καὶ ὕδατι ἀναζέσας
καὶ διηθήσας ἐπιρρυφεῖν χλιαρὸν δίδου.

Οὕτω ταῦτα τὰ νοσήματα θεραπευθέντα ὑγιᾶ γίνεται, ἢν μή τι τοῦ πτύσματος ὑπολειφθὲν ἐν τῷ πλεύμονι πύον γίνηται, ὑφ' οὗ βήσσουσι ξηρὰ βήχια, καὶ
πῦρ καὶ φρίκη ἴσχει καὶ ὀρθοπνοίη, καὶ πυκινὸν
ἀναπνεῖ καὶ ἀθρόον, καὶ ἡ φωνὴ βαρυτέρη ὀλίγῳ καὶ
εὐχροίη σὺν τῇ θέρμῃ τὸ πρόσωπον ἴσχει· προϊόντος
δὲ τοῦ χρόνου μᾶλλον καὶ ἡ νοῦσος σαφὴς δηλοῦται.
152 τοῦτον εἰ ἐντὸς τῶν δέκα ἡμερέων λάβοις, | θερμή-
ναντα χρὴ διαίτῃ καὶ θερμῷ λουτρῷ ἐγχέαι ἐς τὸν
πλεύμονα, ὅ τι πύον ἄξει, καὶ τοῖσιν ἄλλοισι χρῆσθαι
τοῖσι τὸ πύον ἄγουσι, καὶ διαιτᾶν ὡς ἔμπυον, καὶ τὴν
κεφαλὴν ἀποξηραίνειν, ἵνα μὴ ἐπιρρέῃ.

Ἢν δὲ τῷ ἐγχύτῳ μὴ σήπηται καὶ[75] ἀνάγηται τὸ
πύον, ῥήγνυται αὐτῷ ἐκ τοῦ πλεύμονος ἐς τὸν θώρηκα,
καὶ μετὰ τὴν ῥῆξιν δοκέει ὑγιὴς εἶναι, ὅτι ἐκ τῆς
στενοχωρίης ἐς τὴν εὐρυχωρίην ἦλθε τὸ πύον [καὶ τὸ
πνεῦμα, ὃ ἀναπνέομεν, ἕδρην ἔσχεν ἐν τῷ πλεύμονι]·[76]

72 ὀλίγον om. Θ.
73 ὕσσωπον om. M.

44

capers, pepper, and a little soda into honey, vinegar and water; administer warm; the next day have him take hyssop boiled up in vinegar, honey and water. This can also be given to patients with stertorous breathing that are unable to clean themselves. If you want to make something stronger, knead a cheramys of hyssop, mustard, and cress into honey, boil these up in water, sieve, and give warm to drink.

These diseases, if treated in such a way, are cured, unless some of the material that should be coughed up is left behind in the lung and becomes pus; from this patients develop a dry cough, fever, shivering and orthopnoea; the patient breathes frequently and rapidly, his voice is slightly deeper than before, and his face takes on a good colour from the heat. As time passes, the disease reveals itself more clearly. If you take on this patient within ten days from the start, you must warm him by means of a regimen and a hot bath, and infuse into his lung substances that will draw pus; employ the other measures that move pus, prescribe the same regimen as for internal suppuration, and dry his head thoroughly to prevent a flux to the chest.

If, with the infusion, maturation and expulsion of the pus do not occur, the pus breaks out of the patient's lung into his thorax, and after the break he seems to have recovered, since the pus has moved from a confined space into an open one [and the air that we inspire has its seat in the

---

74 Del. Ermerins.

75 σήπηται καὶ om. M.

76 Del. Potter.

προϊόντος δὲ τοῦ χρόνου τὰ στήθεα πύου πληροῦται,
καὶ αἱ βῆχες καὶ οἱ πυρετοὶ καὶ τἆλλα ἀλγήματα
πάντα[77] μᾶλλον πιέζει, καὶ ἡ νοῦσος διαδηλοῦται.
τοῦτον μετὰ τὴν ἔκρηξιν ἐᾶσαι δεῖ δεκαπέντε ἡμέρας,
ὅπως πάλιν πεπανθῇ τὸ πύον· ἅτε γὰρ ἐς εὐρυχωρίην
ἐλθὸν[78] ἀνέψυξέ τε καὶ τὸ ὑπάρχον ὑγρὸν ἐν τῷ θώρηκι
προσηγάγετο πρὸς ἑωυτὸ ὥστε αὐτὸ ἡμισαπὲς εἶναι.
ἢν μὲν οὖν αὐτόματον ἄρξηται πτύεσθαι ἐν τούτῳ τῷ
χρόνῳ, ἢ[79] φαρμάκοισι τιμωρέειν[80] ἢ ποτοῖσιν, ἐν
<δὲ>[81] τῇσι τελευταίῃσιν ἡμέρῃσι τῶν ἡμερέων τῶν
πεντεκαίδεκα σπεύδειν ἀναστῆναι πρὶν μᾶλλον τρύχε-
σθαι τὸ σῶμα, φυλάσσων καθαρὴν τὴν κεφαλὴν τῶν
ἐπιρροῶν εἵνεκεν.

Ἢν δὲ μὴ πτύηται, ἀποσημήνῃ δ' ἐς τὰς πλευράς,
τάμνειν ἢ καῦσαι. ἢν δὲ μήτε πτύηται μήτ' ἀποσημήνῃ ἐς τὰς πλευράς, λοῦσαι πολλῷ καὶ θερμῷ καὶ
νῆστιν καὶ ἄποτον καθίσας ἐπὶ ἕδρης ἀκινήτου ἕτερος
μὲν τῶν ὤμων ἀναλαβέτω, αὐτὸς δὲ σεῖε, τὸ οὖς
παραβαλὼν πρὸς τὰς πλευράς, ἵνα εἰδῇς ὁποτέρωθεν
154 ἀποσημαίνει· βούλου δὲ μᾶλλον πρὸς τὰς | ἀριστε-
ράς· θανατωδέστερον γὰρ καίειν καὶ τάμνειν ἐς τὰς
δεξιάς· ὅσῳ γὰρ ἰσχυρότερά ἐστι τὰ δεξιά, τόσῳ καὶ
τὰ νοσήματα αὐτοῖσιν ἰσχυρότερα γίνεται.

Ἢν δὲ ὑπὸ πάχεος τὸ πύον μὴ κλυδάζηται μηδὲ
ψοφέῃ ἐν τῷ στήθει, πυκινὸν δὲ ἕλκῃ τὸ πνεῦμα καὶ οἱ

[77] πάντα om. Θ.   [78] M adds τὸ πύον.
[79] Θ: ἢν μὴ M.

lung].[6] As time passes, however, the chest fills up with pus and, as coughing, fevers, and all the other evils press the patient more and more, the disease is revealed. You must leave this patient without treatment for fifteen days after the break, in order that the pus can mature anew; for inasmuch as it has moved into an open space, it cools and draws to itself any moisture that happens to be present in the thorax, so that it itself becomes only semi-mature. Now if the patient begins to expectorate spontaneously in this period, assist him with medications or potions, and in the final days of the fifteen urge him to get up before his body becomes any more wasted, and make sure that his head is clean in order to prevent any fluxes to the chest.

If he does not expectorate, but signs point to his side, incise or cauterize. If the patient does not expectorate, but there are also no signs pointing to his side, wash him in abundant warm water and, before he has taken food or drink, set him on a steady chair; let someone else hold him by the shoulders, and you shake him, applying your ear to his sides in such a way as to learn on which side the sign arises. Hope that it is on the left side, for it is more dangerous to cauterize or incise on the right; for in the same proportion that the right parts of the body are stronger, so too are the diseases in them stronger.

If the pus, because of its thickness, does not fluctuate or make any sound in the chest, but the patient draws his

---

[6] This statement seems out of context; it may well be an intruded marginal annotation.

---

[80] τιμωρέειν om. Θ.
[81] Vander Linden.

πόδες ἐποιδέωσι καὶ βηχίον τι προσῇ, μὴ ἐξαπατάτω
ἀλλ᾿ εὖ ἴσθι πλήρη ἐόντα τὸν θώρηκα πύου· ἐς οὖν
Ἐρετριάδα γῆν ὑγρὴν καὶ λ<ε>ίην⁸² τετριμμένην καὶ
χλιαρὴν ἐμβάψας ὀθόνιον λεπτόν, περικάλυψον κύ-
κλῳ τὸν θώρηκα, καὶ ὅπου ἂν πρῶτον ξηραίνηται,
ταύτῃ χρὴ καίειν ἢ τάμνειν ὡς ἐγγυτάτω τῶν φρενῶν,
φυλασσόμενος αὐτῶν τῶν φρενῶν. ἢν δὲ βούλῃ, ἀλείφων τῇ Ἐρετριάδι σκόπει ὁμοίως ὡς ἐν τῷ ὀθονίῳ·
πολλοὶ δὲ ἅμα ἀλειφόντων, ἵνα μὴ τὰ πρῶτα ἀλειφό-
μενα ἀποξηραίνηται.

Μετὰ δὲ τὴν τομὴν ἢ τὴν καῦσιν τῷ μοτῷ τῷ ἐκ τοῦ
ὠμολίνου χρῶ, καὶ ἐξίει κατ᾿ ὀλίγον τὸ πύον, ὅταν δὲ
μέλλῃς καίειν ἢ τάμνειν, ὑποσημαίνου τὸ αὐτὸ σχῆμα
ἔχοντα, ὅπερ ἂν μέλλῃς ἔχοντα τάμνειν ἢ καίειν, ἵνα
μὴ ἐξαπατήσῃ ἀνωτέρω γενόμενον ἢ κατωτέρω τὸ
δέρμα ἐν τῇ μεταβολῇ τοῦ σχήματος· καὶ τὰς βῆχας
φυλάσσειν ἐκ τῆς διαίτης, ὅπως μὴ ἀντισπάσωσι
πάλιν ἐς τὸν πλεύμονα⁸³ τὸ πύον, κακὸν γάρ· ἀλλ᾿
ἐᾶν⁸⁴ μετὰ⁸⁵ τὴν τομὴν ὡς τάχιστα ὑποξηραίνεσθαι.
ἐπειδὰν δὲ δωδεκαταῖος ᾖ κεκαυμένος,⁸⁶ ἅπαν ἀφιέναι
τὸ λοιπὸν πύον, καὶ ἀπὸ τοῦ ὀθονίου μοτοῦν, καὶ
ἀφιέναι δὶς τῆς ἡμέρης τὸ πύον, καὶ τὴν ἄνω κοιλίην
ἐκ τῆς διαίτης ὡς μάλιστα ξηραίνειν.

Οὕτω χρὴ καὶ τὰς ἐκ τῶν τρωμάτων καὶ ἐκ περιπλευμονίης καὶ ἐκ καταρρόων μεγάλων ἐκπυήσιας καὶ

⁸² Littré: λίην ΘΜ.
⁸³ ἐς τὸν πλεύμονα om. Θ.

48

breath rapidly, his feet swell up, and a mild cough is present, do not be deceived, but know well that his chest is full of pus. Soak a piece of fine linen in warm moist finely triturated Eretrian earth; then wrap this all the way around his thorax, and, wherever it first dries, that is where you must cauterize or incise, as close to the diaphragm as possible, but sparing the diaphragm itself. If you prefer, apply the Eretrian earth directly and look for the place the same way you would in the linen; let many people apply the earth simultaneously, in order that the first applied does not become dry.

After the incision or cautery, use a tent of raw linen, and discharge the pus a little at a time. Whenever you are about to cauterize or incise, observe that the patient has the same position he had before, when you first incised or cauterized, in order that you will not be deceived by his skin moving higher or lower as he changes his position. Employ a regimen that minimizes coughing, in order to prevent the patient from drawing pus back into his lung—for that would be bad; after the incision let the patient be dried out as quickly as possible. When twelve days have elapsed after the cautery, remove all the remaining pus and plug the wound with linen; remove pus twice daily, and dry out the upper cavity as thoroughly as possible by means of regimen.

This is also the way you must examine and treat suppurations arising from wounds, pneumonia, and massive

---

84 ἐὰν om. Θ.
85 Ermerins: κατὰ ΘΜ.
86 κεκαυμένος om. Μ.

156 | προσπεσόντος τοῦ πλεύμονος τῇσι πλευρῇσι σκο-
πεῖν καὶ θεραπεύειν.

17. Ψυκτήρια δὲ τάδε δίδου ἐπὶ τοῖσι καύσοισι
πίνειν, ὅταν βούλῃ· πολλὰ δὲ ἀπεργάζεται· τὰ μὲν
γὰρ οὔρησιν ποιέει, τὰ δὲ διαχώρησιν, τὰ δὲ ἄμφω, τὰ
δ' οὐδέτερα, ἀλλὰ ψύχει μοῦνον ὡς ἄγγος ὕδατος ζέον
ἤν τις ἐπιχέῃ ψυχρὸν ὕδωρ, ἢ ψυχρῷ αὐτὸ τὸ ἄγγος
πνεύματι προσαγάγῃ. δίδου δὲ ἄλλα ἄλλοισιν· οὔτε
γὰρ τὰ γλυκέα ἅπασι συμφέρει οὔτε τὰ στρυφνά, οὔτε
ταὐτὰ πίνειν δύνανται.

Τοῦτο μέν· κηρίων ξηρῶν ὅσον δύο κοτύλας βρέχων ὕδατι καὶ ἀνατρίβων γενέσθω, ἕως ἂν ὑπόγλυκυ[87]
γένηται· εἶτα διηθήσας, σέλινα ἐμβαλὼν δίδου πίνειν.

Τοῦτο δέ· λίνου καρποῦ ὀξύβαφον, ὕδατος κοτύλας
δέκα ἐπιχέας, ἕψειν ἐν καινῇ χύτρῃ ἐπ' ἀνθράκων
ἄζεστον, ἵνα ἀναπνέῃ, ἕως ἂν ὁ χυλὸς ἁπτομένῳ
λιπαρὸς γένηται.

Τοῦτο δέ· μελικρήτου ὑδαρέος καθεψήσας τὸ ἥμισυ
λείπειν·[88] ἔπειτα σέλινα ἐμβαλὼν τοῦτο ψύχων κατ'
ὀλίγον δίδου.

Τοῦτο δέ· κριθέων Ἀχιλληΐδων κοτύλην αὐήνας,
ἄρας τὸν ἀθέρα καὶ πλύνας εὖ, ἐπιχέας χοέα ὕδατος,
ἕψε καὶ τὸ ἥμισυ λιπὼν ψύχων δίδου πίνειν.

[87] To judge from the arrangement of words in his glossary, in
Erotian's time the word ἀρτίως stood somewhere in the text of
Diseases III between Littré VII. 122,13 and 156,15 (E. Nachman-
son, *Erotianstudien*, Uppsala, 1917, 404). Two parallel passages

defluxions, and when a lung falls against the side.[7]

17. Give the following cooling agents to drink in ardent fevers whenever you wish; they have many effects: some are diuretic, others laxative, others both, and others neither, merely cooling as if someone were to pour cold water over a vessel of boiling water, or were to move the vessel itself into the cold air. Give different ones to different patients, for the sweet ones do not benefit everyone, nor do the astringent ones, nor are all patients able to drink the same things.

A. Soak about two cotylai of dried honeycomb in water, and stir until the water becomes sweetish to the taste; then sieve, add celery, and give to drink.

B. Pour ten cotylai of water over an oxybaphon of linseed, and simmer in a new pot over a charcoal fire without boiling, in order that it exhales vapour, until the liquid becomes greasy to the touch.

C. Boil dilute melicrat until half is left; then add celery, cool, and give a little at a time.

D. Dry a cotyle of Achilles barley, remove the chaff, wash well, add a chous of water, and boil until half remains; cool and give to drink.

---

[7] See *Diseases II* 59 for a description of this condition.

---

suggest that its original position may have been before ὑπόγλυκυ: *Diseases III* 17 ἕως ἂν ἀτρέμα γλυκανθῇ and *Diseases II* 45 ἄρτι ὑπόγλυκυ ποιέων.

[88] λείπειν om. Θ.

158   Τοῦτο | δέ· Αἰθιοπικοῦ κυμίνου κοτύλης δέκατον μέρος, ἐπιχέας τρία ἡμιχόεα, ἕψε πηλῷ τρηχώδει[89] καταλείψας ἄζεστον, ἕως μέρος τρίτον λίπης, καὶ ψύχων δίδου τοῦτο πρὸς πάντα καῦσον καὶ πυρετόν.

Τοῦτο δέ· ὕδωρ ὄμβριον αὐτὸ καθ᾽ αὑτό.

Τοῦτο δέ· πτισάνης κοτύλῃ χοέα ὕδατος ἐπιχέας, λείπειν τὸ ἥμισυ ἑψῶν· εἶτα διηθήσας,[90] σέλινα ἐμβαλὼν ψυχρὸν δίδου.

Τοῦτο δέ· οἱ σταφίδιοι λευκοὶ οἶνοι ὑδαρέες.

Τοῦτο δέ· τρύγες στεμφυλίτιδες σταφιδευταῖαι ὑδαρέες.

Τοῦτο δέ· ἀσταφίδος λευκῆς[91] ἄνευ γιγάρτων κοτύλην[92] καὶ πενταφύλλου ῥιζέων χεῖρα πλήην φλάσας, εἴκοσι κοτύλας ὕδατος ἐπιχέας, ἀφεψήσας τὸ ἥμισυ[93] ψυχρὸν δίδου κατ᾽ ὀλίγον.

Τοῦτο δέ· κρίμων κριθέων ἀδρῶν[94] ἡμιχοίνικον ὕδατος χοέα ἐπιχέας· ὅταν ἤδη οἰδέῃ τὰ κρίμνα τρίβειν τῇσι χερσίν, ἕως ἂν λευκὸν τὸ ὕδωρ γένηται, καὶ ἀδιάντου δραχμίδα ἐμβαλὼν ἀπαιθριάσας δίδου.

Τοῦτο δέ· ᾠῶν τὸ λευκὸν τριῶν ἢ τεσσέρων κατακυκῶν ἐν ὕδατος χοῒ πινέτω· τοῦτο ψύχει σφόδρα καὶ τὴν κοιλίην ὑπάγει· ἢν δὲ δοκέῃ μᾶλλον[95] ὑπάγειν, τὸν νεοσσὸν προσκατακύκα.

Τοῦτο δέ· καχρύων ἡμιχοίνικον εὖ ἀποπλύνας, ἐν ὕδατος χοῒ ζέσας δὶς ἢ τρὶς ψυχρὸν δίδου.

Τοῦτο δέ· πτισάνης χυλὸν κάθεφθον λεπτὸν καὶ οἶνον γλυκὺν δίδου· τοῦτο δ᾽ οὐκ ἄγει.

E. To the tenth part of a cotyle of Ethiopian cummin add three half choes of water, coat the pot with thick mud, and simmer without boiling until one third remains; cool, and give against every ardent and other fever.

F. Rain-water, pure.

G. Add one chous of water to a cotyle of peeled barley, and boil until half remains; then sieve, add celery, and give cold.

H. Dilute white raisin wine.

I. Dilute wine made from pressed grapes.

J. Crush a cotyle of white raisins without stones and a handful of cinquefoil roots, add twenty cotylai of water, boil off half, and give cold, a little at a time.

K. To a half choinix of ripe barley groats add a chous of water; when the groats have swollen up, knead them with your hands until the water becomes white; add a pinch of maiden-hair, expose to the air, and give.

L. Let the patient beat the white of three or four eggs in a chous of water, and drink; this is very cooling and leads the cavity down. However, if it seems appropriate to evacuate even more strongly, beat in the yolks.

M. Wash a half choinix of parched barley well, and boil it two or three times in a chous of water; give cold.

N. Give thin boiled-down barley-water, and sweet wine; this does not draw.

---

89 Θ: τριχ- M.   90 διηθήσας om. Θ.
91 Ermerins: ἄσταφις λευκὴ ΘM.
92 Ermerins: -λης Θ: -λη M.
93 M adds λείπων.
94 ἁδρῶν om. Θ.
95 ὑπάγει . . . μᾶλλον om. M.

Τοῦτο δέ· σικύου πέπονος ἄνευ τοῦ δέρματος πά-
λην[96] ἐφ᾽ ὕδατι· τοῦτο οὐρέεται καὶ ψύχει καὶ τὴν
δίψαν παύει.

Τοῦτο δέ· ὀρόβους ἐν ὕδατι προεψήσας, εἶτα χύ-
160 τρην καινὴν | ἐν χύτρῃ μέζονι θεὶς πλέῃ ὕδατος,
ἐπιχέας ἕτερον ὕδωρ τοῖσιν ὀρόβοισιν, ἕψε ὀλίγον
χρόνον· εἶτα ἀποχέας τὸ τρίτον μέρος, ἐπειδὰν κάθ-
εφθοι ἔωσιν οἱ ὄροβοι, ψύξας δίδου κατὰ κύαθον
ἐπιπάσσων τῆς τοῦ σικύου πάλης. καὶ ἐκ τῶν ὀρόβων
πάλην· τοῦτο δὲ βεβαίως δίψαν παύει.

Τοῦτο δέ· Θάσιον οἶνον παλαιόν, πέντε καὶ εἴκοσιν
ὕδατος καὶ οἴνου ἕνα δίδου.

Τοῦτο δέ· τρίφυλλον τὸ σικυῶδες ἐν ὕδατι καὶ
κρίμνα κριθέων βρέχων δίδου.

Τοῦτο δέ· σέλινα ὅσον τρὶς τῇ χειρὶ περιλαβεῖν καὶ
γλήχους δραχμίδας δύο ἑψῶν ἐν ὄξους κοτύλῃσι δέκα,
ἕως τρίτον μέρος λείπῃς· τοῦτο μέλιτι καὶ ὕδατι κε-
ραννὺς ὑδαρὲς πινέτω ἀδιάντου δραχμίδα ἐμβαλών·
τοῦτο οὖρον ἄγει καὶ τὴν κοιλίην λύει.

Τοῦτο δέ· μῆλα εὐώδεα γλυκέα φλάσας καὶ ἐν
ὕδατι ἀποβρέξας, δίδου πίνειν τὸ ὕδωρ.

Τοῦτο δέ· μῆλα Κυδώνια ὡσαύτως οἷσιν ἂν καὶ ἡ
κοιλίη λελυμένη ᾖ ἐπὶ πυρετῷ καυσώδει.

Ἰκτέρου δ᾽ ἐπιλαβόντος ἀσταφίδος λευκῆς ἄνευ
γιγάρτων καὶ ἐρεβίνθων λευκῶν, ἡμικοτύλιον ἑκατέ-
ρου, καὶ κριθέων Ἀχιλληΐδων ἴσον, καὶ κνήκου ἴσον,

96 Potter: -ης ΘΜ.

54

O. The finest meal of melon without peel, in water; this is diuretic, cools, and stops thirst.

P. First boil bitter vetches in water; then, setting a new pot in a larger pot full of water, add new water to the vetches and boil for a short time; then pour off one third of the water and, when the vetches are boiled through, cool; give a cyathos at a time, sprinkling it with melon meal. Meal can be prepared from bitter vetches, too; this is very effective in stopping thirst.

Q. Old Thasian wine; give twenty-five parts water and one part wine.

R. Soak cucumber-like[8] clover and coarse barley groats in water, and give.

S. Boil three handfuls of celery and two pinches of pennyroyal in ten cotylai of vinegar until one third remains; mix with honey, and have the patient drink in abundant water, adding a pinch of maiden-hair; this is diuretic and laxative.

T. Crush fragrant sweet apples and, after soaking them in water, give the water to drink.

U. The same with quinces, for patients whose cavity has been evacuated after an ardent fever.

V. When jaundice is present, grind down a half cotyle each of white raisins without stones, white chick-peas, Achilles barley, and safflower in ten cotylai of water with

---

[8] τὸ σικυῶδες has traditionally been taken not as modifying τρίφυλλον but as a separate ingredient, variously identified as cucumber (Calvus, Pylander), cucumber meal (Cornarius, Foes) or melon meal (Littré, Fuchs). I see no reason why this adjective could not have been applied to a particular member of the clover family.

ὕδατος κοτύλας δέκα, καὶ σέλινα, μίνθην, κορίαννον ὀλίγον ἑκάστου ἀνατρίβειν, ἕως ἂν ἀτρέμα γλυκανθῇ, καὶ ἀδιάντου δραχμίδα ὕστερον ἐμβαλὼν αἰθριάσας δίδου.

Τοῦτο δέ, καὶ τὰ τούτοισιν ὅμοια μιμέεσθαι· πάντα δὲ τῷ πυρέσσοντι ἠθριασμένα δίδου, πλὴν οἷσιν ἂν αἱ κοιλίαι μᾶλλον τοῦ δέοντος ῥέωσι.

Τοῦτο δέ· γληχοῦς δραχμίδας τρεῖς, σελίνου διπλάσιον ἐν οἴνῳ κεκρημένῳ ἑψῶν δίδου· τοῦτο καὶ οὐρέεται καὶ διὰ τῆς κοιλίης χολὴν ἄγει.

dashes of celery, mint and coriander, until the mixture becomes slightly sweet; then later add a pinch of maidenhair, expose to the air, and give.

Also imitate the agents described with others that are similar; for the fever patient, expose them all to the air before you give them, except in cases where the cavities have suffered excessive evacuations.

W. Boil three pinches of pennyroyal and twice that amount of celery in wine mixed with water, and give. This is diuretic, and draws bile through the cavity.

INTERNAL AFFECTIONS

INTRODUCTION

Erotian does not mention this treatise in his list of Hippocratic works, and the fact that one word in his glossary may stem from *Internal Affections* 27[1] is less than conclusive evidence that he knew it.[2]

Galen certainly did know *Internal Affections*, albeit under a variety of titles:

> . . . in *Affections the Greater*, which begins "If the bronchial tube of the lung. . . ." Some people give this book the title *Internal Suppuration* (Περὶ ἐμπύων).[3]

> ἄλφιτα: . . . in *Diseases II the Greater* also parched lentils and vetches.[4]

Internal Affections is devoted wholly to the description and, in particular, the treatment of diseases. Each of its

---

[1] B17 βατίδες (Nachmanson p. 29).

[2] See Nachmanson, *Erotianstudien* pp. 411 f.

[3] Kühn XVIII(1). 39; see also Kühn XVIII(2). 512 f.

[4] Kuhn XIX. 76. See also under the words: ἀμαλῶς (XIX. 76), ἀνθίνην οἶνον (XIX. 81), ἀνωργασμένον (XIX. 82), ἀσᾶται (XIX. 86), δίεδρος (XIX. 92), κοτυλίδα (XIX. 114), κρέκειν (XIX. 114). λαμπτήρ (XIX. 117), προσέχει ἡ νοῦσος (XIX. 133), and ῥαγεῖσα (XIX. 134).

fifty-four chapters deals with a specific nosological entity according to the following plan: name or identifying feature; aetiology; symptoms and course: treatment; prognosis.

The overall arrangement of diseases is by anatomy:

Diseases in the Lungs and Sides: 1–12
Diseases in the Abdomen:
    Disease of the Spinal Marrow: 13
    Diseases of the Kidney: 14–17
    Diseases of the Vessels: 18–19
    Diseases of Phlegm: 20–21
    Dropsies: 22–26
    Diseases of the Liver: 27–29
    Diseases of the Spleen: 30–34
General Diseases:
    Jaundices: 35–38
    Typhuses: 39–43
    Ileuses: 44–46
    "Thick" Diseases: 47–50
    Sciatica: 51
    Tetanuses: 52–54

Where there are several varieties of the same disease, an attempt is made to draw significant distinctions. Among the criteria used are the causal agent (e.g. in the four "thick" diseases: phlegm and bile; bile; phlegm; white phlegm), the pathological process (e.g. in the four diseases of the kidney: lithiasis; rupture of the vessels; ulceration; suppuration), the anatomical location (e.g. in the two diseases of the vessels: right; left), the signs (e.g. the patient's colour in the three diseases of the liver: livid; like pomegranate-peel; dark), and the season of occurrence (e.g. in the three ileuses: winter; summer; late autumn).

Internal Affections is present in all the collected editions and translations. Besides, in the seventeenth century it was twice edited and translated into Latin in works devoted to Hippocratic pathology:

Praelectiones in librum Hippocratis . . . De morbis internis auctore M. Ioanne Martino . . . editore M. Renato Morello. . . . Paris. 1637.

Praelectiones in Hippocratis librum De internis affectionibus . . . edente M. Francisco de Saint-André. . . . Caen, 1687.

The work of Jouanna cited in the introduction to *Affections*[5] also contains a newly edited text of several chapters of *Internal Affections*.

[5] See vol. V p. 5.

ΠΕΡΙ ΤΩΝ ΕΝΤΟΣ ΠΑΘΩΝ[1]

1. Ἢν ἡ τοῦ πλεύμονος ἀρτηρίη ἑλκωθῇ, ἤ τι ῥαγῇ τῶν φλεβίων τῶν λεπτῶν τῶν κατακρεμαμένων εἰς τὸν πλεύμονα, ἢ τῶν συρίγγων τῶν διὰ τοῦ πλεύμονος τεταμένων συρραγέωσιν ἐς ἀλλήλας καὶ αἵματος πλησθῶσι—διασπῶνταί τε καὶ καταρρήγνυνται διὰ τάσδε τὰς ἁμαρτίας μάλιστα· διὰ ταλαιπωρίην, διὰ δρόμους, διὰ πτώματα, διὰ πληγάς, δι᾽ ἐμέτους βιαίους γιγνομένους, διὰ πυρετούς—τάδε οὖν πάσχει. τὸ μὲν πρῶτον βὴξ ἴσχει ξηρή, ἔπειτα ὀλίγῳ ὕστερον ἀποπτύει τὸ σίαλον ὕφαιμον, τοτὲ δὲ καθαρόν. οὗτος ἢν μὲν ἐν τάχει παύσηται τῆς νούσου. ἢν δὲ μή, προϊόντος τοῦ χρόνου τὸ αἷμα πλεῖον χωρέει, ἐνίοτε μὲν καθαρόν, ἔστι δ᾽ ὅτε καὶ ὑπόσαπρον. πολλάκις δὲ καὶ ἡ φάρυγξ λανθάνει αἵματος πιμπλαμένη· ἔπειτα θρόμβους αἵματος ἐκβράσσεται κατ᾽ ὀλίγον θαμινά· ἐνίοτε καὶ ὀδμὴ βαρείη ἀπ᾽ αὐτῶν γίνεται. καὶ ὁ φάρυγξ ἔστιν ὅτε ἄχνης πίμπλαται. καὶ ῥῖγος καὶ πυρετὸς ἐπιλαμβάνει, κατ᾽ ἀρχὰς μὲν τῆς νούσου σφόδρα, προϊούσης δὲ βληχρότερον καὶ ἄλλοτε καὶ ἄλλοτε ἐπιλαμβάνει· καὶ ὀδύνη ἐνίοτε ἔγκειται ἐν τοῖσι στήθεσι καὶ ἐν τῷ μεταφρένῳ καὶ ἐν τῇσι

64

INTERNAL AFFECTIONS

1. If the bronchial tube of the lung ulcerates, or one of the narrow vessels leading to the lung tears, or if some of the pipes extending through the lung rupture into one another and are filled with blood—most often such ruptures and tears occur as a result of the following insults: exertion; running; falls; blows; when there is violent vomiting; from fevers—the patient suffers the following: first he has a dry cough; then, a little later, he expectorates sputum charged with blood, and then clear sputum. If this patient gets over the disease quickly, that is all. If not, as time goes on more blood comes up in the sputum, sometimes pure, sometimes somewhat putrid. Often, the throat also fills up with blood, unnoticed; in that case, the patient coughs up clots of blood frequently, a little at a time; sometimes these give off a heavy odour. The throat also sometimes fills up with froth. Chills and fever attack, intensely at the beginning of the disease, but as it advances they become milder and intermittent; pain is sometimes present in the chest, back

---

1 The first and last leaves of the first quire and all 8 leaves of the second quire of Θ are missing. For these parts of the text M is our sole independent ms. authority.

πλευρῇσι. καὶ ὁκόταν τὸ αἷμα παύσηται πτύων, σίαλον πολλὸν ἀποπτύει ὑγρόν, ἐνίοτε δὲ καὶ γλίσχρον.
ταῦτα μὲν οὖν οὕτω πάσχει μέχρι τεσσαρεσκαίδεκα
ἡμέραι παρέλθωσιν. μετὰ δὲ ταύτας ἢν μὴ παύσηται
τὸ νόσημα, λεπίδας ἀπὸ τῆς ἀρτηρίης ἀποβήσσων
ἀποσπᾷ οἵας περ ἀπὸ φλυκταινίδων. καὶ ὀδύνη ἐμπίπτει ἐς τὰ στήθεα καὶ ἐς τὸ μετάφρενον καὶ ἐς τὸ
πλευρόν, καὶ τῶν ὑποχονδρίων ὡς ἕλκος ψαυόμενος
ἀλγέει.

168   Τούτῳ συμφέρει ἡσυ|χίην ὡς μάλιστα τῷ νοσή-
ματι ἔχειν ἔσω, ἢν οὕτως ἔχῃ. ἢν γάρ τι πονήσῃ, ὅ τε
πόνος ὀξύτερος καὶ ἡ βὴξ μᾶλλον ἢ τὸ πρότερον
πιέζει, καὶ τὸ ῥῖγος καὶ ὁ πυρετὸς μᾶλλον ἔχει· καὶ ἢν
πταρῇ, ἡ ὀδύνη ὀξέα ἐπέπεσεν. ἀλγέει δὲ καὶ ἐν τῇ
εὐνῇ, ὁκόταν περιστρέφηται.[2]

    Τούτῳ χρὴ προσφέρειν σιτία μὲν τὰ αὐτά, ἃ καὶ τῷ
ἐμπύῳ, ταῦτα δὲ ὡς πλεῖστα.[3] τῶν δὲ ὄψων τοισίδε
χρήσθω, ἰχθύσι μὲν ῥίνης ἢ φάγρου ἢ γαλεοῦ τοῦ
μεγάλου τοῦ λευκοῦ, ἢ τῶν ἄλλων τῶν τοιούτων,
πᾶσιν ἐν ῥόῳ καὶ ὀριγάνῳ ἠρτυμένοις· κρέας δ᾽ ἐσθιέτω ἀλέκτορος ὀπτὸν ἄναλτον, ἢ αἴγειον ἑφθόν. καὶ
οἴνῳ αὐστηρῷ ὡς παλαιοτάτῳ καὶ ἡδίστῳ μέλανι
χρήσθω. καὶ περιπάτοισι μετρίοισι χρήσθω πυρετοῦ
μὴ ἔχοντος· ἢν δὲ πυρετὸς ἔχῃ, ῥυφήματι ἢ ἀλεύρῳ ἢ
κέγχρῳ χρῆσθαι· ἢν δὲ σιτία προσφέρηται, ὀλίγα
προσφερέσθω, καὶ ὄψα τὰ διαχωρητικά. καὶ ἢν φαρμάκου δοκέῃ σοι δεῖσθαι, ὑποκάθαιρε αὐτὸν τῷ Κνιδίῳ κόκκῳ ἢ τῇ τιθυμαλλίδι, καὶ μετὰ τὴν κάθαρσιν

and sides. When the patient stops expectorating blood, he produces copious moist or sometimes sticky sputum. These things he suffers for fourteen days. After that, if the disease does not go away, in coughing the patient tears off fragments from his bronchial tube as if from blisters; pain occupies his chest, back and side, and on being touched in the hypochondrium, he feels pain as if in an ulcer.

If the case is such, it benefits the patient to rest indoors as much as possible during the disease. For, if he exerts himself in any way, the pain will be sharper, the cough press him more than before, and the chills and fever increase; if he sneezes, it is very painful. He suffers pain even in his bed, whenever he turns over.

You must administer to this patient the same cereals as to one with an internal suppuration, these in very generous amounts. Of main dishes, let him eat the following: of fish angel-fish, braize, large light-coloured dogfish, or others of these kinds, all seasoned in sumach and marjoram; of meats let him eat broiled fowl without salt, or boiled goat. Let him drink dry dark wine that is very old and very pleasant. Have him take moderate walks when he is without fever; if fever is present, give gruel, meal or millet; if cereals are given, let them be given in small amounts, together with laxative main dishes. If the patient seems to you to require a medication, clean him downwards with Cnidian berry or sea-spurge, and after the cleaning give him two

---

2 Θ begins -φηται.

3 ὡς πλεῖστα Θ: οὐ πολλά M.

ἀλεύρου δοῦναι ἑφθοῦ δύο τρυβλία ἐκροφεῖν λιπαροῦ.
καὶ ἔπειτα μετὰ ταῦτα ἀνακομίζειν ὡς μάλιστα, ὅπως
ἂν ὡς ἥκιστα λεπτὸς ᾖ, πρὸς γὰρ τὴν νοῦσον οὐ
συμφέρει λεπτὸν εἶναι.

Καὶ περιπάτους ὀλίγους τὸ πρῶτον ποιείσθω, ὡς
ἂν μὴ κόπος ἐπιλάβῃ. πυριᾶν δὲ ἄλλοτε καὶ ἄλλοτε,
καὶ ᾗ ἂν πυριηθῇ ἡμέρῃ, ἄσιτος ἔστω τὴν ἡμέρην[4]
πλὴν ἀλεύρου ἑφθοῦ τρυβλίον ἐκρυφείτω, ὕδωρ δὲ
πινέτω. τῇ δ' ὑστεραίῃ ἐλάσσονα ἢ ὡς μεμαθήκει
φαγέτω, καὶ πιέτω οἶνον μέλανα, ἡδύν, αὐστηρόν, |
170 ὀλίγον. τὸ δὲ λοιπὸν[5] τρὶς τῆς ἡμέρης τὰ σιτία διδόναι
τούτῳ, μέχρι ἂν καταστήσῃς τὴν κοιλίην, διδοὺς κατ'
ὀλίγα· ἐκ γὰρ τῶν πυρετῶν καὶ τῆς ἀσιτίης,[6] τὸ μὲν
στόμα θέλει,[7] ἡ δὲ κοιλίη οὐκ ἐθέλει δέχεσθαι, ἢν[8] δὲ
ἀθρόον δέξηται, φλεγμαίνει. ἀλλὰ κατ' ὀλίγα διδόναι
χρή· ἢν γὰρ ἀθρόον δῷς καὶ ὀλίγα πονήσῃ τοῖς
περιπάτοισιν, οὐ διαψύχεται ἡ κοιλίη, ἅτε ἀτρέμα
συνεστηκότων τῶν βρωτῶν· ἐνταῦθα δὲ καὶ ὁ πυρετὸς
φιλέει ἐπιγίνεσθαι, καὶ τοῦ μὲν χειμῶνος ἧσσον λυπέει,[9] τοῦ δὲ θέρεος μᾶλλον.[10]

Τοῦτον ἀνακομίζειν ὡς μάλιστα, ὅπως ἂν παχύτατος ᾖ, καὶ τοῖσι περιπάτοισι μετρίως χρήσθω, καὶ
παλαιέτω ἧσσον ἑωυτοῦ, καὶ πονείτω ὀλίγα τὸ πρῶτον, ἔπειτα δὲ πλείω, πολλὰ δὲ οὐδέποτε. ταῦτα ἢν
ποιέῃ, ὑγιὴς ἔσται τάχιστα· ἢν δὲ λεπτὸς γίνηται διὰ

---

[4] τὴν ἡμέρην om. M.   [5] M adds δὶς ἢ.
[6] M adds ἢν.   [7] Θ: μένῃ M.   [8] ἢν om. M.

68

bowls of rich boiled meal to drink. Then after this strengthen him as much as possible, in order that he will definitely not be thin, for against this disease thinness is no help.

At first, let the patient take only short walks, in order that he does not become fatigued. Treat him with vapour-baths from time to time and, on whichever day these treatments take place, let him fast that day, except for taking a bowl of boiled meal and drinking water. On the following day, have him eat less than is his custom, and drink a little pleasant dry dark wine. From then on, give this patient his food three times a day, a little at a time, until you bring his cavity into order; for, as a consequence of the fevers and the fasting, the mouth wants something, but the cavity is not willing to accept it, and if it does receive a large amount all at once, it becomes swollen. Therefore, you must give the food a little at a time; for, if you give it all at once and the patient exerts himself but little in his walks, the cavity is not cooled,[1] inasmuch as the ingesta become congealed and fixed in it, and then fever is inclined to supervene, which produces less stress in winter, but more in summer.

Strengthen this patient thoroughly, in order that he will become very robust; let him take walks in moderation, wrestle less than usual, and exert himself little at first, later more, but never a lot. If he does these things, he will quickly recover; however, if he becomes emaciated be-

[1] Cooling is a phase of normal digestion; cf. *Sacred Disease* 10 and *Diseases IV* 47.

---

[9] λυπέει om. M.
[10] μᾶλλον Θ: κίνδυνος ἐξαμαρτεῖν M.

τὴν ταλαιπωρίην, ἀνιέτω καὶ εὐωχείσθω ἡσυχίην
ἔχων. οὗτος μήτε πρὸς ἄνεμον δράμῃ[11] ὑγιὴς ἐὼν
ὀξέως, μήτε ἐφ' ἵππον μήτ' ἐπὶ ζεύγος ἀναβῇ· φυλασσέσθω δὲ καὶ βοὴν καὶ ὀξυθυμίην.[12] κίνδυνος γὰρ τὴν
νοῦσον πάλιν ἀναλαβεῖν, ἀλλὰ φυλάσσεσθαι χρὴ
τούτων πάντων.

Ἢν δὲ τοῦ σίτου ἀποκλεισθῇ, ὀρόβους φώξας τὰ
κελύφεα[13] ἀποκαθῆραι, εἶτα βρέξαι αὐτοὺς ἐν ὕδατι
τρεῖς ἡμέρας, ἐφ' ἑκάστην δὲ ἡμέρην καὶ ἀπηθέειν τὸ
ὕδωρ καὶ ἄλλο ἐπιχεῖν· ἔπειτα τῇ τετάρτῃ ἡμέρῃ
ἀπηθήσας ξηρῆναι, εἶτ' ἀλέσας διασῆσαι λεπτότατα,
καὶ λίνου καρπὸν φώξας, κόψαι λεῖον, καὶ σήσαμον
φώσας, κόψαι λεῖον, καὶ ἄλφιτα ἄναλτα καθαρὰ
λεπτά· καὶ τῶν μὲν ἀλφίτων καὶ ὀρόβων ἴσον ἑκατέ-
ρου ἔστω, τοῦ δὲ σησάμου τρίτον μέρος μιῆς μερίδος,[14] τοῦ δὲ λίνου ἥμισυ μιῆς μερίδος· ταῦτα ἐν
γάλακτι αἰγείῳ ἑψήσας ὡς ὑγρότατον ῥυφείτω. μετὰ
172 δὲ διδόναι αὐτῷ ἐς ἄριστον σιτία καθαρὰ καὶ ὄψα τῶν
ἰσχυροτέρων· οἶνον δὲ τὸν αὐτὸν πίνειν. διδόναι δ'
αὐτῷ καὶ τῶν ῥιζῶν τῶν πρὸς τὰ ῥήγματα,[15] τῆς
κενταυρίου ἐπ' οἶνον ἐπιχύων· διδόναι δὲ καὶ τοῦ
δρακοντίου ἐπιχύων ἐπ' οἶνον· καὶ τῆς βηχὸς ἕνεκα ἐν
μέλιτι τὸ δρακόντιον ξύων διδόναι λείχειν. καὶ ἢν τὸ
ἕψημα τὸ ἐν τῷ γάλακτι φάσκῃ μὴ δυνατὸς εἶναι
ῥυφεῖν, γάλα βόειον ὡς πλεῖστον πινέτω τὸ τρίτον
μέρος μελικρήτου παραμίσγων.

Καὶ οὕτως τάχιστα ὑγιὴς ἔσται, ἡ δὲ νοῦσος θεραπείης δεῖται πολλῆς, χαλεπὴ γάρ. ἢν δὲ μὴ θεραπεύη-

cause of the exertion, let him give it up, eat heartily, and rest. Let this patient not run against the wind, soon after having recovered, nor ride a horse or in a wagon, and have him avoid shouting and excitement; for there is a danger that the disease will recur, and therefore he must take care in all these matters.

If the patient has no appetite, roast vetches, remove the skins, and then soak them in water for three days, straining off the water each day and pouring in new water; then on the fourth day strain them off and, drying them, grind and sieve them very fine; soak linseed, pound it smooth; do the same with sesame, and with fine white unsalted meal; let there be equal amounts of the meal and the vetches, of the sesame one third that amount, and of the linseed one half; boil these in goat's milk, and let the patient drink this as a very moist gruel. From then on, give him for breakfast fine cereals and main dishes of the heartiest kinds; have him drink the same wine. Also, give him roots effective against tears: grate centaury over wine; grate dragon arum over wine, too, and give it. For the cough, grate dragon arum into honey, and give this to the patient to take. If he says that he is not able to drink gruels boiled in milk, let him drink as much cow's milk as he can, mixing into it a third part of melicrat.

In this way the patient will recover most quickly; the disease requires much treatment, for it is severe. If the

---

11 Θ adds μηδ'.
12 φυλασσέσθω . . . ὀξυθυμίην om. Θ.
13 Ermerins: και λυφα Θ: κέλυφα M.
14 μιῆς μερίδος om. M.
15 M: δήγ- Θ.

ται[16] ὑγιὴς γενόμενος καὶ ἢν μὴ[17] ἐν φυλακῇ ἔχῃ
ἑωυτόν, τοῖς πολλοῖς ὑποτροπάσασα, ἡ νοῦσος ἀπώ-
λεσεν. οὗτος ἢν μὲν ὑπὸ ταύτης τῆς θεραπείης λήξῃ,[18]
ἅλις· εἰ δὲ μή, παχύνας αὐτὸν γάλακτι καῦσαι τὰ
στήθεα καὶ τὸ μετάφρενον· ἢν γὰρ τύχῃς καύσας,
ἐλπὶς ἐκφυγεῖν τῆς νούσου.

2. Ἢν δὲ ἡ ἀρτηρίη σπασθῇ ἢ τῶν φλεβῶν τις[19]
τῶν τεινουσῶν ἐς τὸν πλεύμονα, τάδε πάσχει· κατ᾽
ἀρχὰς τῆς νούσου βὴξ ἴσχει ὀξέη, καὶ ῥῖγος, καὶ
πυρετός, καὶ τὸ σίαλον ἀποπτύει πολὺ καὶ λευκὸν καὶ
ἀφρῶδες, ἄλλοτε δὲ ὕφαιμον, καὶ ὀδύνη τὴν κεφαλὴν
καὶ τὸν τράχηλον ἴσχει. αὕτη ἡ νοῦσος ἰσχυροτέρη
τῆς πρόσθεν· καὶ μέχρι μὲν δέκα ἡμερέων τῶν πρώτων
τοιαῦτα πάσχει· ἔπειτα οἱ πολλοὶ τῇ ἑνδεκάτῃ ἡμέρῃ
πύα ἀποπτύουσι παχέα βιαίως· ἡμέρῃ δὲ καὶ ἡμέρῃ[20]
καθαρώτερα ἀποπτύει, ἢν φύξιμος ᾖ, καὶ τῇ ὀδύνῃ
ἧσσον πονέει, καὶ ἐν τάχει ὑγιὴς γίνεται. ἢν δὲ μέλλῃ
πολυχρόνιος ἡ νοῦσος ἔσεσθαι, τά τε πύα πολλῷ
174 πλείω ἀποπτύει, | καὶ ὁ ἄλλος πόνος ἐν τῷ σώματι
πολλῷ ἔνι πλείων· αἱ δὲ θέρμαι βληχρότερον ἔχουσιν
ἢ τὸ πρίν.

Τοῦτον ἢν λάβῃς κατ᾽ ἀρχάς, ὑποκάθαιρε κάτω
ὀπῷ σκαμωνίης, ἢν ἀπύρετος ᾖ·[21] μετὰ δὲ τὴν κάθαρσιν προσφερέσθω τὰ αὐτὰ ἃ καὶ πρόσθεν. καὶ τὰ
ἄλλα τὰ αὐτὰ προσφερέσθω, ἡσυχίην ἔχων ὡς
μάλιστα τῷ σώματι, καὶ μαλθακῶς κοιμάσθω· ταῦτα

---

[16] μὴ θεραπεύηται Θ: θεραπευθεὶς Μ.

patient is not cared for after he has recovered, and does not keep a watch over himself, in many the disease has returned and killed them. If the patient recovers with this treatment, fine; if not, fatten him on milk, and cauterize his chest and back; for, if your cautery succeeds, there is hope for him to survive the disease.

2. If the bronchial tube is torn, or one of the vessels extending to the lung, the patient suffers the following: the disease begins with a violent cough, chills and fever; he expectorates copious white frothy sputum, sometimes charged with blood, and pain occupies his head and neck. This disease is severer than the preceding one, and the patient suffers these symptoms for the first ten days. Then, on the eleventh day, many cough up thick pus quite violently; day by day the sputum becomes cleaner, if the patient is to escape, he suffers less pain, and he quickly recovers. But, if the disease is to become chronic, he expectorates much more pus, and the suffering in the rest of his body is much greater; the fevers, though, are milder than before.

If you take on this patient at the beginning, clean him downwards with scammony juice, if he is without fever. After the cleaning, let the patient take the same things administered in the preceding case. Otherwise, too, let him take the same things, rest his body as much as possible, and

---

17 καὶ ἢν μὴ Potter: καὶ ἢν Θ: μὴ M.

18 Θ: ἰηθῇ M. 19 ἢ and τις om. M.

20 καὶ ἡμέρῃ Θ: om. M: decimaquarta Cornarius: τεσσαρεσ-καιδεκάτῃ Foes (n. 9), Vander Linden: τετάρτῃ καὶ δεκάτῃ Littré after Mack.

21 ἢν ἀπύρετος ᾖ om. Θ.

73

μὲν κατ᾽ ἀρχὰς ποιείτω μέχρι τῶν δέκα ἡμερέων. ἢν δ᾽
ἔμπυος γένηται, τὰ αὐτὰ ἃ καὶ ὁ ἔμπροσθεν[22] ποιείτω·
ἢν δ᾽ ὑγιὴς γένηται, τῶνδε χρὴ ἀπέχεσθαι, σιτίων μὲν
καὶ ποτῶν ὀξέων καὶ δριμέων καὶ ἁλυκῶν καὶ λιπα-
ρῶν· ταλαιπωρίων δὲ χρὴ ἀπέχεσθαι τῶν αὐτῶν ὧν
καὶ οἱ πρόσθεν. ταῦτα ἢν ποιέῃ, τάχιστα τῆς νούσου
ἀπαλλαγήσεται· ἢν δέ τι τούτων μὴ ποιήσῃ, κινδυνεύσει πάλιν ὑποτροπάσαι, καὶ ἡ νοῦσος κάκιον ἔχειν·
καὶ οἱ πολλοὶ πλευμορρωγέες ἐόντες διατελέουσιν,
ἕως ἂν ἀποθάνωσι.

Τοῦτον ἢν μὴ παραχρῆμά τις ἰήσηται, ὑποτροπα-
σάσης τῆς νούσου, οὐκ ἂν ἔχοις ὠφελῆσαι, εἰ δὲ μὴ
τάδε ποιήσῃς· γάλακτι βοείῳ παχύνας, καῦσαι τὰ
στήθεα καὶ τὸ μετάφρενον· ἢν γὰρ τύχῃς καύσας, ἢ
αὐτὴ ἂν ὠφελίη γένοιτο. ἡ δὲ νοῦσος ἀπὸ τῶν αὐτῶν
ἁμαρτιῶν γίνεται ὥσπερ[23] καὶ ἡ πρόσθεν.

3. Πλεύμονος ἥδε[24] γίνεται μὲν ἡ νοῦσος ἀπὸ τῶνδε
μάλιστα· ὅταν ὁ πλεύμων αἷμα ἑλκύσας ἐφ᾽ ἑωυτὸν ἢ
φλέγμα ἁλμυρὸν μὴ ἀφῇ πάλιν, ἀλλ᾽ αὐτοῦ ξυστρα-
φῇ καὶ ξυσσαπῇ·[25] ἀπὸ τούτων φύματα φιλέει γίνεσθαι ἐν τῷ πλεύμονι καὶ ἐκπυοῦσθαι. οὗτος δὲ τάδε
πάσχει κατ᾽ ἀρχὰς καὶ διὰ παντὸς[26] τοῦ νοσήματος·
βὴξ ὀξέη ἔχει καὶ ξηρή, καὶ ῥῖγος, καὶ πυρετός, καὶ
ὀδύνη ἐν τοῖσι στήθεσι καὶ ἐν τῷ μεταφρένῳ ἔγκειται,
176 | ἐνίοτε δὲ καὶ ἐν τῷ πλευρῷ· καὶ ὀρθοπνοίη σφοδρὴ
ἐπιπίπτει.[27] οὗτος μὲν μέχρι τεσσερεσκαίδεκα ἡμερέων τοιαῦτα πάσχων διατελέει, πολλάκις δὲ καὶ ἐπὶ
πλεῦνας τεσσερεσκαίδεκα ἡμερέων·[28] ἔπειτα ῥήγνυται

sleep in a soft bed. Have him do these things at the beginning, up to the tenth day. If he suppurates internally, let him do the same things that the preceding patient did for internal suppuration. If he recovers, he must refrain from the following: foods and drinks that are acid, sharp, salty and fat; he must also avoid the same exertions that the preceding patients avoided. If he follows these instructions, he will get over the disease most quickly; but, if he does not follow some of them, he runs the risk of a relapse and the disease then being worse. Many patients continue with a tear in their lung until they die.

If someone does not treat this patient right away, when the disease recurs your only means of helping him would be to do the following: fatten him on cow's milk, and then cauterize his chest and back; if your cautery succeeds, the same benefit will result as above. This disease arises as a result of the same insults as the preceding one.

3. This disease of the lung generally arises in the following way: when the lung attracts blood or salty phlegm and does not discharge it again, but it gathers there and grows putrid, from this tubercles are likely to form in the lung and to produce pus. From the beginning and all through the disease this patient suffers the following: a sharp dry cough, chills, and fever; pain in the chest and back, sometimes also in the side; severe orthopnoea. These continue until the fourteenth day, often for even more than fourteen days. Then pus breaks out and the patient coughs much

22 Θ: πρόσθεν M.   23 Θ: ὧν M.
24 Θ: δὲ M.   25 Θ: συμπαγῆ M.
26 καὶ διὰ παντὸς om. Θ.   27 Θ: ἐμ- M.
28 ἐπὶ π. τ. ἡ. Θ: πλείονας δεκατέσσαρας ἡμέρας M.

πύα, καὶ ἀποπτύει πολύ· πολλάκις δὲ καὶ οἷον χιτῶνας
ἀραχνίων ἀποπτύει, πολλάκις δὲ[29] ὕφαιμον. καὶ ἢν μὲν
ἀποκαθαρθῇ καὶ ἀπισχνανθῇ ἐνταῦθα[30] ὁ πλεύμων,
ἐλπὶς[31] ἐκφυγεῖν· ἢν δὲ μὴ προσέχῃ, ἡ νοῦσος ἐπ᾽
ἐνιαυτὸν παρατείνει,[32] καὶ μεταβάλλει ἄλλοτε ἀλλοῖα
πάσχων.

Τούτῳ χρὴ κατ᾽ ἀρχὰς μέν, πρὶν τὰ πύα ῥαγῆναι,
προσφέρειν τάδε· ὅταν ἀνῇ[33] ὁ πυρετός, λούειν θερμῷ
καὶ πολλῷ, καὶ ῥυφήμασι χλιαροῖσι[34] χρήσθω, πτισάνης χυλῷ καθέφθῳ μέλι παραχέας, ὅταν ἕφθῃ·[35]
καὶ οἶνον πινέτω λευκὸν γλυκύν, ἢ μελίκρητον ἐφθόν.
ὅταν δὲ ἅπαξ ἄρξηται τὰ πύα ἀποπτύειν, πινέτω τὰ
αὐτά, ἃ καὶ ὁ πρόσθεν ἔμπυος, καὶ σιτίοισι καὶ ποτοῖσι καὶ ὄψοισι τοῖς αὐτοῖσι χρήσθω, ἀπεχόμενος
ὀξέων καὶ δριμέων καὶ ἁλυκῶν καὶ λιπαρῶν καὶ
λαγνείης καὶ θωρηξίων, ἢν μὴ τῇ νούσῳ πρόσφορον
ᾖ· ἐς χρῆμα[36] δὲ ὁρῶν μελετᾶν ὁποίων ἄν[37] τινων σοι
δοκέῃ δεῖσθαι. τὰ δ᾽ ἄλλα μετὰ ταῦτα ταυτὰ[38] ποιείτω·
πινέτω δὲ καὶ γάλα τὴν ὥρην βοὸς καὶ αἰγός· πρόσθεν
δὲ ὑποκάθαιρε[39] ἑφθῷ ὀνείῳ γάλακτι· πινέτω δὲ καὶ τὸ
ἵππειον γάλα σεσεισμένον ἑκάστης ἡμέρης ἕωθεν
τρικότυλον κύλικα.[40]

Καὶ ἢν μὲν οὕτω μελετώμενος ῥαΐσῃ,[41] καὶ μὴ ῥαγῇ
τὰ πύα ἐς τὰ στήθεα, αὐτὸς ἐφ᾽ ἑωυτοῦ θεραπευέσθω
ἥσυχος[42] ἔχων τῷ σώματι ὡς μάλιστα καὶ τὰ σύμ-

[29] M adds καί.   [30] Θ: ἐν τάχει M.   [31] ἐλπὶς om. M.
[32] ἡ νοῦσος . . . παρατείνει Θ: ἐνιαυτὸν ἡ νοῦσος M.

up; often he also expectorates material that looks like spiders' webs, and often sputum charged with blood. If the lung then becomes clean and the swelling in it goes down, there is hope of recovery; but, if it does not stay that way, the disease stretches out over a year, and the patient suffers different things at different times.

At the beginning before pus breaks out, you must administer the following to this patient: when his fever is in remission, wash him in copious hot water, and let him take warm gruels—boiled-down barley-water to which boiled honey has later been added—and drink wine that is white and sweet, or boiled melicrat. When he once begins to expectorate pus, have him drink the same things as the patient above with internal suppuration, take the same cereals, drinks and main dishes avoiding acid, sharp, salty and fat ones, and refrain from venery and from drinking to intoxication, except where the latter is appropriate to the disease. Pay attention to the patient's condition, and attend to whatever seems to you to require it. After the expectoration, let him do the same things as well; also let him drink cow's and goat's milk in season, after first being cleaned out downwards with boiled ass's milk, and drink a three-cotyle cup of shaken mare's milk early every morning.

If, when cared for in this way, the patient gets better, and pus does not break out into his chest, let him take over his own treatment, keeping his body as quiet as possible

---

33 Θ: ᾗ M.   34 Θ: πολλοῖς M.   35 Θ: ἑφθὸν ᾗ M.
36 Θ: χρῶμα M.   37 Θ: οὖν M.
38 μ. τ. τ. Θ: τὰ αὐτὰ M.   39 Θ: -καθηράσθω M.
40 M adds ἢν ᾗ δυνατός.   41 M adds ἅλις.
42 ἐφ᾽ ἑ. θ. ἥ. Θ: ἑαυτὸν θεραπευέτω ἡσυχίην M.

φορα προσφέρων ἑωυτῷ. ἢν δὲ ῥαγῇ τὰ πύα ἐς τὰ
στήθεα, ὅπῃ ἄν σοι δοκέῃ ἀποσημαίνειν μάλιστα,
178 ταύτῃ ταμὼν ἢ | καύσας, ἀφιέναι τοῦ πύου ὀλίγον τὸ
πρῶτον· τὰ δ᾽ ἄλλα ποιέειν τὰ αὐτά, ἃ καὶ ἐπὶ τοῦ
πρόσθεν ἐμπύου.[43]

4. Ἢν ἐν πλεύμονι κιρσὸς ἐγγένηται, βὴξ ξηρὴ
ἐπιλαμβάνει, καὶ ῥῖγος, καὶ πυρετός, κατ᾽ ἀρχὰς μὲν
τῆς νούσου πάνυ.[44] ἔχει δὲ καὶ ὀρθοπνοίη, καὶ ἐν τῇ
κεφαλῇ ὀδύνη ἔστηκε,[45] καὶ αἱ ὀφρύες δοκέουσιν ἐπι-
κρέμασθαι, καὶ οἴδημα κατέρχεται ἐς τὸ πρόσωπον
καὶ ἐς τὰ στήθεα καὶ ἐς τοὺς πόδας. πολλάκις δὲ καὶ ἐς
τὴν κεφαλὴν ἐρείδει, καὶ ὑπὸ τῆς ὀδύνης, ὅταν ὁ πόνος
ἔχῃ, οὐ δύναται ἀνορᾶν· τὸ δὲ σῶμα ὕπωχρον, καὶ αἱ
φλέβες[46] αὐτοῦ διατείνουσιν ἢ φλόγεαι ἢ μέλαιναι.

Τοῦτον, ὅταν οὕτως ἔχῃ καὶ ὁ πόνος μάλιστα
πιέζῃ, πρῶτον μὲν αἷμα ἀφαιρέειν, ἔπειτα[47] λούειν
πολλῷ καὶ θερμῷ· καὶ ὅταν δίψα ἔχῃ, πίνειν διδόναι
κυκεῶνα ἐν οἴνῳ μέλανι αὐστηρῷ, ὡς ἡδίστῳ, ἴσον
ἴσῳ μίξας· ψυχρὸν δὲ χρὴ ὡς μάλιστα πίνειν· ῥυφήμασι δὲ χρῆσθαι, πτισάνης χυλῷ καθέφθῳ μέλι[48]
παραχέας. ταῦτα χρὴ προσφέρειν ἐν τῇσι πρώτῃσι
τῶν ἡμερέων τεσσερεσκαίδεκα. ἢν δ᾽ ἐπὶ πλέον ἡ
νοῦσος ἔχῃ, ὅ τε πόνος πλείων ἐν τῷ σώματι καὶ ἡ
ἀδυναμίη ἐνῇ, τούτῳ οὕτως[49] ἔχοντι τὰ αὐτὰ προσφέρειν, ἃ καὶ ἐν τῷ ἐμπύῳ τὸν πλεύμονα, ὅταν αἱ
τεσσερεσκαίδεκα ἡμέραι παρέλθωσιν.

[43] M adds γενομένου.

78

and administering to himself what is fitting. But if pus breaks out into his chest, wherever the signs seem to you to point most, incise or cauterize there and at first draw out a small amount of pus; then continue with the same course of treatment as in the case of internal suppuration above.

4. If a varix forms in the lung, a dry cough, chills and fever set in right at the beginning of the disease. There is also orthopnoea, pain occupies the head, the eyebrows seem to overhang, and the face, chest and feet swell up. Often the disease becomes fixed in the head, and from the pain, when it is pressing, the patient cannot look up. The body is pale-yellow, and the vessels show their course through it by their red or dark colour.

This patient, when the case is such and pain is pressing him intensely, first subject to a blood-letting, and then wash him in copious hot water; when thirst is present, give him a cyceon in dry dark very pleasant wine to drink, mixing together an equal amount of each; he must drink this as cold as possible. Also, employ as gruel boiled-down barley-water to which you have added honey. These things you must administer in the first fourteen days. If the disease continues longer, and there is more pain and weakness of the body, after the fourteen days administer to this patient, as long as his condition is such, the same things given for suppuration of the lung.

---

44 M adds σφόδρα.
45 ὁ. ἔ. Θ: ἐνέστηκε M.
46 M adds δ'.
47 πρῶτον . . . ἔπειτα om. Θ.
48 M adds χρηστὸν.
49 Θ: δὲ M.

Αὕτη ἡ νοῦσος μάλιστα⁵⁰ γίνεται ἀπὸ ταλαιπωρίης καὶ αἵματος⁵¹ καὶ χολῆς μελαίνης. (5.) ὅταν τὰ κοῖλα φλέβια τὰ ἐν τῷ πλεύμονι διέχοντα πλησθῇ αἵματος ἢ χολῆς μελαίνης, συνέρρηξεν εἰς ἄλληλα τὰ φλέβια, ἅτε ἐν στενοῖσιν ἐόντα καὶ ἀπειλημμένα καὶ ἔξοδον οὐκ ἔχοντα, ὀδύνην παρέχει καὶ φῦσαν ἐν τῷ πλεύμονι.

180   Αὕτη ἡ νοῦσος χαλεπὴ καὶ θερα|πείης πολλῆς δεῖται· εἰ δὲ μή, οὐκ ἐθέλει ἐκλείπειν, ἀλλὰ τοῖσι πολλοῖσι συναποθνήσκει.

6. Ἢν δ᾽ ἐρυσίπελας ἐν τῷ⁵² πλεύμονι γένηται, ἐγγίνεται δὲ μάλιστα ἀπ᾽ οἰνοφλυγίης καὶ γαστριμαργίης ἰχθύων κεφάλων καὶ ἐγχελύων· ταῦτα γὰρ τὴν πιμελὴν πολεμιωτάτην ἔχει πρὸς τὴν φύσιν τοῦ ἀνθρώπου· ἤδη δὲ τὸ νόσημα ἐγένετο καὶ ἀπὸ φλέγματος, ὅταν μιγὲν τῷ αἵματι ἐπιρρυῇ ἐπὶ τὸν πλεύμονα· προσπίπτει δὲ καὶ ἐκ κρεηφαγίης ἐξ ὕδατος μεταβολῆς.

Τάδε οὖν πάσχει· βήσσει ἰσχυρῶς, καὶ τὸ σίαλον ἀποπτύει ὑγρὸν καὶ πολύ, πολλάκις δὲ λευκὸν καὶ παχύ, οἷον ἀπὸ βράγχου· καὶ ὀδύνη πιέζει ὀξέη ἐς τὰ στήθεα καὶ τὸ μετάφρενον καὶ τοὺς κενεῶνας καὶ τὰ πλευρά, καὶ ἐρεύγεται ὀξύ, καὶ ἐκ τῶν πλευμόνων καὶ στηθέων οἷον γαστὴρ τρυλίζει. καὶ ἐμέει λάμπην ὀξέην, καὶ τὸ ἔμεσμα ἢν ἐκχέῃς χαμάζε, ζύει τὴν γῆν ὥσπερ ὄξος ἐπιχέαντι. καὶ τοὺς ὀδόντας αἱμωδιᾷ, καὶ

⁵⁰ μάλιστα om. M.   ⁵¹ καὶ αἵματος om. M.

This disease usually arises from exertion, blood and dark bile. (5.)[2] For, when the hollow vessels extending through the lung become filled with blood or dark bile, they break into one another and, inasmuch as they are in narrow straits, cut off and without any exit, produce pain and breathlessness in the lung.

This disease is severe and requires much attention; without this, it is not willing to leave off, but clings to many patients until they die.

6. If erysipelas arises in the lung, it is usually from drunkenness or the consumption of too many grey mullets and eels, for these fish contain a fat most harmful to man's constitution. On occasion, the disease has also arisen from phlegm, when, on being mixed with the blood, it flowed to the lung; it may also attack as the result of eating meat, or from a change of water.

The patient suffers the following: he coughs violently and expectorates copious moist sputum, often thick and white like that from a sore throat. Sharp pains press on his chest, back, flanks and sides, he has oxyrygmia, and from his lungs and chest he gurgles as if from a belly. He vomits up a sharp scum, and if you pour out the vomitus onto the earth, it corrodes the earth as vinegar does when it is poured on the earth. His teeth are set on edge, and chills,

[2] From Vander Linden, the first editor to divide the text into chapters, down to Littré, what follows has been designated as a new chapter. I agree with Jouanna (p. 222 n. 3), however, that Littré's chapters 4 and 5 are in fact parts of one and the same disease description.

---

[52] δ᾽ . . . τῷ Θ: φλεγμονὴ ἐν M.

ῥῖγος καὶ πυρετὸς καὶ δίψα ἔχει ἰσχυρή. καὶ ἤν τι
θέλῃ λιπαρὸν φαγεῖν, μύζει πρὸς τὰ σπλάγχνα καὶ
ἔμετον ἄγει· καὶ τὸ σῶμα ἅπαν νάρκα ἔχει. ὅταν δ᾽
ἀπεμέσῃ, ἐπ᾽[53] ὀλίγον δοκέει ῥᾷον εἶναι· ἔπειτα ἐπὴν
τῆς ἡμέρης ὀψίτερον γένηται, βρέμει ἡ κοιλίη καὶ
στρέφει καὶ βορβορύζει.

Τοῦτον ὅταν οὕτως ἔχῃ ὧδε μελετᾶν·[54] μίξας γάλα
καὶ μέλι καὶ ὄξος καὶ ὕδωρ, ταῦτα ἐγχέας εἰς χυτρίδα
χλιαίνειν, καὶ ὀριγάνου κλωνίοισι τῆς κεφαλοειδέος
ταράσσειν· ἔπειτα ὅταν[55] χλιαρὸν ᾖ, δοῦναι ἐκπιεῖν, ἢ
λαβόμενος τῆς γλώσσης, ἔγχει ἡσυχῇ διὰ σύριγγος·
εἶτα κελεύειν συνειληθέντα ἡσυχίην ἔχειν· ἔπειτα ἢν
182   ἔμετος ἔλθῃ αὐτῷ, ἐμείτω προθύμως· ἢν δὲ μὴ | ἐπίῃ,
καταματεύμενος πτερῷ ἐμείτω· καὶ ἤν τι φλέγματος
ἐμέσῃ, ἐπὶ πέντε ἡμέρας ταὐτὰ ποιείτω· ῥάων γὰρ
ἔσται. πινέτω δὲ τοῦτο γυμνασάμενος, ἢν οἷός τε ᾖ,
καὶ λουσάμενος πολλῷ θερμῷ· εἰ δὲ μή, ἀλλὰ λουσάμενος.

Ὅταν δὲ αἱ πέντε ἡμέραι παρέλθωσι, πρώϊος
νῆστις πινέτω ἐν μελικρήτῳ ἢ οἴνῳ καὶ μέλιτι ὀπὸν
σιλφίου ὅσον ὄροβον, καὶ σκόροδον τρωγέτω καὶ
ῥαφανῖδας νῆστις, καὶ οἶνον ἄκρητον ἐπιρρυφανέτω
μέλανα ἢ λευκὸν αὐστηρόν· τρωγέτω[56] δὲ καὶ ἐπὶ σίτῳ
καὶ μετὰ τὸ σῖτον· σιτίοισι δὲ ξηροῖσι καὶ κρέασιν
ὀνείοισιν ἢ[57] κυνείοισι χρῆσθαι ἐφθοῖσιν, ἢν τὸ ῥῖγος
καὶ ὁ πυρετὸς μὴ ἐπιλαμβάνῃ. οὗτος ἢν μὲν ἀπὸ τοῦ

[53] ἐπ᾽ om. Θ.

fever and violent thirst are present. If the patient is willing to eat any substantial food, it rumbles in his inward parts and provokes vomiting; numbness comes over his whole body. When he has vomited, for a short while he seems to be better, but then later in the day his cavity roars and twists and rumbles.

When the case is such treat the patient as follows: pour milk, honey, vinegar and water together into a pot, warm, and stir in twigs of the head-shaped marjoram; then, when this is warm, give it to the patient to drink off, or take hold of his tongue and pour it in gently through a pipe. Next order the patient to cover up and to keep quiet. Then, if vomiting comes on, let him vomit actively; but if vomiting does not occur spontaneously, let him vomit by being tickled with a feather. If he vomits up any phlegm, repeat the same treatment for five days, for he will improve. Have him drink this potion after bathing in copious hot water and, if he is able, exercising; if he is not able to exercise, then at least after bathing.

When the five days are up, early in the morning let the patient drink in the fasting state silphium juice, to the amount of a vetch, in melicrat or in wine and honey, eat garlic and radishes, and on top of that take dry white or dark wine unmixed with water; let him also take these things with his meal, and after it. If chills and fever are not present, give dry cereals and the meat of ass or dog boiled. If this patient is cleaned out to some extent by the instilla-

---

54 ὧ. μ. Θ: καὶ δοκέῃ καιρὸς εἶναι, προσαίρειν ὧδε μελέ-
την M.   55 Θ: ἐπειδὰν δὲ M.
56 Θ: πινέτω M.
57 ὀνείοισιν ἢ om. M.

ἐγχύματος καθαίρηταί τι· εἰ δὲ μή, ἄνω κάθαιρε αὐτὸν
ἐλλεβόρῳ· μετὰ δὲ τὴν κάθαρσιν ἀλεύρου ἑφθοῦ διδόναι δύο τρυβλία ἐκρυφεῖν μέλι παραχέας· οἶνον τὸν
αὐτὸν πινέτω καὶ ὑδαρέα.

Ἢν[58] δὲ μὴ κατ᾽ ἀρχὰς παραγένῃ[59] τῇ νούσῳ,
παχύνας αὐτὸν γάλακτι, καῦσαι τὰ στήθεα[60] καὶ τὸ
μετάφρενον· οὕτω γὰρ ἂν μάλιστα τῆς νούσου ἀπαλ-
λαγείη. ἢν δὲ μὴ καυθῇ, προσέχει καὶ οὐ μάλα ἐκλείπει, ἀλλ᾽ ἐς τὸ γῆρας προσέχει· πολλάκις δὲ καὶ
συναποθνήσκει, ἢν μὴ ἐν τῇσι πρώτῃσι ἡμέρῃσι
τεσσαράκοντα ἀποθάνῃ· ἀλλὰ χρὴ μελεδώνης μάλιστα. καὶ ὀρὸν καὶ γάλα τὴν ὥρην πινέτω βοὸς καὶ
αἰγὸς καὶ ὄνειον καὶ ἵππειον·[61] οὕτω γὰρ ἂν ῥήϊστα
διάγοι· ἡ δὲ νοῦσος χαλεπή.

7. Ἢν πλεύμων [ἀπὸ ἐρυσιπέλατος][62] οἰδήσῃ, τὸ δὲ
οἴδημα μάλιστα ἀπὸ αἵματος γίνεται, ὅταν ἐς ἑωυτὸν
ὁ πλεύμων ἑλκύσῃ αἷμα | καὶ ἔχῃ ἀναλαβών· τὸ δὲ
νόσημα θέρεος [ἢ] ὥρῃ[63] μάλιστα γίνεται. τάδε οὖν
ἀπ᾽ αὐτοῦ πάσχει· βὴξ ξηρὴ ἐμπίπτει, καὶ ῥῖγος, καὶ
πυρετός, καὶ ὀρθοπνοίη, καὶ ὁ πόνος ἰσχυρὸς ἐν τοῖσι
στήθεσι· καὶ τὰς ῥῖνας πίτνα ὡς ἵππος δραμών, καὶ
τὴν γλῶσσαν ἐξίσχει ὡς κύων θέρεος ὑπὸ καύματος.[64]
καὶ οἴδημα τὰ στήθεα κατέχει, καὶ φθέγγεται βρα-
χέως, καὶ ἐρύθημα καὶ κνησμὸς ἐν τῷ σώματι καθέ-

184

---

58 Θ: Εἰ Μ.   59 Θ: -ηται Μ.
60 After στήθεα nine leaves are missing in Θ.
61 Ermerins: -ου καὶ -ου Μ.

84

tion, fine; if not, clean him upwards with hellebore. After the cleaning, add honey to two bowls of boiled meal and give him this to drink off; let him drink the same wine well mixed with water.

If you were not present at the beginning of the disease, first fatten the patient on milk, and then cauterize his chest and back; for this gives him the greater chance of recovering. If he is not cauterized, the disease continues and rarely goes away, tending in most cases to remain into old age; often it even clings to the patient until he dies, if he does not succumb within the first forty days. Treatment is essential. Let the patient drink in season whey and milk of cow, goat, ass and mare, for with this regimen he will fare most easily; the disease is severe.

7. If the lung swells up,[3] the swelling occurs chiefly from blood, when the lung attracts blood to itself and, taking it up, retains it; the disease occurs mainly in summer. This is what the patient suffers from it: there are dry cough, with chills, fever and orthopnoea, and severe pain in the chest; the patient dilates his nostrils like a running horse, and protrudes his tongue as a dog does in summer from the heat. Swelling occupies his chest, the patient speaks little, and redness and itching settle over his body;

---

[3] This disease corresponds to *Diseases II* 58 and *Diseases III* 7 ("if the lung becomes full"), and not to *Diseases II* 55 (erysipelas); the disease just described in *Internal Affections* 6, called by Θ erysipelas but by M φλεγμονή, does correspond to *Diseases II* 55. I therefore delete ἀπὸ ἐρυσιπέλατος here.

---

[62] Del. Potter.   [63] Jouanna (p. 200): ἡ ὥρη M.
[64] M in marg.: ὑπὸ τοῦ πνεύματος M.

στηκε· καὶ ὑπὸ τοῦ πόνου κατακέεσθαι οὐ δύναται,
ἀλλ᾽ αὐτὸς αὑτὸν ῥίπτει ἀλύων. οὗτος θνήσκει ἐν ἑπτὰ
ἡμέρῃσι μάλιστα· ἢν δὲ ταύτας ἐκφύγῃ, οὐ μάλα
θνήσκει.

Τοῦτον, ὁκόταν ὧδε ἔχῃ, ἰῆσθαι τοῖσδεσι· ψύχειν
τὸ σῶμα· τεῦτλα ἐν ὕδατι ψυχρῷ βάπτων προστιθέναι
μάλιστα πρὸς τὸ πονέον⁶⁵ μάλιστα, ἢ ῥάκεα βάπτων
ἐν ὕδατι ψυχρῷ καὶ ἐκθλίβων προστιθέναι. κἢν μὲν
οὕτω ῥηΐσῃ· εἰ μή, κεραμικῇ γῇ ψυχρῇ καταπλάσσειν, καὶ ἐν τῇ αἰθρίῃ κοιμάσθω. οὕτω γὰρ ἂν
μάλιστα μελετώμενος φύγοι ἂν τὰς ἑπτὰ ἡμέρας.
ὁκόταν δὲ αἱ ἑπτὰ ἡμέραι παρέλθωσι καὶ ἡ ὀδύνη
προσέχῃ, χρίων ἐλαίῳ τὸ πονέον μέρος μάλιστα,
χλιάσματα προστιθέναι τὰ αὐτά, ἃ καὶ τῇ πλευρίτιδι.
καὶ τόδε·⁶⁶ πῖσαι αὐτὸν ἐς ὑποκάθαρσιν τοῦ πεπλίου
καὶ τῆς μηκωνίδος καὶ τοῦ κόκκου τοῦ Κνιδίου· καὶ
μετὰ τὴν κάθαρσιν φακῆς τρυβλίον δοῦναι ῥοφῆσαι,
πινέτω δὲ ὕδωρ. τῇ δ᾽ ὑστεραίῃ λοῦσαι αὐτὸν πολλῷ
καὶ θερμῷ πλὴν τῆς κεφαλῆς· ἔπειτα πῖσαι ὀρίγανον
ἐν μελικρήτῳ ἀποβρέξας· ποτοῖσι δὲ ὡς θερμοτάτοισι
χρεέσθω· σιτία δὲ προσφερέσθω τὰ αὐτὰ ταῦτα, ἃ καὶ
186   ὁ ὑπὸ τῆς πλευρίτιδος ἑαλω|κώς, ἢν μὴ πυρετὸς προσ-
ίσχῃ. αὕτη ἡ νοῦσος χαλεπή, καὶ παῦροι ἐκφυγγάνουσι.

8. Ἢν ⟨τὸ⟩⁶⁷ στῆθος καὶ ⟨τὸ⟩⁶⁷ μετάφρενον ἀναρ-
ραγῇ—ῥήγνυται δὲ μάλιστα ἀπὸ ταλαιπωρίης—τάδε
οὖν πάσχει. βὴξ ἴσχει ὀξείη, καὶ τὸ σίαλον ἐνίοτε
ἀποπτύει ὕφαιμον, καὶ ῥῖγος καὶ πυρετὸς ἐπιλαμβάνει

because of the pain, he is not able to lie down, but is distraught and casts himself about. This patient usually dies in seven days; if he survives these, death is rare.

When the case is such, treat the patient by cooling his body with the following: immerse beets in cold water and apply this especially to the most painful areas; or soak rags in cold water, squeeze them out, and apply; if the patient is relieved in this way, fine. If not, plaster him over with cold potter's earth, and have him sleep in the open air. If treated in such a way, the patient is most likely to survive the seven days. When the seven days have passed, if pain is still present, anoint the most painful parts with olive oil, and apply the same fomentations as for pleurisy. Also this: have the patient drink, in order to clean downwards, wild purslane, sea-spurge, and Cnidian berry; after the cleaning, give a bowl of lentil-soup as gruel, and let him drink water. The next day, wash him with copious hot water, except for his head. Then have him drink marjoram well steeped in melicrat; let him take drinks as hot as possible. Administer the same cereals as to a patient with pleurisy, unless fever is present. This disease is severe, and few escape it.

8. If the chest and back are torn apart—in most cases this happens as the result of exertion—the patient suffers the following: he has a sharp cough which sometimes produces sputum charged with blood; in most cases chills and

---

65 Jouanna: τὸ πον νεον M.
66 Jouanna: τῷδε M.
67 Later mss.

τὰ πολλά. καὶ ἐν τῷ στήθεϊ καὶ ἐν τῷ μεταφρένῳ
ὀδύνη ὀξείη ἔνεστι, καὶ ἐν τῷ πλευρῷ δοκέει οἷον
λίθος ἐγκέεσθαι, καὶ κεντέεσθαι ὑπὸ τῆς ὀδύνης διαμ-
περέως, ὡς εἰ βελόνη τις κεντοίη.

Τοῦτον ὁκόταν ὧδε ἔχῃ, παραχρῆμα γάλακτι
πιήνας καῦσαι τὰ στήθεα καὶ τὸ μετάφρενον· καὶ οὕτω
τάχιστα ὑγιὴς ἔσται. τὸ δὲ λοιπὸν ἡσυχάζων τῷ
σώματι μάλιστα διαιτῆσθαι, ἢν γάρ τι πονήσῃ ἢ ἐπὶ
ἄμαξαν ἀναβὰς ἢ ἐφ᾽ ἵππον, ἢ τοῖσιν ὤμοισιν αὐτὸς
ταλαιπωρήσῃ, κινδυνεύσει πάλιν ὑποτροπιάσαι ἡ
νοῦσος, καί, ἢν ὑποτροπιάσῃ, κίνδυνος διαφθαρῆναι·
τὸ γὰρ νόσημα μᾶλλον πιέζει ἢ κατ᾽ ἀρχάς. ἢν δὲ μὴ
καυθῇ, τοῖσι αὐτοῖσι ἰῆσθαι οἷσι καὶ τὸν ἔμπυον,
ῥοφήμασι καὶ σιτίοισι καὶ ποτοῖσι. τὸ δὲ σύμπαν
ἡσυχίην ἔχοντα εὐωχέειν τοῖσι ἐπιτηδείοισι· ἢν γὰρ
οὕτω μελετηθῇ, τάχιστα ὑγιὴς ἔσται. ἡ δὲ νοῦσος
χαλεπή.

9. Ἢν ἐν πλευρῷ φῦμα φύηται καὶ ἔμπυον γένηται,
τάδε πάσχει· ῥῖγος καὶ πυρετὸς ἴσχει, καὶ βὴξ ξηρὴ
πολλὰς ἡμέρας, καὶ ἀλγέει τὸ πλευρόν, καὶ ἐς τὸν
τιτθὸν καὶ ἐς τὴν κληῖδα καὶ ἐς τὰς ὠμοπλάτας ὀδύνη
ἴσχει ἀΐσσουσα.

Οὗτος ὅταν οὕτω ἔχῃ, ἐν μὲν τῇσι πρώτῃσι ἡμέρῃσι ἕνδεκα ῥοφήμασι χρεέσθω, πτισάνης χυλῷ καθέφθῳ μέλι παραχέων, ὁκόταν ἐφθὸν τὸ ῥόφημα ᾖ·
οἴνῳ δὲ χρεέσθω λευκῷ, γλυκεῖ ἢ αὐστηρῷ καὶ[68]
188 ὑδαρεῖ· καὶ ῥοφάνοντα πλεονάκις | τοῦ οἴνου ἐκπτύειν
κελεύειν· καὶ τοῦ ὕπνου κωλύειν, ἔστ᾽ ἂν αἱ ἕνδεκα

fever supervene; in the chest and back there is sharp pain, and in the side there seems to be something like a stone; the patient is pierced through by pain as if a needle were pricking him.

When the case is such, fatten the patient at once on milk, and cauterize his chest and back; if you do this, he will recover most quickly. From then on, let him conduct his life in such a manner as to keep his body very quiet; for, if he strains himself at all by riding in a wagon or on a horse, or if he does hard work with his shoulders, he will run the risk of a relapse, and if this happens there is a danger that he will perish, since the disease then presses more forcefully than it did originally. If the patient is not cauterized, treat him with the same things as in internal suppuration: gruels, cereals and drinks. In general, keep him quiet, and feed him well on suitable foods; for, if he is cared for in this way, he will quickly recover. The disease is severe.

9. If a tubercle forms in the side, and it suppurates, the patient suffers the following: he has chills, fever and a dry cough for many days, he aches in his side, and there are darting pains towards his nipple, collar-bone and shoulder-blades.

When the case is such, for the first eleven days let the patient take as gruel boiled-down barley-water to which honey has been added after the gruel has been boiled, and drink white wine—either sweet or dry—diluted with water; after he has had several drinks of wine, order him to expectorate; prevent sleep until the eleven days have

---

68 Mack (*et* Cornarius): ἤ M.

ἡμέραι παρέλθωσι· μετὰ δὲ ταύτας τὰς ἡμέρας σιτίοισι ὀλίγοισι ὡς μάλιστα χρεέσθω, κρέασι σκυλακεί-
οισι ἢ ἀλεκτρυονείοισι θερμοῖσι· ἐζωμεῦσθαι δὲ χρὴ
καλῶς, καὶ τὸν ζωμὸν ῥυφανέτω· καὶ τοῖσι ῥοφήμασι
πρόσθεν χρεέσθω τοῦ σίτου, καὶ μὴ διψήτω, ἕως ἂν
ἔμπυον γένηται τὸ πλευρόν. πυΐσκεται δὲ μάλιστα ἐν
τεσσεράκοντα ἡμέρῃσι ⟨ἢ⟩[69] ὀλίγῳ πρόσθεν· τούτῳ δὲ
γνώσῃ, ὁκόταν ἔμπυον γένηται τὸ πλευρόν, πύον γὰρ
οὐκ ἀποπτύεται, καὶ οὐκ ἀνεμέεται. τοῦτον, ὁκόταν
οὕτως ἔχῃ, ὅκου ἂν ἀποσημήνῃ, τάμνειν ἢ καίειν.
ἔπειτα ἀφιέναι τὸ πύον κατ᾿ ὀλίγον, καὶ ἐπειδὰν ἀπα-
ρύσῃς, μοτὸν καθιέναι ὠμολίνου· καὶ αὖτις τῇ ὑστεραίῃ ἐξελών, ἀπαρύσαι κατ᾿ ὀλίγον τοῦ πύους, ἔπειτα
μοτῶσαι· καὶ αὖτις τῇ τρίτῃ καὶ τῇσι ἄλλῃσι ἡμέρῃσι
δὶς τῆς ἡμέρης ἀπαρύειν ἕως ἂν ξηρανθῇ. διδόναι δὲ
καὶ τὰ σιτία καὶ τὰ ὄψα, ὁκόταν προσίηται, καὶ πινέτω
ὀλίγον, μὴ πολλόν, ἤν τε οἶνον ἤν τε καὶ ὕδωρ·
τρωγέτω δὲ καὶ τῆς ὀριγάνου τῆς ἀπαλῆς ὡς πλεῖστον, ἐς μέλι ἀποβάπτων. ἢν δὲ μὴ ἀπαλὴν ἔχῃ, ἀλλ᾿
αὔην, λεπτὴν ποιήσας, ἐς τὸ μέλι μίξας λείην, διδόναι
ὡς πλείστην. καὶ μὴ ῥιγούτω, καὶ τοῖσι λουτροῖσι
λούειν, καὶ μαλθακῶς κοιμάσθω.

Οὕτω ταύτην τὴν νοῦσον θεραπεύων, τάχιστ᾿ ἂν
ὑγιέα ποιήσαις· ὁκόταν δὲ ὑγιὴς γένηται, φυλασσέσθω τὸ ψῦχος, τὸ θάλπος, τὸν ἥλιον,[70] καὶ τοῖσι
περιπάτοισι ὀλίγοισι χρεέσθω μετὰ τὸ σῖτον, ὅκως ἂν
μὴ κόπος λάβῃ τὸ σῶμα· ταῦτα ἢν ποιέῃ ὑγιὴς ἔσται.
τούτων τῶν νούσων ἅστινας ἂν καύσῃς, ἐπὶ τὰ[71] καύ-

passed. After that, let the patient take mainly cereals in small amounts, and meats of puppy or fowl warmed; you must boil these well into a soup, and have the patient drink the soup. Let him drink gruels before foods, and prevent thirst until his side has suppurated. Usually suppuration occurs in forty days or a little before; you will be able to tell in this patient when he side has suppurated, for then he no longer coughs or vomits up pus. When the case is such, incise or cauterize the patient wherever the signs point. Then draw off the pus a little at a time and, when you have exhausted it, introduce a tent of raw linen. Draw off pus again the next day a little at a time, and then replace the tent. Again on the third and subsequent days, remove pus twice daily, until it dries up. Give both cereals and main dishes whenever the patient will take them, and let him drink a little, but only a little, wine or water. Let him also eat much fresh marjoram dipped in honey. If there is no fresh marjoram, but only dried marjoram, rub that fine, mix it thoroughly into honey, and give it in a very generous amount. Make sure that the patient does not have a chill, give him baths, and let him sleep in a soft bed.

If you treat this disease in such a way, you will very quickly bring about the patient's recovery. Once he has recovered, have him avoid cold, heat and sun, and take short after-dinner walks such that his body does not become fatigued. If the patient follows these instructions, he will recover. In whichever of these diseases you cauterize, ap-

---

69 Add. I.

70 A later ms: τοῦ ἡλίου M.

71 Later mss: ἔπειτα M.

ματα πράσα τρίψας πολλὰ καταπλάσσειν εὐθὺς μετὰ
τὴν καῦσιν καὶ ἐᾶν μίην ἡμέρην.

190   10. Φθίσιες τρεῖς· αὕτη μὲν γίνεται ἀπὸ φλέγμα-
τος· ἐπὴν ἡ κεφαλὴ φλέγματος πλησθεῖσα νοσήσῃ
καὶ θέρμη ἐγγένηται, συσσήπεται τὸ φλέγμα ἐν τῇ
κεφαλῇ, ἅτε οὐ δυνάμενον κινέεσθαι ὥστε ὑποχωρῆ-
σαι. ἔπειτα ὁκόταν παχυνθῇ καὶ συσσαπῇ καὶ ὑπερ-
πλησθῇ τὰ φλέβια, ῥεῦμα ἐπὶ τὸν πλεύμονα ἐγένετο,
καὶ ὁ πλεύμων ὁκόταν ἀναλάβῃ νοσέει παραχρῆμα,
ἅτε δακνόμενος ὑπὸ τοῦ φλέγματος, ἁλυκοῦ ἐόντος
καὶ σαπροῦ.

Τάδε οὖν πάσχει· πυρετὸς ἄρχεται βληχρὸς ἐπιλαμβάνειν, καὶ ῥῖγος· καὶ πονέει τὰ στήθεα καὶ τὸ
μετάφρενον, ἐνίοτε δὲ καὶ βὴξ πιέζει ὀξείη· καὶ ἀποπτύει τὸ σίαλον πολὺ καὶ ὑγρὸν καὶ ἁλμυρόν. ταῦτα
μὲν κατ᾽ ἀρχὰς τῆς νούσου πάσχει· προϊούσης δὲ τό
τε γυῖον λεπτύνεται, πλὴν τῶν σκελέων· ταῦτα δὲ
οἰδέει, καὶ οἱ πόδες, καὶ οἱ ὄνυχες ἕλκονται· ἐκ δὲ τῶν
ὤμων λεπτὸς καὶ ἀσθενής· ὁ φάρυγξ χνόου πίμπλαται
καὶ συρίζει ὡς διὰ καλάμου· καὶ διψῇ ἰσχυρῶς διὰ
παντὸς τοῦ νοσήματος, καὶ ἀκρασίη πολλὴ τὸ σῶμα
ἔχει.

Οὗτος ὁκόταν οὕτω ἔχῃ, ἐνιαυτῷ φθειρόμενος φαύ-
λως θνήσκει· μελετᾶν δὲ χρὴ ὡς μάλιστα καὶ ἀνακομίζειν. πρῶτον μὲν πῖσαι ἐλλέβορον, κάτω δ᾽ ὑποκαθῆραι ἐπιθύμῳ ἢ τῷ πεπλίῳ ἢ τῷ κόκκῳ τῷ Κνιδίῳ
ἢ τῇ τιθυμαλλίδι· ταῦτα χρὴ τετράκις τοῦ ἐνιαυτοῦ
δοῦναι, ἄνω δίς, κάτω δίς. διδόναι δὲ καὶ ὄνειον γάλα

ply a plaster of many crushed leeks to the site of the cautery immediately after the operation, and leave it there for one day.

10. Three consumptions: the first one arises from phlegm. When the head, on being filled with phlegm, becomes ill and is occupied by burning heat, the phlegm in the head putrefies, inasmuch as it cannot be set in motion to be evacuated. When it has become thick and putrid, and the small vessels are overfilled, a flux to the lung occurs, and the lung, when it takes up the phlegm, immediately becomes ill, being irritated by the salty putrid phlegm.

Thus, the patient suffers the following: at the beginning, there are a mild fever and chills; the patient has pain in his chest and back, and sometimes a violent cough also presses him; he expectorates copious moist salty sputum. These things he suffers at the beginning of the disease. As it goes on, his body grows lean, except for the legs; these swell up, as do the feet, and the nails become curved. In his shoulders the patient is thin and weak; his throat is filled with a film, and whistles as if through a reed-pipe. He has great thirst through the whole course of the disease, and great debility affects his body.

When the case is such, the patient wastes away sorrily for a year, and dies; you must treat him very actively and strengthen him. First have him drink hellebore, and then clean him downwards with dodder of thyme, wild purslane, Cnidian berry, or sea-spurge; these agents must be given four times a year: twice to act upwards, twice to act downwards. To clean downwards give boiled ass's milk,

ἐφθὸν ἐς ὑποκάθαρσιν ἢ βόειον ἢ αἴγειον· πινέτω δὲ
καὶ ὠμὸν τὸ βόειον γάλα, τρίτον μέρος μελικρήτου
προσμίσγων, πέντε καὶ τεσσεράκοντα ἡμέρας, παραμίσγων καὶ τὴν ὀρίγανον. τὴν δὲ κεφαλὴν αὐτοῦ
πρόσθε καθαίρειν, πρὸς τὰς ῥῖνας προστιθεὶς φάρμακον. σιτία δὲ καὶ ὄψα διδόναι μήτε λιπαρὰ μήτε
κνισώδεα μήτε λίην δριμέα. τεκμαιρόμενος δὲ τὸ νόσημα πάντα ποιεῖν χρή. καὶ περιπάτοισι χρέεσθαι
πρὸς τὰ σιτία, τεκμαιρόμενος μὴ ῥιγώῃ· τοῦ δὲ χειμῶνος παρὰ πυρὶ τὴν σίτησιν ποιεέσθω. οἶνον δὲ
πινέτω, αὐστηρόν, μέλανα, ὡς παλαιότατον καὶ ἥδι-
192 στον, ὀλίγον δέ. κἢν | δοκέῃ σοι πρὸ τοῦ φαρμάκου
πυριῆσαι, καὶ οὕτω δοῦναι τὸ φάρμακον· ἢν δὲ μὴ
βούλῃ δοῦναι τὸ φάρμακον, πυριήσας, οὕτω δὲ ἔμετον
ἀπὸ τῶν σιτίων ποιήσασθαι, ὡς τὸ πρόσθε γέγρα-
πται. τούτῳ ἢν ξυμφέρωσι, περιπάτοισι χρεέσθω· ἢν
δὲ μὴ ξυμφέρωσι, ἡσυχίην χρὴ ἔχειν ὡς μάλιστα τῷ
σώματι.

Οὗτος οὕτω μελετώμενος ῥῇστ' ἂν διάγοι ἐν τῷ
νοσήματι. ἡ δὲ νοῦσος θανασίμη, καὶ παῦροι διαφυγ-
γάνουσι.

11. Ἄλλη φθίσις· γίνεται μὲν ἀπὸ ταλαιπωρίης,
πάσχει δὲ πλῆθος τὰ αὐτά, ἃ καὶ ὁ πρόσθεν· ἡ δὲ
νοῦσος διαπαύει αὕτη μᾶλλον τῆς προτέρης, καὶ τοῦ
θέρεος ἀνίησι. τὸ δὲ σίαλον ἀποπτύει, παχύτερον μὲν
τῆς[72] πρόσθεν, καὶ βὴξ πιέζει μάλιστα τοὺς ὄρ-
θρους.[73] καὶ ὁ πόνος ἰσχυρότερος ἐν τοῖσι στήθεσι,
καὶ δοκέει οἷόν περ λίθον ἐν αὐτοῖσι ἐγκέεσθαι· πονέει

94

too, or cow's or goat's; also let the patient drink raw cow's milk, to which one third part of melicrat has been added, for forty-five days, and mix marjoram with it as well; clean the patient's head out beforehand by applying medication to his nostrils. Give cereals and main dishes that are neither rich, nor steaming like roasted meat, nor very sharp. You must do all these things, taking as your guide the state of the disease. Have the patient take walks in conjunction with his meals, but be careful not to have a chill; in winter let him take his food beside the fire. Let him drink wine that is dry, dark, very old and very pleasant, but in small amounts. If it seems advisable to you to treat with vapour-baths before the medication, still give the medication. If you are not willing to give it, apply vapour-baths, and then induce vomiting by means of foods as described above. If walks are beneficial to the patient, let him take them; if they are not, he must keep his body as quiet as possible.

If treated in such a way, the patient will fare best in the disease; the disease is usually mortal, and few escape it.

11. Another consumption: this one arises as the result of exertion, and the person suffers, for the most part, the same things as in the preceding one; this disease, however, makes spontaneous pauses more often than the preceding one, and it remits in summer. The sputum expectorated is thicker than in the preceding disease, and the cough presses most in the mornings. The pain in the chest is more violent, and something like a stone seems to be lying inside

---

[72] Later mss: $\tau\hat{\eta}$ M.
[73] Potter (cf. *Diseases II* 48): ὀρόους M.

δὲ καὶ τὸ μετάφρενον. καὶ ἡ χροιὴ δίυγρος αὐτοῦ
ἐστιν, καὶ ἤν τι πονήσῃ, φύσῃ καὶ ἄσθμα ἴσχει. οὗτος
ἐκ ταύτης τῆς νούσου ἐν τρισὶν ἔτεσι μάλιστα θνήσκει.

Μελετᾶν δὲ χρὴ τοῖσι αὐτοῖσι, οἷσι καὶ τὸν πρόσθεν. αὕτη ἡ νοῦσος προσέχει τοῖσι πολλοῖσι πλὴν
τῶν τριῶν ἐτέων, ἀλλ' ἀποθνήσκουσιν· ἡ γὰρ νοῦσος
χαλεπή.

12. Ἑτέρη φθίσις· ὑπὸ ταύτης τάδε πάσχει· ὁ
μυελὸς αὐτοῦ ὁ νωτιαῖος αἵματος μεστὸς γίνεται·
φθίνει ὁμοίως καὶ ἀπὸ τῶν κοίλων φλεβῶν· αὗται δὲ
φλέγματος ὑδρωποειδέος ἐμπίμπλανται καὶ χολῆς.
πάσχουσι δὲ τὰ αὐτά, ἀφ' ὁποτέρων ἂν φθίνῃ. καὶ ὁ
ἄνθρωπος εὐθὺς μέλας γίνεται καὶ ὑποιδαλέος, καὶ τὰ
194 ὑπὸ τοὺς | ὀφθαλμοὺς ὑπώπια ὠχρά, καὶ αἱ φλέβες αἱ
ἐν τῷ σώματι ὠχραὶ διατέτανται, ἔνιαι δὲ σφόδρα
ἐρυθραί· μάλιστα δὲ δῆλαι αἱ ὑπὸ τῇσι μασχάλῃσι.
καὶ ἀποπτύει ὠχρά, καὶ ὅταν αὐτῷ ἐπίῃ, πνίγεται καὶ
βῆξαι οὐ δύναται ἐνίοτε βουλόμενος. ἐνίοτε δὲ ὑπὸ τοῦ
πνίγματος καὶ τῆς προθυμίης τοῦ βήσσειν ἀθρόον
ἤμεσε χολήν, τότε δὲ λάπην, πολλάκις δὲ καὶ τὰ
σιτία, ὅταν φάγῃ· καὶ ὁκόταν ἀπεμέσῃ, δοκέει κουφό-
τερος εἶναι· εἶτ' αὖτις ὀλίγον χρόνον διαλιπών, ἐν
τοῖσι αὐτοῖσι πόνοισι κέεται. οὗτος καὶ φθέγγεται
ὀξύτερον ἢ ὑγιαίνων, καὶ ῥῖγος καὶ πυρετὸς διαπαύων
ἐπιλαμβάνει ἱδρώδης.

Τοῦτον ὁκόταν ὧδε ἔχῃ, βρωτοῖσι καὶ ῥυφήμασι
καὶ ποτοῖσι καὶ φαρμάκοισι καὶ τοῖσι ἄλλοισι πᾶσι

the chest; the back too is painful. The patient's colour is washed out and, if he exerts himself at all, breathlessness and panting come over him. Death from this disease usually occurs in three years.

You must treat with the same things that you gave to the preceding patient. This disease continues in most patients up to three years, but still they die; for it is severe.

12. Another consumption: from this one the person suffers the following (his spinal marrow becomes filled with blood; or also he may be consumed because the hollow vessels fill with dropsical phlegm and with bile; patients suffer the same symptoms no matter which of these two is the origin of their consumption): he immediately becomes dark and somewhat swollen, the parts of his face below the eyes are pale-yellow, and the vessels through his body are pale-yellow and stretched, or some are very red; especially conspicuous are the ones in the axillae. The patient expectorates pale-yellow sputum, and when an attack occurs he chokes and sometimes cannot cough even though he wants to. Sometimes, because of his choking and eagerness to cough, he all at once vomits bile, then scum, and often even food when he has eaten; after he has vomited, his condition seems to be better; but then after a short time he is again subject to the same distress as before. The patient's voice is shriller than when he was well, and intermittent chills and fever accompanied by sweating occur.

When the case is such, treat this patient with foods, gruels, drinks, medications, and all the other things that

μελετήν, ὥσπερ τὸν πρόσθεν. ἡ δὲ νοῦσος διαφέρει
μάλιστα ἐννέα ἔτεα, ἔπειτα ἀποθνήσκει[74] φθειρόμενος·
παῦροι δὲ φυγγάνουσιν ἐξ αὐτῆς· χαλεπὴ γὰρ ἡ
νοῦσος.

Ἢν δὲ βούλῃ, ὧδε ἰῆσθαι αὐτόν· πρῶτον πυρι-
ῆσθαι· ὁκόταν δὲ πυριηθῇ, τῇ ὑστεραίῃ δοῦναι χρὴ
πιεῖν αὐτῷ μελικρήτου ἡμίχουν καὶ ὄξος παραχέαι
ὀλίγον· τοῦτο δὲ κέλευε ἀπνευστὶ ἐκπιεῖν· ἔπειτα τῶν
ἱματίων ἀμφιέσαι αὐτὸν πολλὰ καὶ ἐᾶν[75] ὡς πλεῖστον
χρόνον. ἢν δὲ μὴ ἀνέχηται, ἀλλ᾽ ἐξεμέσαι βούληται,
ἐξεμεέτω· ἢν δὲ μὴ ἔμετος ἔχῃ χρόνου ἤδη ἐγγενο-
μένου, ἐπιπιὼν ὕδατος χλιεροῦ μεγάλην κύλικα, ἐμε-
έτω καταματτεόμενος πτερῷ· ὁκόταν δὲ ἀπεμέσῃ ὥστε
καλῶς ἔχειν, ἡσυχίην ἐχέτω ταύτην τὴν ἡμέρην. ὁκόταν δὲ δείπνου ὥρη ᾖ, δειπνεέτω μᾶζαν ὀλίγην, καὶ
ὄψον ἐχέτω τάριχος καὶ πράσα, ταῦτα δὲ ἐσθιέτω ὡς
πλεῖστα· οἶνον δὲ πινέτω γλυκύν. τὸν δὲ λοιπὸν χρόνον, λουέσθω τε πᾶσαν ἡμέρην ἅμα ἔωθεν θερμῷ
πλείστῳ, καὶ μετὰ τὸ λουτρὸν φυλάσσειν χρὴ ὡς μὴ
ῥιγώσῃ, ἀλλὰ κατακλιθεὶς εὑδέτω ὡς πλεῖστον χρόνον. ὁκόταν δὲ ἀναστῇ εὕδων, περιελθέτω σταδίους
εἴκοσι τὸ | βραχύτατον ταύτῃ τῇ ἡμέρῃ· τῇσι δὲ
ἄλλῃσι ἡμέρῃσι πέντε σταδίους ἄλλους ὑπερβαλὼν
βαδιζέτω ἑκάστης ἡμέρης ἕως ἂν ἀφίκηται ἐς τοὺς
ἑκατὸν σταδίους. τὴν δὲ κοιλίην ὑποκαθαίρειν δεῖ ἐκ
τῆς ἡμέρης χυλοῖσι τεύτλων καὶ ἀπὸ κράμβης· χωρὶς
ἑκάτερα ἑψήσας ἀπηθῆσαι χοέα ἑκατέρου· εἶτα
συμμίξας ἐς τὠϋτὸ συνεψεῖν· δὲ οἶος στέαρ τὸ ἀπὸ τῶν

196

you gave to the preceding one. Generally the disease continues for nine years, and then, being wasted away, the patient dies. Few escape, for the disease is severe.

If you wish, treat the disease as follows: first administer a vapour-bath; on the day after the patient has had the vapour-bath, give him a half chous of melicrat to which a little vinegar has been added; have him drink this off without taking a breath, and then cover him thickly with blankets, and leave these on for a good long time. If he cannot tolerate this, but wants to vomit, let him vomit. If vomiting does not occur after a certain time has elapsed, let the patient drink, in addition, a large cup of warm water, and vomit by being tickled with a feather. When he has vomited, and feels better, let him rest that day. When dinner time arrives, have him dine on a small barley-cake, and take as main dish salt-fish and leeks, of which he should eat as many as he can; let his wine be sweet. From then on, let the patient bathe every day at dawn in very copious hot water; after the bath, you must make sure that he does not have a chill, by putting him to bed, and having him sleep long. On arising from his sleep, have him walk at least twenty stades[4] that day; on the days that follow, let him walk an additional five stades each day until he reaches one hundred stades. Beginning from the first day, you must clean the cavity downwards with beet- and cabbage-juice: boil these separately, and strain off a chous of each; then mix into each one quarter mina of fat from a sheep's kid-

4 See table of measures p. 290.

---

74 Potter (cf. chs. 19, 32): διαφέρει M.
75 Potter: ἢν M.

νεφρῶν τεταρτημόριον μνᾶς ἐν ἀμφοτέροις· ἑψεῖν διελών·
ὁκόταν δὲ μέλλῃ πιεῖσθαι, πρὸς μὲν τῆς κράμβης τὸν
χυλὸν ἅλα παραβαλεῖν, πρὸς δὲ τὸν τῶν τευτλίων
μέλι παραχέειν, εἶτα[76] χωρὶς ἑκάτερον πίνειν· ἢ μέλι
παραχέας παρὰ τὴν ἑτέρην κύλικα [καὶ][77] πίνειν, παρὰ
δὲ τὴν ἑτέρην ἅλας· ἐκπιεῖν δὲ χρὴ πάντα τὸν χυλόν.

Ταῦτα μὲν ποιέειν χρὴ τριήκοντα ἡμέρας. τῷ δὲ
δευτέρῳ μηνὶ ἐσθιέτω ἄρτον καὶ κρέα πίονα ὑὸς ἑφθά,
ἄλλο μηδέν· οἶνον δὲ πινέτω λευκόν, αὐστηρόν, καὶ
ὁδὸν ὁδοιπορεέτω μὴ ἐλάσσω σταδίων τριήκοντα πρὸ
τοῦ δείπνου, μετὰ δὲ τὸ δεῖπνον δέκα· καὶ μὴ ῥιγούτω,
ἀλλ᾽ ἐσκεπάσθω. ταῦτα ἢν ποιέῃ ῥήιον οἴσει τὴν
νοῦσον.

Τῷ δὲ τρίτῳ μηνὶ κυκεῶνα ἀνθινὴν πινέτω· σελίνου
ῥίζας καὶ ἄνηθον καὶ πήγανον καὶ μίνθην καὶ κορίαννον καὶ μήκωνας ἁπαλὰς καὶ ὤκιμον καὶ φακὸν καὶ
ῥοιῆς γλυκείης καὶ οἰνώδεος χυλόν· εἶναι δὲ χρὴ τὰς
γλυκείας διπλασίας· εἶναι δὲ χρὴ συναμφοτέρων τοῦ
χυλοῦ ἡμικοτύλιον καὶ οἴνου μέλανος ἡδέος αὐστηροῦ
ἡμικοτύλιον καὶ ὕδατος κοτύλης ἥμισυ· ἔπειτα ἄνθεα
τρίψας λεῖα, διῆναι τούτῳ τῷ συγκεκρημένῳ, καὶ
198 ἐγχέαι ἐς κύλικα· ἔπειτα ἐπι|βαλεῖν ἄλευρα ὀρόβων,
ὁκόσον ὀξύβαφον, καὶ ἄλφιτον ἴσον, καὶ τυροῦ παλαιοῦ αἰγείου ξέσας τὸ ἴσον τοῖσι ὀρόβοισι· ταῦτα
συγκυκήσας ἐκπιέτω· ἔπειτα διαλιπὼν ὀλίγον χρόνον
ἀριστάτω ἄρτον, καὶ ὄψον ἐχέτω τέμαχος νάρκης ἢ
ῥίνης ἢ γαλεοῦ ἢ βατίδος, καὶ κρέα οἰὸς ἐσθιέτω
ἑφθά· καὶ παχυνέτω ἑωυτὸν ἡσυχίην ἄγων ὡς μά-

ney, and boil them, still separately; when the patient is ready to drink, into the cabbage-juice sprinkle salt, and into the beet-juice pour honey; then have him drink each separately; alternatively add honey to one cup and have him drink it, and then salt to the next cup; he must drink all the juice.

These things you must do for thirty days. In the second month, let the patient eat bread and boiled fat pork, but nothing else; let him drink dry white wine, and walk a distance of not less than thirty stades before dinner, and after dinner ten stades; let him keep himself covered in order to avoid a chill. If he does these things, he will bear the disease more easily.

In the third month, let the patient drink a cyceon flavoured with plants: celery roots, dill, rue, mint, coriander, fresh poppies, basil, lentil, and the juices of sweet and vinous pomegranates, the amount of the sweet being double that of the vinous; there must be a half cotyle of the two juices together, a half cotyle of pleasant dry dark wine, and a half cotyle of water. Grind the plants fine, soak them in this mixture, and pour into a cup. Then add vetch-meal to the amount of an oxybaphon, an equal amount of barley-meal, and grate in an amount of aged goat's cheese equal to the vetches. Stir all these into the cyceon, and have the patient drink it off. Then, leaving a short interval, let him breakfast on bread, take as main dish a slice of torpedo, angel-fish, dogfish or skate, and eat boiled mutton; let him

---

76 Potter: ἢν δὲ M.
77 Del. later mss.

λιστα· καὶ πυριὴν διὰ δεκάτης ἡμέρης [ἐς ἑωυτὸν ἡσυχῇ].⁷⁸

Τῷ δὲ τετάρτῳ μηνὶ πυριὴν διὰ πέμπτης ἡμέρης ἀτρέμα, καὶ ἐσθιέτω ὄψον ὡς πλεῖστον· ὄψῳ δὲ χρεέσθω τυροῖσι καὶ κρέασι ὀλίγοισι οἰὸς ἑφθοῖσι. ὁδοιπορεέτω δὲ σταδίους οὗτος τῷ τετάρτῳ μηνὶ ἀρξάμενος τῇ πρώτῃ ἡμέρῃ ἀπὸ δέκα σταδίων, μέχρις ὀγδοήκοντα αὐτῷ στάδιοι γένονται· περιπατεέτω δὲ τῆς ἡμέρης ὀγδοήκοντα σταδίους, μετὰ τὸ δεῖπνον εἴκοσιν, ὄρθρου δὲ τριήκοντα.

Τὸ δὲ λοιπὸν τοῦ χρόνου διαιτάσθω μᾶζαν καὶ ἄρτον ἐσθίων ἀμφότερα· καὶ ὄψον ἐχέτω σελάχια, καὶ κρέα πάντα ἐσθιέτω πλὴν βοείων καὶ χοιρείων· ἰχθύων δὲ τῶνδε ἀπεχέσθω, κεστρέος καὶ ἐγχέλυος καὶ μελανούρου· ἐσθιέτω δὲ νάρκην καὶ ῥίνην καὶ βατίδα καὶ γαλεὸν καὶ τρυγόνα καὶ βατράχους, τῶν δὲ λοιπῶν μηδένα. ἢν δὲ δοκέῃ ἀσινέα εἶναι, καὶ κυκεῶνα, ἐπειδὰν μέλλῃ καθευδήσειν, πινέτω ἀπὸ οἴνου μέλανος, ἡδέος, παλαιοῦ, δικότυλον κύλικα· καὶ μεθ᾽ ἡμέρην τῷ αὐτῷ οἴνῳ χρεέσθω ἐπὶ σιτίῳ. καὶ ὁδοιπορεέτω τῆς ἡμέρης, ἑκατὸν πεντήκοντα σταδίους, μετὰ δεῖπνον εἴκοσι, ὄρθρου δὲ τεσσεράκοντα. οὗτος γίνεται ὑγιὴς μάλιστα ἐνιαυτῷ οὕτω θεραπευόμενος.

13. Ἢν μυελὸς ὁ κατὰ τὴν ῥάχιν αὐαίνηται· αὐαίνεται δὲ μάλιστα, ὁκόταν τὰ | φλέβια ἀποφραχθῇ τὰ ἐς τὸν μυελὸν τείνοντα καὶ ἡ ἐκ τοῦ ἐγκεφάλου ἔφοδος. διὰ κάκωσιν δὲ τοῦ σώματος τάδε πάσχει καὶ νοσέει—αὐαίνεται δὲ μάλιστα καὶ ἀπὸ λαγνείης—

200

102

fatten himself by keeping very quiet; administer a vapour-bath every tenth day.

In the fourth month, administer mild vapour-baths every fifth day, and let the patient eat main dishes as much as he can; as such employ cheeses and small amounts of boiled mutton. Let this patient walk stades in the fourth month, beginning on the first day with ten stades, and increasing until he reaches eighty stades; then let him walk eighty stades daily, of which twenty should be after dinner and thirty in the morning (sc. the other thirty between his meals).

From then on, let the regimen include eating both barley-cakes and bread; let the patient take as main dish selachians and eat all meats except beef and pork; of fish, let him abstain from grey mullet, eel and black-tail, and eat torpedo, angel-fish, skate, dogfish, sting-ray and fishing-frog, but no others. If he seems quite well, let him also drink a cyceon, when he is about to retire, a two-cotyle cup made from pleasant old dark wine; by day, let him use this same wine with his meal. Have him walk one hundred and fifty stades a day, of which twenty should be after dinner and forty in the early morning. When treated in this way, the patient usually recovers in a year.

13. If the marrow of the spine becomes dry (it generally becomes dry when the small vessels extending into the marrow are blocked, and also the passage out of the brain), as a result of the body's insult the patient suffers the following and is ill (the marrow also often becomes dry from

---

[78] Del. Potter.

τάδε οὖν πάσχει· ὀδύνη ὀξέη ἐμπίπτει αὐτῷ ἐς τὴν κεφαλήν, καὶ ἐς τὸν τράχηλον καὶ τὴν ὀσφῦν, καὶ ἐς τοὺς μύας τῆς ὀσφύος, καὶ ἐς τὰ ἄρθρα τῶν σκελέων, ὥστε ἐνίοτε οὐ δύνανται ξυγκάμπτειν. καὶ ἡ κόπρος οὐ διαχωρέει, ἀλλ' ἵσταται, καὶ δυσουρέεται. οὗτος κατ' ἀρχὰς μὲν τῆς νούσου ἡσυχαίτερον διάγει· ὁκόσῳ δ' ἂν ὁ χρόνος τῇ νούσῳ ἀπομηκύνηται, πονέει ἅπαντα μᾶλλον. καὶ τὰ σκέλεά τε οἰδέει ὡς ἀπὸ ὑδέρου, καὶ ἔκλεα ἐκφλυνδάνει ἐκ τῆς ὀσφύος, καὶ τὰ μὲν ἄλλα ὑγιαίνεται, τὰ δὲ ἄλλα παραγίνεται.

Τούτῳ, ὁκόταν οὕτω ἔχῃ, καθῆραι τὴν κεφαλὴν τῷ ἱππόφεω ὀπῷ ἢ τῷ Κνιδίῳ κόκκῳ πυριήσας πρῶτον τὸ σῶμα εὖ μάλα. τῆς δὲ ἑσπέρης μετὰ τὴν κάθαρσιν πτισάνης δύο τρυβλία ῥοφεέτω μέλι παραχέας· οἶνον δὲ λευκὸν πινέτω μαλθακόν. τῇ δ' ὑστεραίῃ ὀνείου γάλακτος διδόναι αὐτῷ ἑφθοῦ, μέλι παραχέας, ὀκτὼ κοτύλας ἐκπιεῖν· ἢν δὲ μὴ ὄνειον ἔχῃς, βοείου ἢ αἰγείου ἑφθοῦ τρία ἡμίχοα, παραχέας μέλι. καὶ τὴν ὤρην γαλακτοποτεέτω ἐν ὀρῷ⁷⁹ καὶ γάλακτι πέντε καὶ τεσσεράκοντα ἡμέρας. σιτίοισι δὲ καὶ ὄψοισι χρεέσθω ὡς διαχωρητικωτάτοισι· οἶνον δὲ πινέτω λευκόν, μαλθακόν, Μένδαιον.⁸⁰ ὁπόταν δὲ παχύτατος ᾖ, καῦσαι αὐτοῦ ἐς τὴν ὀσφῦν ἑκατέρωθεν τῶν σπονδύλων τέσσαρας ἐσχάρας, καὶ ἐς τὸ μετάφρενον δεκαπέντε ἑκατέρωθεν, καὶ εἰς τὸν αὐχένα δύο μεταξὺ τῶν τενόντων· ἢν γὰρ τύχῃς καύσας, [καὶ]⁸¹ ὑγιέα ποιήσεις· ἡ δὲ νοῦσος χαλεπή.

venery); he suffers, then, as follows: sharp pain occupies the head, neck, loins, lumbar muscles, and the joints of the legs, so that sometimes patients cannot flex their hip. The stools do not pass off, but are stopped, and the patient has dysuria. At the beginning of the disease, this patient goes along quite peacefully, but as the period of the illness lengthens, his sufferings increase in every way. His legs swell up as if from dropsy, and ulcers break out in the region of the loins; while some of these ulcers are healing, others develop.

When the case is such, clean out the patient's head with hippopheos juice or Cnidian berry, after first applying very thorough vapour-baths to his body. In the evening after the cleaning, let him take two bowls of barley-gruel to which honey has been added; let him drink mild white wine. On the next day, give him eight cotylai of boiled ass's milk with honey, to drink off. If you do not have ass's milk, then give three half choes of boiled cow's or goat's milk with honey. In season, also let him drink whey and milk for forty-five days. Of cereals and main dishes have him eat the most laxative, and drink mild white Mendean wine. When the person is in a state of corpulence, burn four eschars in the lumbar region on both sides of the vertebrae, fifteen on each side in the back, and in the neck two between the tendons; for, if your cautery succeeds, you will cure him. The disease is severe.

---

79 Littré (*sero* Cornarius): ὀροβίῳ M.
80 Potter (cf. chs. 16–18, 24): Μενδήσιον M.
81 Del. Cornarius.

14. Ἀπὸ νεφρῶν αἵδε νοῦσοι γίνονται τέσσαρες.

Ἀπὸ τῆς πρώτης τάδε πάσχει· ὀδύνη ὀξείη ἐμ-πίπτει ἐς τὸν νεφρὸν καὶ ἐς τὴν ὀσφῦν καὶ ἐς τὸν κενεῶνα καὶ ἐς τὸν ὄρχιν τὸν κατὰ τὸν νεφρόν· καὶ οὐρέει πυκινά, καὶ στάζει[82] κατ' ὀλίγον τὸ οὖρον· καὶ ἅμα τῷ οὔρῳ προέρχεται[83] ψάμμος, καὶ ὁκόταν ἐξίῃ διὰ τῆς οὐρήθρης ἡ ψάμμος, ὀδύνην παρέχει ἰσχυρὴν ἐν τῇ οὐρήθρῃ. ὁκόταν δὲ διεξουρήσῃ, ⟨ἡ⟩[84] ὀδύνη ἀνίησι· ἔπειτα αὖτις ἐν τοῖσι αὐτοῖσι ἄλγεσι κέεται· ὁκόταν δὲ οὐρέῃ, καὶ τὸν καυλὸν ὑπὸ τῆς ὀδύνης τρίβει.

Πολλοὶ δὲ τῶν ἰητρῶν οἱ μὴ συνιέντες τὴν νοῦσον, ὁκόταν ἴδωσι τὴν ψάμμον, δοκέουσι λιθιῆν τὴν κύ-στιν, ⟨ἣν⟩[85] μὲν οὐ λιθιῇ, τὸν δὲ νεφρὸν λιθιᾷ. αὕτη ἡ νοῦσος γίνεται ἀπὸ φλέγματος, ὁκόταν ὁ νεφρὸς ἐς ἑωυτὸν ἀναλαβὼν φλέγμα μὴ ἀφίῃ πάλιν, ἀλλ' αὐτοῦ ξυμπωρωθῇ· τοῦτο γίνεται λίθοι λεπτοὶ οἷον ψάμμος.

Τοῦτον ὁκόταν οὕτως ἔχῃ, τῷ ὀπῷ τῆς σκαμμωνίης ἢ αὐτῇ τῇ ῥίζῃ, πυριήσας[86] πρόσθεν ἅπαν τὸ σῶμα, ὑποκαθῆραι. τῇ δ' ὑστεραίῃ ἀπὸ ἐρεβίνθων λευκῶν τῷ χυλῷ ὑποκαθῆραι δύο χοεῦσι· ἅλα[87] δὲ παρεμβαλὼν διδόναι πίνειν. μετὰ δὲ ταῦτα ποτοῖσι καὶ βρωτοῖσι καὶ λουτροῖσι μελετᾶν, διδοὺς τὰ αὐτὰ ⟨ἃ⟩[88] καὶ τῷ στραγγουριῶντι δίδοται φάρμακα. ὁκόταν δὲ ἡ ὀδύνη πιέσῃ, λούειν πολλῷ καὶ θερμῷ, καὶ χλιάσματα προσ-τιθέναι ὅπῃ πονέει μάλιστα. ὁκόταν δὲ ἀποιδήσῃ καὶ

---

[82] Potter: στύφει M.   [83] Cornarius: προσ- M.

14. From the kidneys these four diseases arise.

In the first one, the patient suffers the following: a sharp pain attacks his kidney, loin, flank, and his testicle on the same side as the kidney; he urinates frequently, and drips urine a little at a time; together with the urine sand too is passed, and when the sand discharges through the urethra, it produces violent pain in it. When the patient has finished urinating, the pain stops; later, though, he labours under the same distress again. When he is passing urine, he rubs his penis because of the pain.

Many physicians that do not understand the disease, when they see the sand, think the patient is suffering from stones of the bladder, which he is not, but rather from stones of the kidney. This disease arises from phlegm, when the kidney takes up phlegm into itself and does not release it, but it solidifies there; this forms fine stones like sand.

When the case is such, clean the patient downwards with scammony juice or the root itself, first applying vapour-baths to the whole body. On the following day, clean downwards with juice from white chick-peas to the amount of two choes; add salt, and give this to drink. After that, treat with drinks, food and baths, giving the same medications that are given to a patient with strangury. When pain is pressing, wash with copious hot water, and apply fomentations where the pain is worst. When the

---

84 Later mss.   85 Potter.

86 Littré: -ῆσαι M.   87 Later mss: ἄλλα M.

88 Grmek and Wittern, "Die Krankheit des attischen Strategen Nikias und die Nierenleiden im Corpus Hippocraticum", *Arch. int. hist. sci.* XXVI. 100 (1977) 10.

ἐξαρθῇ, ὑπὸ τοῦτον τὸν χρόνον παραχρῆμα[89] τάμνειν
κατὰ τὸν νεφρόν, καὶ ἐξελὼν τὸ πύος, τὴν ψάμμον
διουρητικοῖσι ἰῆσθαι· ἢν μὲν γὰρ τμηθῇ, ἐλπὶς ἐκφυ-
γέειν· ἢν δὲ μή, ἡ νοῦσος τῷ ἀνθρώπῳ συναποθνήσκει.

15. Ἄλλη νεφροῦ· αἱ μὲν ὀδύναι ἰσχυρῶς πιέζουσι
204 ὡς | ἐν τῇ πρόσθεν· γίνεται δὲ τὸ νόσημα ἀπὸ ταλαι-
πωρίης, ὁκόταν ῥαγῇ τὰ φλέβια ἐς τὸν νεφρὸν τείνοντα, ἔπειτα ὁ νεφρὸς αἵματος πλησθῇ. οὗτος ὁκόταν
ταῦτα πάθῃ, ἐξουρέει ἅμα τῷ οὔρῳ αἷμα κατ᾽ ἀρχὰς
τοῦ νοσήματος, ἔπειτα πύα[90] προϊόντος δὲ τοῦ χρόνου.
οὗτος ἢν ἡσυχίην <ἔχῃ>[91] τῷ σώματι, τάχιστα ὑγιὴς
ἔσται· ἢν γάρ τι πονήσῃ αἱ ὀδύναι πολλῷ μᾶλλον
ἕξουσι. ὁκόταν οὖν ἔμπυος ᾖ ὁ νεφρός, ἀποιδέει παρὰ
τὴν ῥάχιν.

Τοῦτον, ὅταν οὕτω ἔχῃ, τάμνειν κατὰ τὸ ἀποιδέον,
μάλιστα μὲν βαθείην τομὴν κατὰ τὸν νεφρόν· κἢν
μὲν τύχῃς ταμών, παραχρῆμα ὑγιέα ποιήσεις· ἢν δὲ
ἁμάρτῃς, κίνδυνος ἕλκος ἔμμοτον γενέσθαι. ἢν δὲ
συμφυῇ τὸ ἕλκος, ἐκπυοῦται εἴσωθεν ἡ κοιλίη [ἡ][92]
ἀπὸ τοῦ νεφροῦ· κἢν μὲν ῥαγῇ ἔσωθεν καὶ χωρήσῃ
κατὰ τὸν ἀρχὸν τὰ πύα, ἐλπὶς ἐκφυγέειν· ἢν δὲ ψαύσῃ
τοῦ ἑτέρου νεφροῦ, κινδυνεύσει καταφθαρῆναι.[93] μελετᾶν δὲ χρὴ[94] φαρμάκοισι τοῖσιν αὐτοῖσι πᾶσιν, ὡς[95]

---

[89] Potter: παρασχῇ M.
[90] Grmek and Wittern, 16: πύει M.   [91] Later mss.
[92] Del. Potter.   [93] Θ resumes with -ρῆναι.

affected area swells up and becomes raised, incise then immediately over the kidney, draw out the pus, and attend to the sand with diuretics. For, if this person is incised, he has a hope of survival, but if he is not, the disease clings to him until he dies.

15. Another disease of the kidney: pains press violently just as in the preceding disease. This disease arises from exertion, when the vessels extending to the kidney rupture, and the kidney is then filled with blood. When a patient suffers this, he passes blood with his urine, at the beginning of the disease, and then, as time goes on, pus. If he rests his body, he will recover very quickly; however, if he exerts himself in any way, the pains increase greatly. Now, when the kidney suppurates, swelling appears beside the spine.

When the case is such, incise the patient at the site of the swelling, making an especially deep cut over the kidney. If your incision succeeds, you will quickly bring about recovery, but, if you fail, there is the danger that an ulcer requiring tents will arise. If this ulcer unites, the cavity is made to suppurate from within by the kidney; if the pus formed breaks into the intestine and passes off through the rectum,[5] there is hope of survival, but if it comes to involve the other kidney, the patient will be in danger of perishing. You must treat with all the same medications as in the pre-

---

[5] I.e. escapes from the abdominal cavity by passing into the gastro-intestinal tract and out through the anus.

καὶ τὸν πρόσθεν, καὶ τὴν δίαιταν τὴν αὐτὴν ἐχέτω.
αὕτη ἡ νοῦσος χαλεπή, καὶ πολλοὶ ἐκ ταύτης τῆς
νούσου ἐς φθίσιν⁹⁶ νεφρίτιδα κατέστησαν.

16. Ἄλλη νεφροῦ· τὸ μὲν οὖρον προέρχεται οἷον
ἀπὸ κρεῶν βοείων ὀπτῶν χυλός· γίνεται δὲ τὸ νόσημα
ἀπὸ χολῆς μελαίνης, ὅταν χολὴ ἐς τὰ φλέβια συρρυῇ
τὰ τείνοντα ἐς τὸν νεφρόν· καὶ ὅταν στῇ, ἑλκοῖ τὰ
φλέβια καὶ τὸν νεφρόν· ὑπὸ οὖν τῆς ἑλκώσιος τοιοῦ-
τον ὑποχωρέει ἅμα τῷ οὔρῳ. αἱ δὲ ὀδύναι ἔχουσιν ἐν
τῇ ὀσφύϊ καὶ τῇ κύστει, καὶ ἐν τῷ περινῷ καὶ ἐν αὐτῷ
τῷ νεφρῷ ἐπ' ὀλίγον χρόνον· ἔπειτα ἀνῆκεν ὁ πόνος,
καὶ αὖτις ἐπέλαβεν ὀξὺς δι' ὀλίγου· καὶ ἐς τὸ λεπτὸν
τῆς γαστρὸς ἔστιν ὅτε ὀδύνη ἐμπίπτει.

Τοῦτον ὅταν οὕτως ἔχῃ, ὑποκαθῆραι τὴν κοιλίην τῇ
σκαμωνίῃ ῥίζῃ· πίνειν διδόναι ταῦτα ἃ καὶ τῷ στραγ-
206 γουριῶντι· | καὶ ὅταν ἡ ὀδύνη ἔχῃ, λούειν θερμῷ⁹⁷ καὶ
πολλῷ, καὶ χλιάσματα προστιθέναι πρὸς τὸ πονέον
μάλιστα. καὶ ῥυφήματι χρήσθω ἀλεύρῳ ἐφθῷ, μέλι
παραχέων, καὶ τῇ ἄλλῃ διαίτῃ χρήσθω ὡς διαχωρη-
τικωτάτῃ, καὶ οἶνον πινέτω λευκὸν Μένδαιον μελι-
χρόν, ἢ ἄλλον⁹⁸ τὸν ἥδιστον καλὸν κεκρημένον. αὕτη
ἡ νοῦσος μάλιστα⁹⁹ ἐκλείπει. καὶ τὴν ὥρην¹⁰⁰ γαλακτοποτείτω ἐς κάθαρσιν τῷ ὀρῷ· τὴν δὲ γαλακτοπωσίην ἐν θέρει¹⁰¹ ποιείσθω πέντε καὶ τεσσεράκοντα
ἡμέρας. ταῦτα ἢν ποιέῃ, ῥᾷστα τὴν νοῦσον διάξει.

⁹⁶ Θ adds ἢ.
⁹⁷ θερμῷ om. Θ.

110

ceding case, and have the patient follow the same regimen. This disease is severe, and many patients have gone on from it into a nephritic consumption.

16. Another disease of the kidney: the urine passed is like the juice of roasted beef. This disease arises from dark bile, when it collects in the vessels extending to the kidney; when it has come to rest there, it ulcerates the vessels and the kidney; thus, because of the ulceration, material of the kind described passes off with the urine. For a short time, pains are present in the loins, bladder, perineum and the kidney itself; then the attack relents, but after a short period it presses sharply again; sometimes pain also attacks the narrow part of the belly.

When the case is such, clean the cavity downwards with scammony root, and give the same things to drink that are given to a patient with strangury. When pain is present, wash in copious hot water, and apply fomentations where it is worst. Let the patient drink boiled meal as gruel, adding honey, and otherwise follow a regimen that is strongly laxative; let him drink white Mendean wine sweetened with honey, or some other very pleasant one, well mixed with water. This disease usually goes away. In season, also let the patient drink whey, in order to bring about cleaning; in summer, let him do this for forty-five days. If the patient does these things, he will get through the disease most easily.

---

98 M adds λευκὸν.
99 Θ: οὐ μάλα M.
100 M adds ὀροποτεέτω καὶ.
101 Potter: θρει Θ: ὥρῃ M.

17. Ἄλλη νεφροῦ· τὸ μὲν νόσημα γίνεται ἀπὸ χολῆς καὶ φλέγματος, τοῦ δὲ[102] θέρεος μάλιστα γίνεται·[103] γίνεται δὲ καὶ ἀπὸ λαγνείης ἡ νοῦσος. οὗτος τάδε[104] πάσχει· ὀδύναι πιέζουσιν αὐτὸν ἐς τὴν λαπάρην καὶ ἐς τὸν κενεῶνα καὶ ἐς τὴν ὀσφὺν καὶ ἐς τοὺς μύας τῆς ὀσφύος, καὶ πάσχει οἷα γυνὴ ὠδίνουσα. καὶ οὐκ ἀνέχεται ἐπὶ τοῦ ὑγιέος κατακείμενος· ἃ γὰρ πονέει[105] τοῦ κενεῶνος δοκέει ἀποκρέμασθαι ὡς ἀπορρησσόμενα· ἐπὶ δὲ ἃ πονεῖ ἢν[106] κατακέηται, οὐκ ἀλγέει. οἱ δὲ πόδες καὶ αἱ κνῆμαι αἰεὶ ψυχραὶ αὐτοῦ. τὸ δὲ οὖρον μόγις προέρχεται ὑπὸ τῆς θερμασίης[107] καὶ παχύτητος τοῦ οὔρου, καὶ ἢν ἐάσῃς αὐτὸ ὀλίγον χρόνον καταθεὶς τέως ἂν καταστῇ, ὄψῃ τὸ ὑπεστηκὸς παχύ, οἷόν περ ἄλευρον· καὶ ἢν μὲν χολὴ ἐπικρατέῃ, ὑπόπυρρον αὐτὸ ὄψῃ· ἢν δὲ ἀπὸ φλέγματος ᾖ τὸ νόσημα, λευκὸν καὶ παχὺ ἔσται.

Καὶ τὸ μὲν πρῶτον ἐς ἐνιαυτὸν ἢ[108] ὀλίγῳ ἐλάσσω ἢ ὀλίγῳ πλείω χρόνον τοιαῦτα πάσχων διατελέει· ἢν δ' ὁ[109] χρόνος πλείων τῇ νούσῳ ἀπομηκύνηται,[110] 208 πονέει μᾶλλον καὶ ἐκπυοῦται, καὶ ὅταν ἔμπυος | ᾖ, ἀποιδέει. καὶ ὅπου ἂν μάλιστα ἀποιδέῃ τάμνειν ἐς τὸν νεφρόν, καὶ ἀφιέναι τὰ πύα· καὶ ἢν τύχῃς ταμών, παραχρῆμα ὑγιῆ ποιήσεις.

Τοῦτον ὅταν οὕτως ἔχῃ, τοῖσιν αὐτοῖσι πᾶσι θεραπεύειν κατὰ τὸ πρόσθεν. κατ' ἀρχὰς τῆς νούσου πρὸ τῆς ὑποκαθάρσιος πυριᾶσαι. καὶ τοῖσι λουτροῖσι

102 δὲ Θ: δ' ἔτεος Μ.   103 γίνεται om. Μ.

17. Another disease of the kidney: this disease arises from bile and phlegm, and occurs mainly in summer; it also arises from venery. The person suffers the following: pains press him in the side, flank, loins, and the lumbar muscles, and he experiences the same as a woman in labour. He cannot tolerate lying on his healthy side, for the part of the flank in which he feels pain seems to hang down as if it were being torn away; if, however, he lies on the affected side, he has no pain. His feet and his legs below the knees become permanently cold. Hardly any urine passes, because it is so hot and thick, and if you leave it a short while, setting it down until it deposits its sediment, you will see that the precipitate is thick, just like meal; if bile predominates, the sediment will appear reddish; if the disease is from phlegm, the sediment will be thick and white.

First, the patient goes through a year, more or less, like this. If the illness goes on longer, his pains increase and he suppurates internally; when this happens, he swells up. Wherever the swelling is greatest, incise the kidney, and draw off pus; if your incision succeeds, you will quickly bring about recovery.

When the case is such, treat with all the same measures as above. At the beginning of the disease, before cleaning downwards, apply vapour-baths; let the patient not take

104 M: ταχέως Θ.

105 ἃ γὰρ πονέει Θ: ἀλλ' ἄγαν πονέει καὶ M.

106 ἐπὶ δὲ απονειην Θ: ἔπειτα πονέει ἐπὶ δὲ τὰ πρηνέα ἦν M.   107 Θ: φλεγμασίης M.

108 ἢ om. Θ.   109 δ' ὁ Θ: ὧδε M.

110 Θ: καὶ ὑπο- M.

μὴ πυκινὰ λούσθω, ἀλειφέσθω δὲ μᾶλλον, μηδὲ ῥιγούτω, καὶ τοῦ ἡλίου ἀπεχέσθω, μηδὲ λαγνευέτω.
ταῦτα ἢν ποιέῃ,[111] τάχιστα ὑγιὴς ἔσται· ἡ δὲ νοῦσος
χαλεπή.

Ἢν δὲ βούλῃ ἄνευ φαρμάκων ὑγιᾶ ποιῆσαι,[112] πα-
χὺν ποιῆσαι ἀπὸ τῆς διαίτης, ἤν τε ταύτην τὴν νοῦσον
κάμνῃ, ἤν τε τῶν προτέρων τινά. τὰ σιτία διελὼν ἃ
μεμαθήκει ἐσθίειν δέκα μερίδας, ἔπειτα μίαν ἀφελὼν
μερίδα, τὰ λοιπὰ καταφαγέτω, ὄψον δ᾽ ἐχέτω κρέας
οἰὸς τετρυμένον· καὶ περιπατησάτω δέκα σταδίους
ταύτης τῆς ἡμέρης. τῇ δὲ ὑστεραίῃ καὶ τῇ τρίτῃ καὶ
μέχρι τῶν δέκα ἡμέρων ὑποβαίνων μερίδα, ἐλάσσω
ἐσθιέτω, καὶ περιπατείτω δέκα σταδίους αἰεὶ πλείω
ἑκάστης ἡμέρης. ὅταν δὲ ἐς τὴν ἐσχάτην τοῦ σιτίου
μερίδα ἀφίκηται καὶ ἐς τοὺς ἑκατὸν σταδίους, ἐσθιέτω
μίαν μόνην μερίδα, καὶ ταύτης τῆς ἡμέρης περιπατησάτω ἑκατὸν σταδίους· μετὰ δὲ τὸ δεῖπνον εἴκοσιν,
ὄρθρου δὲ τεσσεράκοντα. οἶνον δὲ πινέτω Μένδαιον,[113]
λευκόν, αὐστηρόν. ταῦτα ποιείτω τρεῖς ἡμέρας.[114]
ἔπειτα τὸν λοιπὸν χρόνον ὑποβαίνων τῶν περιπάτων
καὶ τῶν σιτίων πλείω ἐσθίων, τὸν αὐτὸν τρόπον
210 ὥσπερ ἀφῆρει, οὕτω | προστιθέτω· τῶν δὲ περιπάτων
ἀφαιρείτω μέχρι τῶν δέκα ἡμερέων. ἔπειτα ἡσυχίην
ἐχέτω ὡς μάλιστα, καὶ εὐωχείσθω σιτία τε καθαρὰ
καὶ ὄψα ὡς πλεῖστα, ἔχων[115] καὶ τὰ γλυκέα πάντα
σύμφορα· λαχάνων δὲ ἀπεχέσθω καὶ τῶν ὀξέων καὶ
τῶν δριμέων καὶ ὁπόσα φῦσαν παρέχει ἁπάντων· καὶ
λούσθω πολλῷ καὶ θερμῷ, καὶ μὴ ῥιγούτω.

many baths, but rather be anointed with oil; he must avoid chills, sun and venery. If he does these things, he will quickly recover. The disease is severe.

If you wish to cure the patient without medications, whether he is suffering from this particular disease or from one of the ones above, first fatten him by means of his regimen; then have him divide the cereals he is accustomed to eat into ten portions, and on the first day subtract one portion and eat the rest; as main dish let him eat ground mutton; also have him walk ten stades. On the next day, the third, and up to the tenth, have him each day reduce the amount of cereal he eats by one portion, and each day walk ten stades more. When he has arrived at the final portion of his cereal and at one hundred stades, let him eat only the one portion, and on that day walk the hundred stades: after dinner twenty, and early in the morning forty (sc. the other forty between his meals). Also give him dry white Mendean wine to drink. Let him do this for three days. Then have him decrease his walks for ten days, and eat more cereals, adding a portion a day in the same way as he subtracted. Then, let him keep as quiet as possible, be well fed on fine cereals and very generous main dishes, and also have all the suitable sweets; let him abstain from vegetables, foods that are acid and sharp, and everything that produces flatulence; let him bathe in copious hot water, but be careful to avoid a chill.

---

111 M adds καὶ μὴ.
112 ὑ. π. Θ: ἰῆσθαι M.
113 M: μὲν παλαιόν Θ.
114 H, Littré: μῆνας ΘM.
115 Θ: πιότατα, ἐχέτω M.

Ταῦτα ἢν ποιέῃ, τάχιστα ὑγιὴς ἔσται.

18. Ἐκ τῆς νεφρίτιδος ἐπιλαμβάνει ἥδε ἡ νοῦσος, καί ἐστι[116] μεγάλη τῶν φλεβῶν τῶν κοίλων, αἳ τείνουσιν ἐκ τῆς κεφαλῆς παρὰ τὰς σφαγὰς διὰ τῆς ῥάχιος ἐς τὸ σφυρὸν τὸ ἐκτὸς τοῦ ποδὸς καὶ ἐς τὸ μεταξὺ τοῦ μεγάλου δακτύλου. γίνεται δὲ τὸ νόσημα ἀπὸ φλέγματος καὶ χολῆς, ὅταν ἐς τὰς φλέβας συρρυῇ· αἱ δὲ φλέβες αὗται αἵματός εἰσι πλήρεις· ἢν οὖν τι παρέλθῃ ἀλλοῖον ἐς αὐτάς,[117] νοσέουσι.

Τάδε πάσχει, ἢν μὲν ἐπὶ δεξιὰ νοσέῃ· ἄρχεται ἡ ὀδύνη ἔχουσα ἐκ τῆς κοτυληδόνος ἐς τὸ ἰσχίον[118] κατ' ἀρχάς. ὅσῳ δ' ἂν πλείω χρόνος προΐῃ καὶ ἀπομηκύνηται, ἥ τε ὀδύνη ὀξυτέρη καὶ κατέρχεται κατωτέρω· καὶ ὅταν ἐς τὸ σφυρὸν ἀφίκηται τὸ ἐκτὸς τοῦ ποδὸς τότ'[119] ἐς τὴν ῥάχιν πάλιν ἀνέρχεται καὶ ἐς τὴν κεφαλήν·[120] καὶ ὅταν ἐν τῇ κεφαλῇ στῇ τὸ ἄλγος,[121] πιέζει ἰσχυρῶς, καὶ δοκέει διαρρήσσειν τὴν κεφαλήν· καὶ οἱ ὀφθαλμοὶ αἵματος[122] πίμπλανται.[123]

Τοῦτον, ὅταν οὕτως ἔχῃ, ἐλατήριον πῖσαι ἢ θαψίης
212 ῥίζαν ἢ ἐλλέβορον ἢ ὀπὸν | σκαμωνίης. μετὰ δὲ τὴν κάθαρσιν ταῦτα προσφέρειν, ἃ καὶ τοῖσι πρόσθεν. ἢν δὲ μὴ ὑπὸ ταύτης τῆς θεραπείης παύηται, γάλακτι παχύνας καῦσαι παρὰ[124] τὴν ὠμοπλάτην τὴν δεξιὴν τέσσερας ἐσχάρας, καὶ ἐς τὴν κοτυληδόνα τοῦ ἰσχίου τοῦ δεξιοῦ τρεῖς, καὶ ὑπὸ τὸν γλουτὸν δύο, καὶ ἐν μέσῳ τοῦ μηροῦ δύο, καὶ ὑπὲρ τοῦ γούνατος μίαν, καὶ ὑπὲρ

---

116 καί ἐστι om. M.   117 M: ταύτας Θ.

116

If he does these things, he will quickly recover.

18. This disease develops out of nephritis and is serious, involving the hollow vessels that extend down from the head past the throat, along the spine, to the outer malleolus of the foot and the middle of the large toe; it arises when these vessels are invaded by phlegm and bile, for inasmuch as the vessels are normally filled with blood, if anything else enters them, they become ill.

If the disease is on the right side, the patient suffers the following: at first pain is present between the acetabulum and the hip-joint. The more time goes by and the longer the illness becomes, the sharper is the pain and the further down it moves; when it arrives at the outer malleolus of the foot, it ascends again to the back and the head, and when it comes to rest in the head, it presses violently, and seems to be splitting the head apart; the eyes fill with blood.

When the case is such, have the patient drink squirting-cucumber juice, thapsia root, hellebore, or scammony juice. After the cleaning, administer the same things as to the patients above. If the disease does not go away with this treatment, fatten the patient on milk, and burn four eschars beside his right shoulder-blade, three into the acetabulum of his right hip-joint, two under his buttock, two in the middle of his thigh, and one each above his knee

---

118 Θ: τοῦ ἰσχίου M.
119 Potter: τὸ Θ: om. M.
120 M adds ἔρχεται.
121 Θ: ἕλκος M.
122 Θ: φλέγματος M.
123 M adds καὶ τὸ σῶμα.
124 Θ: κατὰ M.

τοῦ σφυροῦ μίην.[125] οὗτος, ἢν οὕτω καυθῇ, οὐ παρήσει
οὔτε ἄνω οὔτε κάτω τὴν νοῦσον διαχωρέειν. ἢν δέ που
ἡ ὀδύνη φθῇ[126] ῥαγεῖσα, ἢν μὲν στηρίξῃ πρὶν καυθῆ-
ναι ἐς τὸ σκέλος, χωλὸς ἔσται· ἢν δὲ ἐς τὴν κεφαλήν,
κωφὸς ἢ τυφλός· ἢν δ᾽ ἐς τὴν κύστιν, προχωρέει ἅμα
τῷ οὔρῳ καὶ τοῦ αἵματος[127] τεσσεράκοντα ἡμέρῃσιν.
ἀλλὰ χρή, ἢν ἐς τὴν κύστιν ῥαγῇ, διδόναι τὰ αὐτὰ
φάρμακα, ἃ καὶ τῷ στραγγουριῶντι· καὶ ἤν που ἄλλῃ
ἡ ὀδύνη στῇ, καῦσαι· καίειν δὲ χρὴ τὰ μὲν σαρκώδεα
σιδηρίοισι, τὰ ὀστώδεα καὶ νευρώδεα μύκησι.

Τάδε δὲ τούτων πρότερον χρὴ ποιῆσαι, ἢν κατ᾽
ἀρχὰς τῇ νούσῳ παραγένῃ. οἶνον λευκὸν Μένδαιον
διδόναι πίνειν ὀλίγῳ ὑδαρέστερον ὡς πλεῖστον μεθ᾽
ἡμέρην[128] τέως ἂν αἱμορραγήσῃ κατὰ τὰς ῥῖνας· ὅταν
δὲ ἄρξηται, ἐᾶν ῥυῆναι ἡμέρας τὸ ἐλάχιστον[129] τρεῖς
καὶ δέκα· ὅταν δὲ αὗται αἱ ἡμέραι διαγένωνται, μηκέτι
μεθυσκέσθω, μηδ᾽ ὅταν ἄρξηται ἅπαξ ῥεῖν· πινέτω
μέντοι ὀλίγῳ πλέονα τὸν οἶνον ἐπὶ τῷ σίτῳ, ὅπως ἂν
ῥέῃ τὸ αἷμα. ἤδη δὲ παυσθέντος τοῦ αἵματος ἐρράγη
τισὶν ἐς τὴν κύστιν καὶ ἐχώρησεν αἷμα καὶ πύα· ἢν
214 οὖν ῥαγῇ, διδόναι τὰ αὐτὰ | φάρμακα, ἃ καὶ τῷ
στραγγουριῶντι, καὶ τοῦ οἴνου διδόναι τοῦ αὐτοῦ
πίνειν πολύν. οὗτος οὕτω μελετώμενος καὶ σιτία

[125] καὶ ὑπὲρ τοῦ σφυροῦ μίην om. Θ.
[126] Littré: ὀφθῇ ΘΜ.
[127] καὶ τοῦ αἵματος Θ: αἵματος μάλιστα Μ.
[128] Μ adds καὶ μεθυσκέσθω.   [129] ἐᾶν . . . ἐλάχιστον
Θ: ῥέειν, τὸ ἐλάχιστον ἡμέρας ῥέει Μ.

and his ankle. If a person is cauterized in this way, it will not allow the disease to migrate either upwards or downwards. If, however, pain breaks out first, and, before you can cauterize, it becomes fixed in the leg, the patient will become lame; if it becomes fixed in the head, he will become deaf or blind, if in the bladder, blood will be passed along with the urine for forty days; if pain occupies the bladder, give the same medications as to a patient with strangury. If the pain settles somewhere else, cauterize: burn fleshy parts with irons, osseous and fibrous ones with fungi.[6]

First, however, you must do the following, if you attend the disease at its beginning: by day give very large quantities of white Mendean wine to drink, slightly more dilute than normal, until the patient bleeds through his nostrils; when this begins, allow the flow to continue for at least thirteen days. When these days have passed, the patient need no longer be drunk, nor actually even when the flow has once begun, but do have him drink somewhat more wine than usual with his meals, in order that the blood flow will continue. In some patients, after the epistaxis had already stopped, blood and pus have broken into the bladder and passed with the urine; if such a break occurs, give the same medications as to a patient with strangury, and the same wine to drink, in large amounts. This patient, if he

[6] I.e. moxibustion; cf. Caelius Aurelianus, *Chronic Diseases* V (Drabkin 916–18) and J. S. Milne, *Surgical Instruments in Greek and Roman Times*, Oxford, 1907, 120.

προσφερόμενος διαχωρητικὰ καὶ τὰ ὄψα, τάχιστ' ἂν
ὑγιὴς γένοιτο. ἡ δὲ νοῦσος χαλεπή.

19. Ἄλλη δὲ ἥδε[130] ἀπὸ τῆς ἀριστερῆς φλεβός· τὰ
μὲν ἄλλα πλῆθος τὰ αὐτὰ πάσχει, ἃ καὶ ὁ πρόσθεν· ἐς
δὲ τὸν σπλῆνα ὀδύνη ἐνστηρίζει ὀξέη εὐθὺς κατ'
ἀρχὰς τοῦ νοσήματος. καὶ ἢν μὲν συνίῃς[131] παραχρῆμα πρὶν καταστηρίξῃ ἐς τὸν σπλῆνα,[132] μύκησι
καῦσαι ὀκτὼ ἐσχάρας, τὰς κεφαλὰς ἀπολαβὼν τοῦ
σπληνός, ὡς τάχιστα· καὶ ὅπου ἂν ἄλλη ἡ ὀδύνη
στηρίξῃ, καῦσαι καὶ οὕτω παραχρῆμα ὑγιής.[133] ἢν δὲ
μὴ καυθῇ, ὑγιὴς δὲ γένηται ἀπὸ τοῦ αὐτομάτου, τοῖσι
πολλοῖσι δωδεκάτῳ ἔτει ἡ νοῦσος αὖτις ὑπετρόπασε,
καὶ ἢν λάβηται τοῦ σπληνός, τοῖσι πολλοῖσιν ὕδερον
ἐποίησεν. ἀλλὰ χρὴ παραχρῆμα θεραπεύειν ὡς τὸν
πρόσθεν, καὶ ἢν δοκέῃ, καῦσαι ὥσπερ τὸν ἕτερον,[134]
ἢν ἡ ὀδύνη καθεστήκῃ ἐν τοῖσιν αὐτοῖσιν ἄρθροισιν.
ἢν δὲ μὴ οὕτω μελετηθῇ, τὸ λοιπὸν φθειρόμενος θνή-
σκει· ἡ γὰρ νοῦσος χαλεπή.

20. Περὶ τοῦ φλέγματος τὰς αὐτὰς γνώμας ἔχω, ἃς
καὶ περὶ[135] χολῆς· ἰδέας[136] πολλὰς εἶναι.

Καὶ ἐπιδήμιον μέν ἐστι τὸ νεώτατον, ἑωυτοῦ[137] καὶ
ἡ ἴησις ῥᾴστη. ἐμέτους γὰρ χρὴ ποιέεσθαι μετὰ τὸ
σιτίον ἡμέρας δύο ἢ τρεῖς, προαριστῶντα καὶ ἡσυ-
χάζοντα, ἢν εἰώθῃ[138] τὰς πρόσθεν ἡμέρας μονοσιτέειν
216 τε καὶ ταλαιπωρέεσθαι· ἢν δὲ μή, | τῇ αὐτῇ διαίτῃ

---

130 ἥδε om. M.   131 Θ: μὴ ξυνίῃ M.
132 ἐς τὸν σπλῆνα Θ: ἀλλ᾽ ἢ ἐς τὸν πλεύμονα M.

120

is treated in such a way, and he takes laxative cereals and main dishes, will very quickly recover. The disease is severe.

19. The corresponding disease arising from the left vessel: generally this patient suffers the same things as the preceding one, except that right at the onset of the disease a sharp pain becomes fixed in his spleen. If you discover this at once, before the pain is firmly established in the patient's spleen, very quickly burn eight eschars with fungi, holding their heads away from the spleen; wherever else the pain settles, cauterize; if you do this, there is immediate recovery. However, if the patient is not cauterized, but recovers spontaneously, in many cases the disease recurs in the twelfth year and, if it involves the spleen, frequently produces dropsy. Therefore, you must treat at once just as you did the preceding patient, and, if it seems advisable, cauterize too if pain becomes fixed in the same organs. If the patient is not treated in this way, in the time that follows he wastes away and dies; for the disease is severe.

20. About phlegm I have the same views as about bile, that there are several forms.

The common variety is present for only a very short time, and its cure is easiest: the patient must induce vomiting after his meals for two or three days, and, if it has been his habit in the days that have gone before to eat but one meal and to exert himself strenuously, now take breakfast

133 M adds ἔσται.   134 Θ: πρότερον M.
135 ἃς and περὶ om. Θ.   136 M adds φημὶ.
137 ἐπιδήμιον . . . ἑωυτοῦ Θ: τὸ μὲν ἐπιδήμιόν ἐστι, τὸ δὲ νεώτατον M.   138 M: εἴωθεν Θ.

χρήσθω, λούσθω δὲ πολλῷ καὶ θερμῷ ὅταν μέλλῃ
ἔμετον ποιέεσθαι. καὶ μᾶζαν τ᾽ ἐσθιέτω ψαιστήν, καὶ
ἄρτον ἔωλον ἔξοπτον, ἕλκοι γὰρ ἂν μᾶλλον τὸ φλέγ-
μα· ὄψοισι δὲ χρήσθω καὶ λαχάνοισι δριμέσι, καὶ τὰ
λιπαρὰ καὶ τὰ ὀξέα καὶ τὰ γλυκέα, ταῦτα ἅπαντα
ἐπιτήδεια ξυμμεμιγμένα προσφέρεσθαι· καὶ πᾶσι
χλωροῖσι τοῖσι λαχάνοισι χρήσθω. καὶ ἐπιπινέτω ἐπὶ
τῷ σίτῳ ὀλίγον πυκνὰ οἶνον γλυκύν, καὶ πλακοῦντα
ἐπιφαγέτω ἐπὶ τελευτῆς καὶ μέλι καὶ σῦκα. ὅταν δὲ
δειπνήσῃ, πινέτω λαύρως τὰς κύλικας, καὶ ὅταν ἤδη
πλήρης ᾖ, κατακοιμηθήτω ὀλίγον, ἔπειτ᾽ ἐπεγερθεὶς
ἐμείτω πίνων οἴνου μεγάλην κύλικα χλιερῷ[139] κεκρη-
μένην· ἕλκοι γὰρ ἂν μᾶλλον τὸ φλέγμα[140] τῶν σαρκῶν
καὶ τὸν χυμόν, καὶ ξηραίνοιτο ἂν μᾶλλον τὸ σῶμα.
ἐμείτω δὲ ἐς ὃ[141] ἂν τὰ σῦκα ἐξεμέσῃ, ὕστατα γὰρ
ἐξεμέεται ταῦτα.[142] τῇ ὑστεραίῃ ἐν ἡσυχίῃ ἐχέτω[143]
ἑωυτὸν μέχρι δείπνου, δειπνήτω δὲ ἄρτον αὐτοπυρίτην, ὄψον δ᾽ ἐχέτω τῶν ἰσχυροτέρων· οἶνον δὲ πινέτω
μέλανα αὐστηρόν. αὕτη μὲν[144] τοῦ ἐπιδημίου φλέγματος ἴησις.

Ἢν δὲ δυνατὸς ἐὼν ἐσθίειν καὶ πίνειν ἤδηται τοῖσι
σιτίοισι, εἶτα τὰ σκέλεα βαρύνηται, καὶ ἡ χροιὴ
μετηλλαγμένη ᾖ, τούτου[145] φάναι ἐν τῇ κοιλίῃ φλέγμα
τὸ λυπέον εἶναι. ἀλλὰ χρή, ὁπόταν οὕτως ἔχῃ, κλύζειν
μέλιτι καὶ οἴνῳ γλυκεῖ καὶ ἐλαίῳ λίτρου παραμίξας
ὅσον οἶος ἀστράγαλον· ταῦτα γὰρ τῇ φύσει εὐμε-

---

139 Θ: καὶ χλιερῷ ὕδατι Μ.

and keep himself quiet; if this was not his habit, he can continue with his normal regimen, but bathe in copious hot water when he is about to induce vomiting. Let him eat ground barley-cake and day-old well-baked wheat bread, for these will draw the phlegm very well; also employ sharp main dishes and vegetables, all the suitable rich, acid and sweet ones mixed together, and all green vegetables. Let the patient drink a little thick sweet wine with his meals, and at the end eat a flat-cake, honey and figs. After he has had his dinner, have him drink his cups rapidly and, when he is full, lie down a little to sleep; then rouse him, give him a large cup of wine mixed with warm water to drink, and have him vomit; for this will better draw the phlegm and fluid out of the tissues, and his body will be better dried; let him continue to vomit until the figs come up, for these are vomited up last. On the following day, have the patient rest until dinner, and then dine on whole-wheat bread and main dishes from among the stronger ones; also let him drink dark dry wine. This is the treatment for the common variety of phlegm.

If the patient is able to eat and drink, and he enjoys his meals, but then his legs become weighed down, and his colour altered, you may suppose that the phlegm of the grievous kind is present in his cavity. When the case is such, you must clean him out below, by using soda to the amount of a sheep's vertebra mixed with honey, sweet wine and olive oil, for these things are the most agreeable to

---

140 M adds ἐκ.   141 ἐς ὃ Θ: ἕως M.
142 ἐ. τ. Θ: τὰ σῦκα ἐξεμέεται. ταῦτα μὲν τῇδε M.
143 ἐν ᾗ. ἐ. Θ: συνεχέτω M.
144 M adds οὖν.   145 Θ: -ῳ M.

νέστατα τοῦ ἀνθρώπου ἐς τὸν κλυσμόν· μέτρον γὰρ
χρὴ ἑκάστου εἶναι, τοῦ μὲν οἴνου κοτύλην, ἡμικοτύλιον δὲ τοῦ ἐλαίου, καὶ τοῦ μέλιτος ἴσον. ἢν δὲ μὴ
218 κλύζειν βούλῃ, | ὑγρὸν χρὴ τὸν ἄνθρωπον ποιῆσαι,
πυριάσαντα ἐν ὑγρῇ τῇ πυριήσει.[146] τάχα γὰρ ἂν
οὕτως ὑποκενωθείη ἡ κόπρος. ὑπὸ γὰρ τῆς ὑπερξηρα-
σίης τῶν σιτίων τοῦτο πάσχει· εἰ μὲν οὖν τις ἐσθίοι
σιτία λίαν ἔγχυλα, οὐκ ἂν πάσχοι ταῦτα οὕτω σφό-
δρα· εἰ δὲ καὶ πάσχοι ποτέ, ὀλίγης ἂν ἰήσιος δέοιτο.
τοῦτον οὕτως ἰώμενος τάχιστ' ἂν ὑγιέα ποιήσαις.

21. Ἢν δὲ τύχῃ παλαιότερον ἐὸν τὸ φλέγμα—λευκὸν δὲ καλέεται τοῦτο τὸ φλέγμα[147]—πάσχει[148] τάδε·
βαρύνει τὸν ἄνθρωπον μᾶλλον, καὶ ἰδέην ἔχει ἀλλοίην
τοῦ ἐπιδημίου. ὠχρότερός τέ ἐστι, καὶ οἰδέει διὰ παντὸς[149] τὸ σῶμα, καὶ τὸ πρόσωπον ἐρεύθει, καὶ τὸ
στόμα ξηρόν, καὶ δίψα ἔχει, καὶ ὅταν φάγῃ, τὸ πνεῦμα
πυκινὸν ἐπιπίπτει αὐτῷ. οὗτος αὐτῆς τῆς ἡμέρης τότε
μὲν γίνεται ῥάων, τότε δὲ πονέει ἐξαπίνης καὶ δοκέει
ἀποθανεῖσθαι. τούτῳ ἢν μὲν ἡ γαστὴρ αὐτομάτη
ταραχθῇ, ἐγγυτάτω ὑγιὴς προβαίνει.

Ἢν οὖν μὴ ταραχθῇ αὐτομάτη ἡ κοιλίη, καθαίρειν
χρὴ διδόντα τοῦ κνεώρου ἢ τοῦ Κνιδίου κόκκου ἢ τοῦ
ἱππόφεω ἢ τῆς Μαγνησίης λίθου· καὶ μετὰ τὴν κάθαρσιν ῥυφεῖν δοῦναι φακῆς τρυβλίον ἓν ἢ δύο, συν-
εψέσθω δὲ τῇ φακῇ σκόροδα· καὶ σεύτλου λιπαροῦ
ἀνηδύντου, ἐπ' ἀλφίτων περιπάσαντα, δοῦναι τρυ-

---

[146] Θ: πυρίῃ Μ.

man's constitution as an enema; let the amounts be one cotyle of wine and a half cotyle each of olive oil and honey. If you do not want to use an enema, then you must moisten the person by applying moist vapour-baths; for, in this way, the faeces will be rapidly emptied below. A person is affected by this condition as the result of excessive dryness of his foods; thus, if someone eats foods that are very succulent, he will not be as likely to suffer from it, and, even if he were to, it would require little treatment. If you treat this patient as indicated, you will very quickly make him well.

21. If the phlegm happens to be of longer duration—this phlegm is called white—the person suffers the following: he is afflicted more intensely, and has different signs than in the common variety of the disease; he is paler, his body swells up all over, his face becomes red, his mouth is dry, he is thirsty, and when he eats anything rapid breathing comes over him. In the course of the same day, this patient is at one time better, but at another time suddenly suffers an attack and seems about to die. If his belly has a spontaneous movement, he proceeds to health very quickly.

However, if the cavity does not have a spontaneous movement, you must clean it out by giving spurge-flax, Cnidian berry, hippopheos or magnetic stone; after the cleaning, give one or two bowls of lentil-soup with boiled garlic, to drink; also give a bowl of unseasoned beets boiled in grease, over which you have sprinkled meal. Let the

---

147 λευκὸν . . . φλέγμα om. Θ.
148 M adds οὖν.
149 δ. π. Θ: οἰδήματι πᾶν M.

βλίον. πινέτω δὲ οἶνον μέλανα αὐστηρὸν ἰσχυρόν. τῇ
δ᾽ ὑστεραίῃ περιπατησάτω σταδίους εἴκοσι τὸ ἑωθι-
νόν· ἐλθὼν δὲ φαγέτω ἄρτον μικρὸν[150] ἔξοπτον, καὶ
ὄψον ἐχέτω σκόροδα ὀπτά· καὶ πινέτω τοῦ αὐτοῦ οἴνου
ὀλίγον ἀκρητέστερον. ἔπειτα βαδιζέτω σταδίους τρι-
ήκοντα, καὶ ὅταν ὥρη ᾖ δείπνου, δειπνήτω[151] ὅσον |
220 περ καὶ ἠριστήκει, ὄψον δ᾽ ἐχέτω μάλιστα μὲν πόδα
ὑὸς καὶ κεφάλαια. εἰ δὲ μή, ἀλεκτρυόνος κρέα ἢ ὑός·
τετρυμένοισι δὲ καὶ ἐφθοῖσι[152] χρήσθω· ἰχθύων δὲ
σκορπίῳ ἢ δράκοντι ἢ κόκκυγι ἢ καλλιωνύμῳ ἢ κωβιῷ ἢ τῶν ἄλλων ἰχθύων ὅσοι τὴν αὐτὴν δύναμιν
ἔχουσι· λαχάνων δὲ σκορόδοισι χρήσθω καὶ ἄλλῳ
λαχάνῳ μηδενὶ χρήσθω.[153] ταῦτα δὲ ὡς πλεῖστα τρωγέτω καὶ ὠμὰ καὶ ὀπτὰ καὶ ἐφθά. καὶ ἐσθιέτω αἰεὶ
πλείω ἑκάστης ἡμέρης, καὶ ταλαιπωρείτω πρὸς τὰ
σιτία τεκμαιρόμενος καὶ ὀλίγῳ πλείω.

Τοῦτο τὸ νόσημα γίνεται μάλιστα θέρεος ὥρῃ[154]
ἀπὸ ὑδροποσίης[155] καὶ ὕπνου· κρίνεται δ᾽ ἐν τριήκοντα
ἡμέρῃσιν, εἰ θανάσιμον ἢ οὔ. ταῦτα μὲν ποιείτω ὅταν
αἱ τριάκοντα ἡμέραι παρέλθωσιν. ἐν δὲ τῇσι πρώτῃσι
τῶν ἡμερέων ῥυφήματι διαχρήσθω φακῇ λεπτῇ[156]
ἐπωκεστέρῃ[157] τῷ ὄξει, καὶ πτισάνῃ ὀξέῃ· πινέτω δὲ
χλιαρὸν μελίκρητον, ἄλφιτα ἐπιβάλλων[158] ὀλίγα, ἵνα
ἀνωργασμένον[159] τὸ σῶμα ᾖ πρὸς τὴν φαρμακοπω-
σίην· καὶ εὑδέτω ὑπαίθριος ταύτας τὰς ἡμέρας. καὶ ἤν
σοι δοκέῃ τοῦ αἵματος ἀφελεῖν ἀπὸ τῆς ὀσφύος,

150 Θ: πικρὸν Μ.   151 δ., δ. Θ: δειπνεέτω Μ.

patient drink strong dry dark wine. On the next day, have
him walk twenty stades early in the morning, and, on coming back, eat a small loaf of well-baked bread, as main dish
have baked garlic, and drink a little of the same wine,
mixed with a very little water. Then let him walk thirty
more stades, and, when it is dinner time, eat as much for
dinner as he had for breakfast; as main dish let him best
have feet and head-parts of a swine; if not that, the meat of
fowl or swine—let him employ these ground and boiled;
also of fish scorpion fish, weever, piper, star-gazer, goby or
others of the same nature; of vegetables let him employ
only garlic, but of these eat as many as possible, raw, baked
and boiled. Each day have him eat more, and also exert
himself a little more, determining how much according to
his food.

This disease occurs mainly in summer from drinking
water and from sleeping; thirty days are the critical period
that decide whether or not the patient will die. Do the following when the thirty days have passed: on the first days,
have the patient regularly take as gruel thin lentil-soup
well acidified with vinegar, and acid barley-water; let him
drink cool melicrat over which a little meal has been sprinkled, in order that his body will be relaxed for a medication; also let him sleep outdoors on these days. If it seems
advisable to you to draw blood from the loins, apply a cup-

---

152 δὲ καὶ ἑφθοῖσι om. M.   153 χρήσθω om. M.
154 Later mss: ὥρης ΘΜ.   155 M adds ἔτι δὲ.
156 Θ: ἑφθῇ Μ.
157 Cornarius: ἐπιεικεστέρη ΘΜ.
158 Θ: -πάσων Μ.
159 Littré: ἂν ὀργισμένον Θ: ἂν ὠργισμένον Μ.

σικύην προσβάλλειν,[160] καὶ τὰς ἐν ὄσχη φλέβας σχά-
σον τὰς παχυτάτας. οὗτος οὕτω θεραπευόμενος τάχιστα ὑγιὴς ἔσται.

22. Ἀπὸ φλέγματος μάλιστα περιίσταται ἐς ὕδε-
ρον· καὶ ἡ πιμελὴ συντήκεται καὶ γίνεται ὕδωρ ὑπὸ
τοῦ καύματος τοῦ ἐν τῷ φλέγματι ἐνεόντος. γνώσῃ δὲ
222   τούτῳ, ὅστις δυνατός ἐστιν ἰηθῆναι | καὶ ὅστις μή·
ἕως ἄν τινι[161] ἐπὶ τῷ ἤτρῳ ἐπῇ ἡ πιμελή,[162] δυνατὸς[163]
ἰηθῆναί ἐστι. γνώσῃ δὲ τοισίδε μάλιστα, εἰ ἔστι
πιμελὴ ἐπὶ τῷ ἤτρῳ ἢ οὔ· ἢν μὲν[164] πυρετοὶ ἐπιγένωνται καὶ μὴ δύνηται ἀνίστασθαι καὶ ὁ ὀμφαλὸς
ἔξω ἐξέχῃ πεφυσμένος, φάναι μηκέτι ἐπεῖναι πιμελήν.
ἢν δὲ πυρετὸς[165] μὴ ἐπιγένηται, καὶ δυνατὸς[166] ᾖ ἀνίστασθαι, καὶ ὁ ὀμφαλὸς μὴ ἐξέχῃ, φάναι ἐπεῖναι
πιμελὴν καὶ ἰήσιμον εἶναι.

Τούτῳ ξυμφέρει τὴν κοιλίην ξηραίνειν, διδόντα
ἄρτον μέλανα[167] αὐτοπυρίτην, θερμόν, μὴ ἔωλον, ὄψον
δὲ ὄνου[168] κρέας καὶ κυνὸς τελείου, καὶ ὑὸς καὶ οἰὸς ὡς
πιότατα ἐφθά,[169] καὶ ἀλεκτρυόνος ὀπτὰ καὶ θερμά· καὶ
πουλύποδας ἐσθιέτω ἑψῶν ἐν οἴνῳ μέλανι αὐστηρῷ·
οἶνον δὲ πινέτω μέλανα ὡς παχύτατον καὶ στρυφνό-
τατον· ἰχθύων δὲ χρήσθω κωβιῷ, δράκοντι, καλλιωνύμῳ, κόκκυγι, σκορπίῳ καὶ ἄλλοισι τοῖσι τοιούτοισι
πᾶσιν ἐφθοῖς ἑώλοισι καὶ ψυχροῖσι· ξηρότατοι γὰρ
οὗτοι μάλιστά εἰσι· καὶ ἐς τὸν ζωμὸν μὴ ἐμβαπτέσθω,
καὶ ἄναλτοι ἔστωσαν οἱ ἰχθύες. λαχάνων δὲ χρήσθω

---

[160] Potter: -βάλλων Θ: -βαλεῖν Μ.

ping instrument and slit the widest vessels of the scrotum. If this patient is treated in such a way, he will very quickly recover.

22. From phlegm the most frequent change is to dropsy: fat melts, from the burning heat of the phlegm, and becomes water. You will know by the following who can be healed and who not: as long as fat is present on the patient's abdomen, he can be healed; and you will know, especially by the following, whether or not fat is present on the abdomen: if fevers supervene, if the patient is unable to stand up, and if his navel is distended and protrudes, then assume that fat is no longer present; but if fever does not supervene, if the patient can stand up, and if his navel does not protrude, assume that fat is present, and the patient curable.

It benefits this patient if you dry out his cavity by giving him fresh warm dark whole-wheat bread, and as main dish the meat of ass, mature dog, swine and sheep, these very fat and boiled, or meat of fowl, roasted and warm; also let him eat polyp boiled in dry dark wine; let him drink dark wine that is very thick and sour. Of fish let him take goby, weever, star-gazer, piper, scorpion fish and others of the same sort, all boiled and eaten cold on the following day; for these are generally the driest; let the patient not dip them in sauce; the fish must not be salted. Of vegetables

---

161 Θ: γὰρ ἄν τις Μ.   162 ἐ. ἡ π. Θ: ἔχει πιμελήν Μ.
163 Θ: ἀδύνατος Μ.   164 Μ: μὴ Θ.
165 δὲ π. Μ: τε πυρετός τε Θ.   166 Θ: ἀδύνατος Μ.
167 Θ: μὲν Μ.   168 Θ: λαγωοῦ Μ.
169 π. ἐ. Θ: ὀπτά Μ.

ῥαφανίσι καὶ σελίνοισιν· ἑψήσθω[170] δὲ καὶ φακὴν τῷ
ὄξει ἐπωκεστέρην· καὶ περιπατείτω καθ᾽ ἡμέρην καὶ
μετὰ τὸ δεῖπνον καὶ ὄρθρου, καὶ ὄψιος εὑδέτω, καὶ
πρώϊος ἐξεγειρέσθω. καὶ ἢν μὲν ὑπὸ τούτων καθίστη-
ται· εἰ δὲ μή, πῖσαι αὐτὸν κνέωρον ἢ ἱππόφεω ὀπὸν ἢ
Κνίδιον κόκκον, καὶ μετὰ τὴν κάθαρσιν φακῆς δύο
τρυβλία ῥυφείτω καὶ ἄρτον σμικρὸν καταφαγέτω·
οἶνον δὲ πινέτω μέλανα, στρυφνόν, ὀλίγον· πινέτω δὲ
τὸ φάρμακον δὶς τῆς ἡμέρης, τέως ἂν λαπαρὸς γένη-
ται. ἢν δὲ οἴδημα καθεστήκῃ ἐν τῇ ὄσχῃ καὶ τοῖσι |
224   μηροῖσι καὶ τῇσι κνήμῃσι, κατασχᾶν χρὴ ὀξέῃ τῇ
μαχαίρῃ πολλὰ καὶ πυκινά. ταῦτα ἢν ποιέῃς, τάχιστα
ὑγιᾶ ποιήσεις.

    23. Ὁ δὲ ὕδερος ἀπὸ τῶνδε γίνεται· ὅταν θέρεος
ὥρῃ διψήσας ὕδωρ πολὺ πίῃ ἐπισπάδην,[171] ἐκ τούτου
γίνεσθαι φιλέει μάλιστα· ὁ γὰρ πλεύμων πλησθεὶς
αὖθις ἀφίησιν ἐς τὰ στήθεα, καὶ ὅταν ἐν τοῖσι στήθεσι γίνηται, καῦμα παρέχει σφόδρα ὥστε τήκειν τὸ
στέαρ, τὸ[172] ἐπὶ τῇσιν ἀρτηρίῃσιν ἐπεόν·[173] καὶ ἢν
ἅπαξ ἄρξηται τήκεσθαι τὸ στέαρ πολλῷ πλέον, ἐν
ὀλίγῳ χρόνῳ τὸν ὕδερον ἐνεποίησε. γίνεται δὲ καὶ ἢν
φύματα ἐν τῷ πλεύμονι ἐμφυῇ καὶ πλησθῇ ὕδατος καὶ
ῥαγῇ ἐς τὰ στήθεα. ὡς δὲ γίνεται καὶ ἀπὸ φυμάτων ὁ
ὕδερος, τόδε μοι μαρτύριον καὶ ἐν βοῒ καὶ ἐν ὑῒ καὶ ἐν
κυνί· μάλιστα γὰρ τῶν τετραπόδων τούτοισι γίνεται
φύματα ἐν τῷ πλεύμονι ἃ ἔχει ὕδωρ, διαταμὼν δὲ ἂν
γνοίης τάχιστα, ῥεύσεται γὰρ ὕδωρ. δοκέει δὲ καὶ ἐν
ἀνθρώπῳ ἐγγίνεσθαι τοιαῦτα πολλῷ μᾶλλον ἢ ἐν

have him eat radishes and celery; also boil lentil-soup well acidified with vinegar. Let the patient take walks each day, after dinner and early in the morning, go to bed late, and be awakened early. If, with this treatment, he settles down, fine. If not, have him drink spurge-flax, hippopheos juice, or Cnidian berry, and, after the cleaning, take two bowls of lentil-soup and eat a small loaf of bread; let him drink a small portion of sour dark wine, and take a medication twice a day until he has been opened. If swelling occurs in the scrotum, thighs, and legs below the knees, you must make repeated incisions with a sharp scalpel. If you do these things, you will very quickly make the patient well.

23. Dropsy arises in the following way: when, in summer, a person that is thirsty drinks a large amount of water at one draught, it is most likely to arise from this; for when the lung becomes full, it sends water back into the chest, and when this enters the chest, it produces great burning heat, so that the fat present on the bronchial tubes melts; once this fat begins to melt very much, it soon gives rises to dropsy. The disease also arises if tubercles form in the lung, fill up with fluid, and rupture into the chest. (That dropsy truly does also arise from tubercles, here is my proof from the cow, swine and dog: tubercles containing fluid occur most frequently among quadrupeds in the lungs of these animals, as you would very quickly discover by cutting through one, for water will run out; and it seems probable that in man such things are present much more

---

170 Some later mss, Littré: ὀψάσθω Θ: ὀψήσθω M.
171 Littré: -στάδην Θ: ἐπὶ πάσῃ διήν M.
172 τὸ σ., τὸ Θ: τὴν πιμελήν, τὴν M.
173 Θ: ἐπεοῦσαν M.

προβάτοισιν, ὅσῳ καὶ τῇ διαίτῃ χρώμεθα ἐπινούσῳ
μᾶλλον. ἐγένοντο δὲ πολλοὶ καὶ ἔμπυοι φυμάτων ἐγ-
γενομένων.

Τάδε οὖν κατ᾽ ἀρχὰς τῷ νοσήματι ἐπιγίνεται· βὴξ
ξηρή, καὶ ἡ φάρυγξ δοκέει κρέκειν,[174] καὶ ῥῖγος καὶ
πυρετὸς ἐπιγίνεται καὶ ὀρθοπνοίη, καὶ ὁ χρὼς ἐποι-
δαλέος, καὶ οἱ πόδες μάλιστα ἐποιδέουσι,[175] καὶ οἱ
ὄνυχες ἕλκονται. καὶ ἕως μὲν ἐν τῇ ἄνω κοιλίῃ ὁ
ὕδερος ἐνῇ, ὁ πόνος ὀξύς· ὅταν δ᾽ ἐς τὴν κάτω κοιλίην
ἔλθῃ, δοκέει ῥάων εἶναι· ἔπειτα πάσχει ταῦτα προϊ-
226 όν|τος τοῦ χρόνου οἷά περ ὁ πρόσθεν, πιμπραμένης[176]
τῆς κοιλίης. ἔστι δ᾽ ὅτε ἀποιδέει πρὸς τὸ πλευρόν, καὶ
δηλοῖ ᾗ χρὴ τάμνειν.

Ἢν δὲ μὴ ἀποδηλοῖ,[177] λούσας πολλῷ καὶ θερμῷ,
τῶν ὤμων λαβόμενος σεῖσον· εἶτα ἀκροᾶσθαι ἐν ὁποτέρῳ ἂν τῶν πλευρέων μᾶλλον κλυδάζηται.[178] συνεὶς
δὲ τάμνειν κατὰ[179] τὴν πλευρὴν τὴν τρίτην ἀπὸ τῆς
νεάτης μέχρι τοῦ ὀστέου. εἶτα τρυπῆσαι[180] πέρην
τρυπάνῳ περητηρίῳ,[181] καὶ ὅταν τρυπηθῇ, ἀφεῖναι τοῦ
ὕδατος[182] ὀλίγον· καὶ ὅταν ἀφῇς, μοτῶσαι ὠμολίνῳ,
καὶ ἄνωθεν ἐπιθεῖναι σπόγγον μαλθακόν· εἶτα κατα-
δῆσαι ὡς μὴ ἐκπέσῃ ὁ μοτός. ἀφιέναι δὲ[183] δώδεκα
ἡμέρας τὸ ὕδωρ,[184] ἅπαξ τῆς ἡμέρης. μετὰ δὲ τὰς
δώδεκα ἡμέρας τῇ τρισκαιδεκάτῃ ἅπαν ἀφιέναι τὸ
ὕδωρ, καὶ τὸν λοιπὸν χρόνον ἢν ὑπογίνηται ὕδατός τι,
ἀφιέναι. καὶ ὑποξηραίνειν τὴν κοιλίην.

[174] Θ: κέρχειν Μ.   [175] μ. ἐ. Θ: οἰδέουσι Μ.

than in animals, since we employ a more unhealthy regimen.) Many patients also suppurate internally when tubercles are present.

At the beginning, then, this disease includes the following: a dry cough; the throat seems to whistle; chills and fever set in, and orthopnoea; the skin is puffy; the feet are very swollen; the nails become curved. As long as the dropsy occupies the upper cavity, the distress is acute, but when it moves to the lower cavity, the patient seems better. Then, with the passage of time, his cavity becomes distended, and he suffers the same things as the preceding patient. Sometimes swelling appears in the side and indicates where you must incise.

If there is no indication, wash the patient in copious hot water, take hold of him by the shoulders, and shake him; then listen in which of the two sides there is more fluctuation. When you have discovered this, incise down to the bone at the third lowest rib; then pierce right through to the inside with a straight-pointed trephine and, after boring, draw off a little of the fluid. When you have done this, plug the wound with a tent of raw linen and apply a soft sponge from above; then tie the sponge tight in order that the tent does not fall out. Draw off fluid once a day for twelve days. On the thirteenth day, remove all the fluid, and from then on, if any new fluid forms, draw it off. Also dry out the cavity below.

---

176 Θ: πιμπλαμένης M.   177 μὴ ἀ. M: ἀποιδήσῃ Θ.
178 Θ: κλύζ- M.   179 κατὰ om. M.
180 M: -ήσας Θ.   181 Littré (περιτ- Mack): πειρητ- Θ:
τρυγλητ- M.   182 Later mss: τὸ ὕδωρ Θ: τοῦ ὕδρωπος M.
183 M adds χρὴ.   184 Θ: τὸν ὕδρωπα M.

Τάδε δὲ[185] διδόναι μετὰ τὴν τμῆσιν· σκευάσας ὁποῦ
σιλφίου δραχμὴν σταθμόν, καὶ ἀριστολοχίης[186] κνή-
σας[187] ὅσον ἐλάφειον ἀστράγαλον, καὶ φακῶν καὶ
ὀρόβων πεφρυγμένων ἄλφιτα καθήρας ὅσον ἡμιχοίνικον ἑκατέρων· εἶτα ταῦτα συμφυρῆσαι μέλιτι καὶ ὄξει·
εἶτα πλάσαι κόλλικας ἑξήκοντα. τούτων τρίβων ἕνα
ἑκάστης ἡμέρης διεῖναι ἐν οἴνου ἡμικοτυλίῳ μέλανος
αὐστηροῦ ὡς ἡδίστου·[188] εἶτα διδόναι πίνειν νήστει.
τὴν δὲ ἄλλην δίαιταν καὶ ταλαιπωρίην τὴν αὐτὴν
κελεύειν διαιτᾶσθαι ἣν καὶ ὁ πρόσθεν.[189] καὶ ἢν οἰδήσῃ τὰ αἰδοῖα καὶ τοὺς μηρούς, θαρσῶν κατασχάσαι.

Τοῦτον ἢν οὕτως μελετᾷς, τάχιστα ὑγιέα ποιήσεις.[190]

24. Ὅδε δὲ ὁ[191] ὕδερος ἀπὸ τοῦ ἥπατος γίνεται, |
228   ὅταν ἐς τὸ ἧπαρ φλέγμα ἐπινεμηθῇ,[192] καὶ ἀναλάβῃ τὸ
ἧπαρ καὶ ὑγρανθῇ. εὐθὺς οὖν[193] καῦμα παρέχει τὸ
ἧπαρ, καὶ φῦσαν ἐμποιέει, ἔπειτα χρόνῳ ὕδατος ἐμ-
πίμπλαται· κἄπειτα δηγμὸς ἐς τὸ σῶμα ἐμπίπτει, καὶ
οἴδημα ἐν τῇσι κνήμῃσι καὶ τοῖσι ποσίν ἐστι. καὶ τὸ
ἧπαρ σκληρὸν καὶ οἰδέει, καὶ αἱ κληῗδες λεπτύνονται.

Τοῦτον ὅταν οὕτως ἔχῃ, κατ᾽ ἀρχὰς τοῦ νοσήματος
διδόναι αὐτῷ, ἢν ἀλγέῃ τὸ ἧπαρ, τρίβων ὀρίγανον καὶ
ὀπὸν σιλφίου ὅσον ὄροβον, διδόναι δὲ διεὶς[194] πίνειν
ἐν οἴνου ἡμικοτυλίῳ λευκοῦ·[195] πινέτω δὲ καὶ γάλα
αἰγός, τρίτον μέρος μελικρήτου παραμίσγων, τετρα-

185 Θ: Τὰ δὲ χρὴ Μ.   186 Μ: -ίαν Θ.
187 Jouanna (p. 216): κνῆσαι ΘΜ.

134

After the incision, give the following: prepare silphium juice a drachma in weight, grate aristolochia to the amount of a deer's vertebra, and sift a half-choinix each of the meals of parched lentils and vetches; then knead these together with honey and vinegar, and form into sixty trochisci. Grind one of these down each day, soak it in a half-cotyle of dry dark very pleasant wine, and give to the fasting patient to drink. Order him to conduct the rest of his regimen and his exercise the same as the preceding patient. If he swells up in the genital organs and the thighs, make incisions without hesitation.

If you treat this patient in such a way, you will very quickly make him well.

24. The next dropsy arises from the liver, when phlegm encroaches into the liver, and the liver takes it up and becomes moist: the liver immediately produces burning heat, and gives rise to tympanites, and then, after a time, it fills up with fluid. After that, gnawing pains attack the body, and swelling occupies the legs below the knees, and the feet; the liver is hard and swollen, and the collar-bones become lean.

When the case is such, if at the beginning of the disease the patient has pain in his liver, give him ground marjoram and silphium juice to the amount of a vetch, soaked in a half-cotyle of white wine, to drink. Let him also drink a four-cotyle cup of goat's milk, to which one third part of

---

188 ἐν οἴνῳ . . . ἡδίστου Θ: οἴνου μέλανος ἡμικοτυλίῳ αὐστηρῷ ὡς ἡδίστῳ M.   189 ὁ π. Θ: τὸν π. χρόνον M.
190 Θ: ὑγιὴς ἔσται M.   191 Θ: Ὁ δὲ M.
192 Θ: -γένηται M.   193 οὖν om. Θ.   194 δὲ δ. Potter: διεὶς M: δὲ δὶς Θ.   195 Ermerins: οἴνῳ . . . λευκῷ ΘM.

κότυλον κύλικα. σιτίων δὲ ἀπεχέσθω τὰς πρώτας
ἡμέρας δέκα· αὗται γὰρ κρίνουσιν, εἰ θανάσιμος ἢ οὔ.
ῥυφανέτω δὲ πτισάνης χυλὸν κάθεφθον μέλι παραχέ-
ων· οἶνον δὲ πινέτω Μένδαιον λευκὸν ἢ ἄλλον τινὰ[196]
ἥδιστον ὑδαρέα. ὅταν δὲ αἱ δέκα ἡμέραι παρέλθωσι,
σιτία προσφερέσθω καθαρά, καὶ ὄψον ἐχέτω ἀλεκτρυόνος κρέα ἑφθὰ καὶ ὀπτά·[197] ἐχέτω δὲ καὶ σκύλακος
ἑφθά. ἰχθύϊ δὲ γαλεῷ καὶ νάρκῃ χρήσθω ἑφθοῖσιν·[198]
οἶνον δὲ τὸν αὐτὸν πινέτω.

Καὶ ἢν μὲν ὑπὸ τούτων παύηται·[199] ἢν δὲ μή, ὅταν
αὐτὸς ἑωυτοῦ παχύτατος ᾖ καὶ τὸ ἧπαρ μέγιστον,
καῦσαι μύκησιν· οὕτω γὰρ ἂν τάχιστα ὑγιέα ποιή-
σαις· καῦσαι δὲ χρὴ ὀκτὼ ἐσχάρας.

Ἢν δὲ ὁ ὕδερος ἐγγένηται καὶ ῥαγῇ ἐς τὴν κοιλίην,
τοῖς αὐτοῖσιν ἰᾶσθαι οἷσι καὶ τὸν πρόσθεν, φαρμάκοισι καὶ ποτοῖσι[200] καὶ βρωτοῖσι καὶ ταλαιπωρίῃσιν·
οἶνον δὲ πινέτω μέλανα αὐστηρόν. ἢν δέ σοι δοκέῃ
ἀφίστασθαί που τὸ ἧπαρ, καῦσαι σιδηρίῳ καὶ ἀφιέναι τοῦ ὕδατος κατ' ὀλίγον ὡς τὸ πρόσθεν· καὶ τἆλλα
ἰᾶσθαι τὸν αὐτὸν τρόπον. ἢν δὲ μὴ ὑπὸ τούτων ὑγιὴς
230 γένηται, φθειρόμενος χρόνῳ[201] θνῄ|σκει· ἡ γὰρ νοῦ-
σος χαλεπή, καὶ παῦροι ἐκφυγγάνουσιν.

25. Ὅδε δὲ ὁ ὕδερος ἀπὸ τοῦ σπληνὸς γίνεται[202]
ἀπὸ τῆσδε τῆς προφάσιος μάλιστα· ὅταν ὀπώρη ᾖ καὶ
αὐτῆς[203] φάγῃ πολλὴν σύκων χλωρῶν καὶ μήλων.

196 Θ: τὸν Μ.
197 ἑ. καὶ ὀ. Θ: ὀπτὰ θερμά Μ.

melicrat has been added. Have him abstain from foods for the first ten days—this is the critical period that decides whether or not he will die—take gruels of boiled barley-water with honey, and drink a white Mendean or some other very pleasant wine diluted with water. When the ten days have passed, let him take fine cereals, have as main dish the meats of fowl boiled or grilled, and also boiled puppy, of fish employ boiled dogfish and torpedo, and drink the same wine.

If this brings an end, fine. If not, when the patient has spontaneously become very robust and his liver is at its maximum, cauterize with fungi; for in this way you will most quickly make him well; you must burn eight eschars.

If in dropsy there is a break into the cavity, treat with the same things used for the preceding patient: medications, drinks, foods, and exercises; let the patient drink dry dark wine. If the liver seems to you to stand out at some point, cauterize it with an iron, and draw off fluid a little at a time as above; for the rest, apply the same treatment. If the patient does not recover with this treatment, in time he will waste away and die; for the disease is severe, and few escape it.

25. The next dropsy arises from the spleen, most often in the following way: when in late summer a person eats a lot of the season's fruits such as green figs and apples; many

---

198 Θ: ὀπτοῖσι M.
199 Θ: παύσηται, ἅλις M.
200 καὶ ποτοῖσι om. Θ.
201 χρόνῳ om. M.
202 Θ adds δὲ.
203 Θ: -ὸς M.

πολλοὶ δὲ ἤδη βότρυας πολλοὺς καταφαγόντες καὶ
γλεῦκος πιόντες τὴν νοῦσον ἔλαβον.

Ἦν οὖν μέλλῃ[204] ἐς τὸ νόσημα ἐμπεσεῖσθαι, παραχρῆμα ἐν τοῖσι πόνοισίν ἐστιν· ὀδύναι τε γὰρ ὀξέαι ἐν
τῷ σπληνὶ καθεστᾶσι, μεταπίπτουσι δὲ καὶ ἐς τὸν
ὦμον καὶ ἐς τὴν κληῖδα καὶ ἐς τὸν τιτθὸν καὶ ἐς τὴν
λαγόνα· καὶ πυρετοὶ ἰσχυροὶ ἔχουσι· καὶ ἢν φάγῃ τι,
ἡ γαστὴρ πίμπραται·[205] καὶ ὁ σπλὴν ἀείρεται καὶ
ὀδύνην παρέχει. τούτῳ ἢν χρονισθῇ τὸ νόσημα, τὸν
μὲν ἄλλον χρόνον οὐ[206] πονέει, ὅταν δὲ ἡ ὀπώρη ᾖ καὶ
φάγῃ αὐτῆς, τότε[207] πονέει μάλιστα.

Τοῦτον, ὅταν οὕτως ἔχῃ, κατ᾿ ἀρχὰς μελετᾶν, ἄνω
μὲν ἐλλέβορον διδούς, κάτω δὲ κνέωρον ἢ ἱππόφεω
ὀπὸν ἢ κόκκον Κνίδιον· διδόναι δὲ καὶ ὄνειον γάλα
ἑφθὸν[208] ὀκτὼ κοτύλας μέλι παραχέας. καὶ ἢν μὲν ὑπὸ
τούτων καθίστηται, ἅλις· ἢν δὲ μή, ὅταν μέγιστος ᾖ ὁ
σπλὴν καὶ οἰδέῃ μάλιστα, καῦσαι μύκησι, τὰς κεφαλὰς πολλὰς ἀπολαβών, ἢ σιδηρίοισι, φυλασσόμενος
ὅπως μὴ πέρην διακαύσῃς. ταῦτα μὲν κατ᾿ ἀρχὰς
ποιέειν τοῦ νοσήματος· καὶ δίαιταν τήνδε προσφέρε-
σθαι, πυρετοῦ μὴ ἔχοντος· ἄρτῳ μὲν χρήσθω πυρί-
232   νῳ·[209] | ὄψον δ᾿ ἐχέτω τάριχον Γαδειρικὸν ἢ σαπέρδην,
καὶ κρέας τετρυμένον οἰός, καὶ τὰ ὀξέα καὶ τὰ[210] ἁλμυρὰ πάντα ἐσθιέτω, καὶ πινέτω οἶνον Κῷον αὐστηρὸν
ὡς μελάντατον· τῶν δὲ γλυκέων ἀπεχέσθω. ἢν δ᾿

---

204 Θ adds παραχρῆμα.
205 Θ: -πλαται Μ.

persons have also taken the disease after eating too many
grapes and drinking grape-juice.

If, then, a person is about to fall into the disease, he is at
once subject to its sufferings; for sharp pains occupy his
spleen, and also migrate to his shoulder, collar-bone, nipple and flank, there are violent fevers and, if the patient
eats anything, his belly is distended; his spleen swells up
and is painful. If the disease becomes prolonged in this patient, at other times of the year he is free of pain, but, when
late summer arrives and he eats its fruits, his suffering is
great.

When the case is such, at the beginning treat the
patient by giving him hellebore to clean upwards, and
spurge-flax, hippopheos or Cnidian berry to clean downwards; also give him eight cotylai of boiled ass's milk with
honey. If, with this treatment, the disease settles down,
fine. If not, when the spleen is at its greatest size and most
swollen, cauterize it with fungi, holding their many heads
away from the spleen, or with irons, taking care not to burn
right through. Do these things at the beginning of the disease. If there is no fever, prescribe the following regimen:
let the patient eat wheat bread; as main dish have Cadiz
salt-fish or saperdes, and ground mutton; eat all acid and
salty foods; drink dry very dark Coan wine, and abstain
from sweets. If he gets up, and is able, let him wrestle with

---

206 Θ: ἧσσον M.
207 α., τ. Θ: ἀντὶ τοῦ πρόσθεν M.
208 ἑφθὸν om. M.
209 M adds ὀπτὸν ἢ τῶν σκληρῶν πυρῶν διπυρίτην.
210 Θ adds ἄλλα.

ἐξανίστηται καὶ δυνατὸς ᾖ, παλαιέτω ἀπ' ἄκρων[211] τῶν
ὤμων, καὶ ταλαιπωρείτω περιόδοισι πολλῇσι δι' ἡμέρης, καὶ εὐωχείσθω ἃ προείρηται μάλιστα.

Ἢν δὲ ὕδερος ἐπιγένηται,[212] ἰᾶσθαι κατὰ ταὐτὰ καὶ
τοῖς αὐτοῖσι καθάπερ τοὺς πρόσθεν.

26. Ὅδε δὲ ὁ[213] ὕδερος ἀπὸ τῶνδε γίνεται· θέρεος
ὥρην, ἢν ὁδοιπορέων ὁδὸν μακρὴν ἐπιτύχῃ ὀμβρίῳ
ὕδατι καὶ στασίμῳ καὶ πίῃ αὐτοῦ ἐπισπάδην πολλόν,
καὶ[214] αἱ σάρκες ἀναπίωσι καὶ ἐν ἑωυτῇσιν ἴσχωσι τὸ
ὕδωρ, ὑποχώρησις δὲ μὴ γένηται μηδαμῇ.

Τάδε οὖν πάσχει· ἐν μὲν[215] τῇ σαρκί, καῦμα
παρέχει ἐν τῇ κοιλίῃ καὶ τῷ σώματι, ὥστε τὸ στέαρ τὸ
ἐὸν ἐπὶ[216] τῇ κοιλίῃ τήκει. οὗτος τέως μὲν ἂν βαδίζῃ,
οὐδὲν δοκέει κακὸν ἔχειν, ὅταν δὲ παύσηται βαδίζων
καὶ ὁ ἥλιος δύῃ,[217] παραχρῆμα τὸν πόνον ἔχει.[218]
προϊούσης δὲ τῆς νούσου λεπτύνεται σφόδρα· ἢν δὲ
καὶ ἀσιτίη ἐπιγένηται, καὶ πολλῷ μᾶλλον λεπτύνεται.
ἢν δὲ τὰ σιτία μὲν ἐσθίῃ,[219] ταλαιπωρέειν δὲ ἀδύνατος
ᾖ, τοῖσι πολλοῖσι οἴδημα καθίσταται ἐς ἅπαν τὸ
σῶμα, καὶ τῷ μὲν λεπτῷ ἡ χροιὴ πελιδνὴ γίνεται, καὶ
ἡ γαστὴρ μεγάλη, καὶ δίψα ἔχει ἰσχυρή· τὰ γὰρ
σπλάγχνα αὐτοῦ θερμαίνεται[220] ὑπὸ τῆς θερμασίης.
τοῦ δὲ χρόνου προϊόντος, αὐτὸς μὲν πρόθυμός ἐστιν
234   ἐσθίειν ὅσα[221] τις διδοῖ, καὶ | πίνειν, καὶ ἀλγέει οὐδέν.

---

[211] Θ: ἐπ' ἄκρον Μ.   [212] Θ: γέν- Μ.
[213] Θ: Ὁ δὲ Μ.   [214] Θ: ἢν οὖν Μ.
[215] ἐν μ. Θ: ἢν μὲν ᾖ ἐν Μ.   [216] ἐὸν ἐ. Potter: σὸν ἐ.
Θ: ἐπιὸν Μ.   [217] Θ adds εὐθὺς.

140

the tips of his shoulders,[7] and exert himself through the day with frequent walks; also feed him well, mainly on what has been mentioned above.

If dropsy is present, treat it according to the same principles and with the same things used for the preceding patients.

26. The next dropsy arises in the following way: if, in summer, a person on a long journey happens upon some stagnant rain water, and drinks a large amount of it at one draught, if his tissues drink up the water and hold it within themselves, and if no evacuation at all occurs.

The patient, then, suffers the following: the water in the tissues produces burning heat in the cavity and the body, so that the fat present in the cavity melts. As long as the person keeps walking, he does not seem to suffer any harm, but when he stops and the sun goes down, he immediately has an attack. As the disease progresses, he becomes very lean, and, if he loses his appetite as well, even leaner. If he eats his meals, but is unable to exert himself, in most cases swelling occupies his whole body; the emaciated patient's skin becomes livid, his belly large, and he has a violent thirst, for his inward parts are heated by the burning heat. As time passes, the patient is eager to eat and drink as much as anyone will give him, and his pains go

---

[7] Wrestlers are sometimes depicted on Greek vases standing with arms extended and hands grasping each other's shoulders. It may be that a type of skirmishing in this position was considered safe enough for persons in a state of convalescence.

ПЕΡΙ ΤΩΝ ΕΝΤΟΣ ΠΑΘΩΝ

ἢν δὲ τὸ οἴδημα κατέχῃ, ἡ χροιὴ αὐτοῦ γίνεται ὠχρή,
καὶ διὰ τοῦ σώματος φλέβες μέλαιναι διατέτανται
πυκναί· θυμαίνει δὲ καὶ λυπεῖται ἐπὶ παντός, οὐδενὸς
ἐόντος νεωτέρου· ἡ δὲ γαστὴρ μεγάλη καὶ δίυδρος
καὶ²²² ὥσπερ λαμπτήρ· καὶ τοῦ χρόνου προϊόντος τὰ
σιτία οὐ προσίεται, ἀλλὰ δοκέει αὐτῷ ὄζειν σικύου
ἀγρίου ὑπὸ τῆς βδελυρίης.

Τούτῳ, ὅταν οὕτως ἔχῃ, διδόναι τοῦ κνεώρου ἢ τοῦ
ἱππόφεω τὸν ὀπὸν ἢ τὸν Κνίδιον κόκκον· ταῦτα δὲ τὰ
φάρμακα διδόναι ὧδε χρή· τὸ μὲν κνέωρον δι' ἕκτης
ἡμέρης, τὸν δὲ τοῦ ἱππόφεω ὀπὸν δι' ὀγδόης, τὸν δὲ
Κνίδιον κόκκον διὰ δεκάτης ἡμέρης· διδόναι δὲ χρὴ
ταῦτα, ἕως ἂν ἐκκαθαρθῇ καὶ λαπαρὸς γένηται· τὰς δὲ
μεταξὺ τῶν ἡμερέων εὐωχέειν τοῖς αὐτοῖς οἷσι καὶ
τοὺς πρόσθεν. μάλιστα δὲ τοῦ ὕδατος τοῦ αὐτοῦ πίνειν
διδόναι, ὑπ' ὅτευ καὶ τὸ νόσημα ἔλαβεν, ὡς πλεῖστον,
ὅπως ἀναταράξῃ αὐτοῦ τὴν κοιλίην καὶ ὑποχωρήσῃ
σφόδρα· οὕτω γὰρ ἂν τάχιστα ὑγιέα ποιήσαις.²²³ ἢν
δέ σοι δοκέῃ, κλύζειν θαμινά· χρὴ τοῦ κνεώρου τρίψας
ἥμισυ πόσιος, μέλιτος παραμίξας τρίτον μέρος κοτύλης, σεντλίου τέσσερσι κοτύλησι²²⁴ διεῖναι, εἶθ' οὕτω
κλύζειν. καὶ ὀνείου γάλακτος ἐφθοῦ τῇ δ' ὑστεραίῃ
δοῦναι ὀκτὼ κοτύλας, μέλι παραχέων ἢ ἅλας παραβάλλων, πίνειν. καὶ μετὰ τὴν κάθαρσιν τοῖσιν αὐτοῖσι
χρῆσθαι οἷσι καὶ οἱ πρόσθεν· καὶ τὰς μεταξὺ τῶν

²²² καὶ om. M.
²²³ ἂν . . . ποιήσαις Θ: μάλιστα ὑγιέα ποιήσεις M.

away. However, if the swelling prevails, his skin becomes pale-yellow, and throughout his body extend numerous dark vessels; he is angry and vexed with everything, even when nothing new happens; his belly is large, full of water, and like a lantern.[8] With the passage of time, the patient no longer accepts food, since it seems to him to smell like squirting-cucumber, because of his nausea.

When the case is such, give the patient spurge-flax, hippopheos juice or Cnidian berry; you must give these medications as follows: the spurge-flax every sixth day, the hippopheos juice every eighth day, and the Cnidian berry every tenth day; you must give them until the patient is cleaned out and loosened. On the days between, feed him well on the same things given to the previous patients. In particular, give him the same water to drink from which he took the disease, and in large amounts, in order that it stirs up his cavity and passes off powerfully below; for in this way you will most quickly make him well. If it seems advisable to you, administer several enemas: you must grind one half draught of spurge-flax, mix in one third cotyle of honey, soak this in four cotylai of beets, and then employ as an enema. The next day, give eight cotylai of boiled ass's milk with honey or salt, to drink. After the cleaning, let this patient employ the same things as the preceding ones and,

---

[8] I.e. with a watery, yellow, translucent appearance; cf. chapter 43 below.

---

224 Θ: τέταρτον κοτύλης M.

ἡμερέων σιτίοισι καὶ ποτοῖσι τοῖς αὐτοῖσι χρήσθω
καὶ τοῖσι περιπάτοισιν.

Οὗτος οὕτω θεραπευόμενος τάχιστα τῆς νούσου
ἀπαλλαγήσεται τρίμηνος ἢ ἑξάμηνος· ἢν δὲ ἀμελείη
τις ἐγγένηται καὶ μὴ παραχρῆμα μελετηθῇ, ἐν τάχει
ἀποθνήσκει. καὶ τὸν καταλεπτυνόμενον δὲ τοῖς αὐτοῖ-
σιν ἰᾶσθαι· προϋγρῆναι δὲ χρὴ πρότερον αὐτοῦ τὸ |
236 σῶμα πυριάσαντα, ὅπως ἂν μᾶλλον τῷ φαρμάκῳ
ὑπακούσῃ. ἀλλὰ χρὴ παραχρῆμα μελετᾶν· εἰ δὲ μή,
τοῖσι πολλοῖσι συγγηράσκει ἡ νοῦσος. κλύζειν δὲ
χρὴ καὶ ἄλλοισι[225] ὁποτέρην ἂν βούλῃ τῶν νούσων·
οἴνου λευκοῦ δύο κοτύλας λαβὼν καὶ μέλιτος ἡμικοτύλιον, καὶ ἐλαίου ἡμικοτύλιον, λίτρου τεταρτημόριον
Αἰγυπτίου ὀπτοῦ, σικύου ἀγρίου τῶν φύλλων κόψας
καὶ ἐκπιέσας[226] τοῦ χυλοῦ κοτύλην, ταῦτα μίξας πάντα, ἐγχέαι ἐς χυτρίδιον, κἄπειτα ζέσας[227] οὕτω κλύζειν.

27. Ἡπατῖτις ἡ νοῦσος γίνεται ἀπὸ χολῆς μελαίνης, ὡς[228] ἐπιρρυῇ ἐπὶ τὸ ἧπαρ· προσπίπτει δὲ
μάλιστα φθινοπώρου καὶ ἐν τῇσι μεταβολῇσι τοῦ
ἐνιαυτοῦ. τάδε οὖν πάσχει· ἐς τὸ ἧπαρ ὀδύνη ὀξέη
ἐμπίπτει,[229] καὶ ὑπὸ τὰς νεάτας πλευρὰς καὶ ἐς τὸν
ὦμον καὶ ἐς τὴν κληῖδα καὶ ἐς[230] τὸν τιτθόν, καὶ πνὶξ
ἔχει ἰσχυρή, καὶ ἐνίοτε ἀπεμέει πελιδνὴν χολήν, καὶ
ῥῖγος, καὶ πυρετὸς τὰς μὲν πρώτας ἡμέρας σφόδρα,
ἔπειτα μέντοι βληχρότερος ἔχει· καὶ ψαυόμενος ἀλγέει κατὰ τὸ ἧπαρ, καὶ ἡ χροιὴ ὑποπέλιος αὐτοῦ· καὶ
τὰ σιτία ἃ πρόσθεν ἐβεβρώκει πνίγει[231] προσπίπτον-

on the days between cleanings, the same foods, drinks and walks.

This patient, if treated in such a way, will be relieved readily of the disease within three or six months; but if there is any negligence, and he is not treated at once, he soon dies. Also treat the very emaciated patient with these same measures; first, though, you must moisten his body in advance by applying vapour-baths, in order that he will respond better to the medication. Treatment must be immediate; if it is not, in most cases the disease accompanies the person into old age. Administer this other enema in whichever disease you wish: take two cotylai of white wine, a half-cotyle each of honey and oil, and a fourth-cotyle of burnt Egyptian soda, and cut off the leaves of a squirting-cucumber plant and squeeze out a cotyle of juice; mix all these ingredients together, pour them into a small pot, and then boil and administer as an enema.

27. This hepatic disease arises from dark bile, when it flows to the liver; the disease generally attacks in fall and at the year's changes. The patient suffers the following: sharp pain befalls his liver, the region beneath his lowest ribs, his shoulder, his collar-bone, and his nipple; there is violent choking, and sometimes the patient vomits livid bile; chills and fever, intense on the first days, but then milder, set in. On being touched in the region of his liver, the patient feels pain; his colour is somewhat livid; the foods he had eaten previously provoke choking now, when they come

---

225 καὶ ἄ. Potter: *ἀλλον Θ: καὶ τοισίδε M.
226 Θ adds κεκρημένου.   227 Θ: καὶ ἐπιζέσας M.
228 Θ: ὁκόταν M.   229 M adds αὐτῷ.
230 Θ: ὑπὸ M.   231 M adds αὐτὸν.

τα καὶ καίει καὶ στρέφει[232] τὴν κοιλίην. ταῦτα μὲν
πάσχει κατ᾽ ἀρχὰς τῆς νούσου· προϊούσης δὲ τῆς
νούσου[233] οἵ τε πυρετοὶ ἀφιᾶσι καὶ ἀπ᾽ ὀλίγων σιτίων
πίμπλαται, καὶ ἐν τῷ ἥπατι ἡ ὀδύνη μοῦνον λείπεται,
καὶ αὐτὴ ποτὲ μὲν ἰσχυρή, ποτὲ δὲ ἥσσων διαπαύουσα. ἐνίοτε δὲ ὀξέη ἐπιλαμβάνει,[234] καὶ πολλάκις
ἐξαπίνης τὴν ψυχὴν ἀφῆκε.

    Τούτῳ ξυμφέρει, ὅταν μὲν ἡ ὀδύνη ἔχῃ,[235] χλι-
άσματα προστιθέναι ταῦτά, ἃ καὶ τῇ πλευρίτιδι· ὅταν·
238 δὲ ᾖ[236] ὁ πόνος λούειν αὐτὸν πολλῷ καὶ θερμῷ, | καὶ
μελίκρητον δίδου πίνειν καὶ οἶνον λευκὸν γλυκὺν ἢ
αὐστηρὸν ᾖ[237] ὁπότερος ἂν συμφέρῃ, καὶ ῥυφήματα
τὰ αὐτὰ καὶ ὅσα τῷ[238] ὑπὸ πλευρίτιδος ἑαλωκότι. τῆς
δὲ ὀδύνης ἕνεκα[239] τάδε χρὴ διδόναι πίνειν· ἀλεκτορίδος ᾠοῦ ἑφθοῦ τὸ ὠχρὸν τρίψας παραχέαι στρύχνου
χυλοῦ ἡμικοτύλιον, καὶ μελίκρητον ἐπιχέαι[240] ἐν ὕδατι
πεποιημένον ἥμισυ ἡμικοτυλίου, τούτοισι διεὶς δοῦναι
πίνειν, καὶ παύσεις τὴν ὀδύνην· διδόναι δὲ ἑκάστης
ἡμέρης, τέως ἂν ἡ ὀδύνη παύσηται· πινέτω δὲ καὶ
σιλφίου ὀπὸν ὅσον ὄροβον·[241] καὶ ὀρίγανον τρίβων
διεῖναι οἴνῳ λευκῷ καὶ οὕτως πίνειν νῆστιν· ἐπιπινέτω
δὲ καὶ τὰ ἐν τῇ πλευρίτιδι διδόμενα τῆς ὀδύνης φάρμακα. πινέτω δὲ καὶ γάλα αἴγειον τρίτον μέρος μέλιτος παραμίσγων, τοῦ δὲ γάλακτος ἔστωσαν τέσσερες
κοτύλαι· τοῦτο ἔωθεν πινέτω ὅταν τἆλλα μὴ πίνῃ.
σίτων δὲ ἀπεχέσθω, τέως ἂν κριθῇ ἡ νοῦσος· κρίνεται

---

[232] Θ adds ἐς.

into contact with the cavity, and burning and colic. These are the things the person suffers at the beginning of the disease; as it progresses, though, the fevers slacken, he becomes full on little food, and pain persists only in the liver; there it is sometimes severe, but at other times intermittently milder. Sometimes the pain attacks sharply, and many a patient has suddenly given up the ghost.

It benefits this patient, when the pain is present, to apply the same fomentations as for pleurisy. When an attack is occurring, wash him in copious hot water, and give him melicrat to drink, sweet or dry white wine or whichever other one benefits him, and the same gruels in the same amounts as you give to a patient with pleurisy. Against the pain you must give the following to drink: mash the yolk of a boiled hen's egg, pour in a half-cotyle of nightshade juice, and add a quarter-cotyle of melicrat made with water; mix these together and give to the patient to drink; you will stop the pain; give this each day until the pain ceases. Also let the patient drink silphium juice to the amount of a vetch; also grind marjoram, soak it in white wine, and have him drink it thus in the fasting state; afterwards, let him also drink the medications given in pleurisy against pain. Also let him drink four cotylai of goat's milk to which one third part of honey has been added; let him drink this at dawn, when he is not taking the other drinks. Have the patient abstain from foods until the disease reaches its cri-

---

233 τῆς νούσου· . . . νούσου Θ: τῆς δὲ ν. π. M.
234 Θ: τε λαμβάνει M.   235 M adds τά τε ἄλλα καὶ τὰ.
236 ᾗ Θ: ἀνῇ M.   237 ᾗ and ἢ om. M.   238 τὰ . . .
τῷ Potter: τὰ om. Θ: ταῦτα ἃ καὶ ὡς M.   239 ἕνεκα om. M.
240 ἐπιχέαι om. M.   241 M: ὄβολον Θ.

δὲ μάλιστα ἐν ἑπτὰ ἡμέρῃσιν, ἐν[242] ταύτῃσι γὰρ
ἀποδηλοῖ εἰ θανάσιμος ἢ οὔ.

Ἢν δὲ πνῖγμα προσίστηται, τάδε χρὴ διδόναι,
ἕως[243] ἂν ἀπεμέσῃ· μέλι καὶ ὕδωρ, ὄξος καὶ[244] ἅλας,
ταῦτα μίξας ἐγχέαι ἐς χυτρίδιον καινόν· εἶτα χλιαίνειν, καὶ ταράσσειν ὀριγάνου κλωνίοισι τῆς κεφαλοειδέος σὺν τῷ καρπῷ· ὅταν δὲ χλιανθῇ, ἐκπιεῖν δοῦ-
ναι·[245] εἶτα ἐπιβαλὼν ἱμάτια ἐᾶν, περιστείλας ὅπως[246]
ἂν ἱδρῷ μάλιστα. καὶ ὅταν ἔμετος ἔχῃ αὐτόν, ἐμείτω
προθύμως καταματεύμενος τῷ πτερῷ. ἢν δὲ μὴ ἔμετος
ἔχῃ,[247] ἐπιπιὼν μελικρήτου χλιαροῦ κύλικα δικότυλον,
οὕτως ἐμείτω. καὶ ἤν τι ἀπεμέσῃ χολῆς ἢ φλέγματος
αὖτις ταὐτὰ χρὴ ποιέειν ἐπὶ τέσσερας ὥρας· ὠφελήσει
γάρ.

Μετὰ δὲ τῆς νούσου τὴν κρίσιν μελετᾶν, σιτία
διδοὺς ὀλίγα, ταῦτα δὲ καθαρά· καὶ ἢν μὲν ἄρτον
ἐσθίῃ, θερμὸν ὡς μάλιστα ἐσθιέτω· ἢν δὲ μᾶζαν, |
240 ἄτριπτον ἐσθιέτω, πρότερον προφυρήσας. ὄψον δὲ
ἐχέτω σκυλακίου ἑφθὰ ἢ πελειάδος ἢ ἀλεκτορίδος
νεοσσοῦ, χρήσθω δὲ[248] ἑφθοῖσι πᾶσιν· ἰχθύων δὲ
γαλεῷ, νάρκῃ, τρυγόνι καὶ βατίσι τῇσι σμικρῇσι,
πᾶσιν ἑφθοῖσι. καὶ λούσθω ἑκάστης ἡμέρης, καὶ τὸ
ψῦχος φυλασσέσθω, καὶ περιπατείτω ὀλίγα ἕως ἐν[249]
ἀσφαλεῖ.[250] ταῦτα ἢν φυλάσσηται, οὐχ ὑποτροπιάσει
πάλιν ἡ νοῦσος. ἡ γὰρ νοῦσος χαλεπὴ καὶ χρονίη.

28. Ἄλλη ἥπατος· αἱ μὲν ὀδύναι πιέζουσι κατὰ τὰ
αὐτὰ κατὰ[251] τὸ ἧπαρ, καὶ ἡ χροιὴ διαφέρει τῆς
πρόσθεν, σιδιοειδὴς γάρ ἐστι. τοῦ δὲ ἔτεος θέρεος

sis; usually the crisis occurs in seven days, for in these the patient reveals whether or not he will die.

If choking comes on, you must give the following medication until the patient vomits: honey and water, vinegar and salt; mix these together and pour them into a new pot; then warm, and add twigs of the head-shaped marjoram with their seeds; when it is warm give to the patient to drink off; then cover him with blankets, wrapping him so that he will sweat heavily, and leave him. When vomiting occurs, let him vomit actively by being tickled with a feather. If vomiting does not occur, make the patient provoke it by drinking, in addition, a two-cotyle cup of warm melicrat. If he vomits up any bile or phlegm, he must do the same every four hours, for this will help.

After the disease's crisis, treat by giving fine cereals in small amounts; if the patient eats bread, let him take it very hot, if barley-cake, let it be unpounded but mixed a while before it is baked. As main dish, let him have boiled puppy, pigeon or chicken, taking all these boiled, and of fish dog-fish, torpedo, sting-ray and small skates, also all boiled. Have him bathe each day, guarding against cold, and take short walks until he is in safety. If the person takes care in these matters, the disease will not recur. The disease is severe and lasts a long time.

28. Another disease of the liver: pains press in the same way over the liver, and the patient's colour is different from what it was before, being like pomegranate-peel. This dis-

---

242 ἐν om. M.   243 Θ: ὅκως M.   244 καὶ om. M.
245 M adds χλιερόν.   246 Θ: ἕως M.   247 ἔ. ἔ. Θ: ἐμέσῃ M.   248 χρήσθω δὲ om. M.   249 M: ἂν Θ.
250 Potter: -λη Θ: -λείη M.   251 Θ: ἐς M.

μάλιστα ἡ νοῦσος ἐπιπίπτει· γίνεται δὲ μάλιστα ἐκ
κρεηφαγίης βοείων κρεῶν καὶ ἐξ οἰνοφλυγίης· ταῦτα
γὰρ πάντα[252] πολεμιώτατα ταύτην τὴν ὥρην τῷ ἥπατι,
καὶ χολὴ μάλιστα προσίσταται[253] πρὸς τὸ ἧπαρ.

Τάδε οὖν πάσχει· ὀδύναι ὀξέαι ἐμπίπτουσι, καὶ οὐκ
ἐκλείπουσιν οὐδεμίην ὥρην,[254] ἀλλ' αἰεὶ μᾶλλον ἐμπίπτουσιν.[255] ἔστι δ' ὅτε καὶ ἐμέει χολὴν ὠχρήν, καὶ
ὅταν ἐμέσῃ, ἐπ' ὀλίγον δοκέει ῥᾴων εἶναι. ἢν δὲ μὴ
ἀπεμέσῃ, ἐς τοὺς ὀφθαλμοὺς ἡ χολὴ καθίσταται, καὶ
ὠχροὶ γίνονται σφόδρα, καὶ οἱ πόδες οἰδέουσι. ταῦτα
πάσχει τότε[256] μὲν σφόδρα, τότε δὲ ἧσσον. ὅταν δὲ αἵ
τε[257] ἡμέραι παρέλθωσιν ἐν ᾗσι κρίνεται τὸ νόσημα, ὅ
τε πόνος ἐλάσσων ἔχῃ, ἀναμάρτητον[258] διαιτᾶσθαι
χρὴ τῇ τοιαύτῃ διαίτῃ, ᾗ καὶ πρόσθεν· ἢν γὰρ μεθ-
υσθῇ παρὰ καιρὸν ἢ λαγνεύσῃ ἢ ἄλλο τι ποιήσῃ μὴ
ἐπιτήδειον, τὸ ἧπαρ παραχρῆμα γίνεται αὐτοῦ σκλη-
ρόν, καὶ οἰδέει, καὶ σφύζει ὑπὸ τῆς ὀδύνης, καὶ ἢν τι
σπεύσῃ, πονέει ἐξαπίνης τὸ ἧπαρ καὶ τὸ σῶμα ἅπαν.

Τοῦτον, ὅταν οὕτως ἔχῃ, ὅταν[259] αἱ πρῶται ἡμέραι |
242 παρέλθωσι, πυριᾶσαι, εἶτα ὑποκαθῆραι τῇ σκαμωνίῃ·
ἢν δὲ ἡ κοιλίη ξυγκεκαυμένη ᾖ, κλύσαι τοῖς αὐτοῖς,
οἷσι καὶ τοὺς πρόσθεν, ὅ τι ἄξει καλῶς. καὶ μετὰ τὸν
κλυσμὸν ὀνείῳ γάλακτι ἑφθῷ ὑποκαθῆραι ὀκτὼ κοτύ-
λαις· μέλι δὲ παραχέας διδόναι πίνειν· διδόναι δὲ καὶ
τὸ αἴγειον, τρίτον μέρος μελικρήτου παραμίσγων, τὸ
ἑωθινόν, τετρακότυλον κύλικα· διδόναι δὲ καὶ ἑφθοῦ

[252] πάντα om. Θ.   [253] Θ: χολὴν . . . προσίστησι Μ.

ease attacks most often in summer, and it arises mainly from eating beef and from drunkenness; for in that season both of these are most harmful to the liver, and bile in particular assails the liver.

The patient suffers the following: sharp pains attack and, without relenting for so much as an hour, become more and more intense. Sometimes the patient vomits up pale-yellow bile, and, after he has, for a short time he seems to be better. If he does not vomit, bile settles into his eyes, which become very yellow, and his feet swell up. These things the patient suffers at one time more, at another time to a lesser degree. When the days in which this disease has its crisis arrive and the attack relents, the patient must follow strictly the regimen laid down in the preceding case; for, if he becomes drunk at an inopportune time, or engages in venery, or does anything else that is inappropriate, his liver immediately becomes hard, swells up, and throbs with pain, or if he exerts himself, he immediately has pain in his liver and in his whole body.

When the case is such, once the first days have passed administer a vapour-bath to the patient, and then clean him downwards with scammony; if his cavity is burnt up, employ an enema containing the same substances used in the enemas of the patients above, choosing whichever will draw well. After the enema, clean downwards with eight cotylai of boiled ass's milk: add honey and give this to drink. Also give a four-cotyle cup of goat's milk early in the

---

254 οὐδεμίην ὥρην om. M.

255 Θ: πιέζουσι M.   256 Θ: ποτὲ M.

257 τε om. M.   258 Θ: ἢν ἀναμάρτητος ᾖ M.

259 Θ: καὶ M.

τοῦ αἰγείου δύο χοέας,[260] τρίτον μελικρήτου παραμίσγων, ἢ αὐτὸ μοῦνον μέλι παραχέων·[261] διδόναι δὲ
καὶ τὸ ἵππειον γάλα τὸν αὐτὸν τρόπον τῷ ὀνείῳ.

Καὶ ἢν μὲν ὑπὸ τούτων μελετώμενος, ἡ νοῦσος
ἐξέλθῃ, ἅλις· εἰ δὲ μή, τάμνειν τοῦ ἀγκῶνος τοῦ δεξιοῦ
τὴν ἐντὸς φλέβα καὶ ἀφιέναι τοῦ αἵματος. ἢν δέ σοι
μὴ[262] δοκέῃ ὀνείῳ γάλακτι ὑποκαθῆραι, βοείου γάλακτος ὠμοῦ διδόναι δύο κοτύλας, τρίτον μελικρήτου
παραμίσγων, ἑκάστης ἡμέρης ἐφ' ἡμέρας δέκα· εἶτα
ἄλλας δέκα, ἕκτον μέρος τοῦ μελικρήτου παραμίσγων,[263] πίνειν διδόναι· τὸ δὲ λοιπὸν αὐτὸ τὸ
γάλα[264] διδόναι, ἕως ἂν πιανθῇ, δύο κοτύλας. ἢν δὲ
μηδ' οὕτω παύηται, καῦσαι χρὴ ὅταν μέγιστον τὸ
ἧπαρ ᾖ καὶ ἐξεστήκῃ μάλιστα, πυξίνοις ἀτράκτοισι
βάπτων ἐς ἔλαιον ζέον, προστιθέναι τέως ἂν δοκέῃ
σοι καλῶς ἔχειν καὶ κέκαυσθαι εὖ·[265] ἢ[266] μύκησιν
ὀκτὼ ἐσχάρας καῦσαι. ἢν γὰρ τύχῃς καύσας, ὑγιᾶ
ποιήσεις, καὶ τὸν λοιπὸν τοῦ χρόνου ῥάων διάξει. ἢν
δὲ μὴ τύχῃ καυθεὶς ἢ[266] ὑπὸ τῶν ἄλλων ὑγιὴς γενόμενος, τὸ λοιπὸν φθειρόμενος χρόνῳ[267] ἀποθνήσκει.

29. Ἄλλη ἥπατος· τὰ μὲν ἄλλα πλῆθος τὰ αὐτὰ
πάσχει τοῖσι δὲ[268] πρόσθεν· ἡ δὲ χροιὴ μέλαινα· τοῦ
δὲ ἥπατος ὡς λογιζόμεθα ἡ χολὴ φλέγματος καὶ
αἵματος πλησθεῖσα, διαρρήγνυται· καὶ ὅταν διαρρα-
γῇ, τάχιστα μαίνεται, καὶ ἀναΐσσει,[269] καὶ διαλέγεται

---

[260] Θ: κοτύλας Μ. [261] ἢ . . . παραχέων om. Θ.
[262] μὴ om. Θ. [263] Θ: -χέων Μ.

morning, adding one third part of melicrat; also two choes
of boiled goat's milk, adding one third-cotyle of melicrat or
just plain honey; also mare's milk administered in the same
manner as the ass's.

If, on being treated with these measures, the disease
passes off, fine; if not, incise the inner vessel of the bend of
the right arm, and draw blood. If it does not seem advisable to you to clean downwards with ass's milk, give two
cotylai of uncooked cow's milk daily for ten days, adding
one third-cotyle of melicrat; then give the same for another ten days with one sixth-cotyle of melicrat; from then
on, give two cotylai of pure milk, until the patient becomes
fat. If the disease does not come to an end thus, you must
cauterize, when the liver is largest and most protuberant,
with boxwood spindles; dip these in boiling oil, and apply
them until you think you have accomplished what you
wanted and the cautery is adequate; or burn eight eschars
with mushrooms. For, if your cautery succeeds, you will
make the patient well, and he will pass the time from then
on more easily. But if he is neither cauterized successfully
nor brought to health by any of the other treatments, he
wastes away after that, and in time dies.

29. Another disease of the liver: generally this patient
suffers the same things as the preceding one, except that
his colour is dark, and the hepatic bile, as we infer, becomes filled with phlegm and blood, and breaks out; when
this bile breaks out, the patient very soon rages, casts him-

---

264 λ. α. τὸ γ. Θ: γ. α. τὸ λ. M.   265 εὖ om. M.
266 ἢ om. M.   267 ὑγιὴς . . . χρόνῳ Θ: λοιπὸν ὑγιὴς
μὴ γενόμενος, φθειρόμενος M.   268 τοῖσι δὲ Θ: τῇσι M.
269 Potter: ἀν ἐς σει Θ: ἀγανακτεῖ M.

244 ἀσύνετα. καὶ ὑλακτέει ὡς κύων, καὶ οἱ ὄνυχες φοι-
νίκεοί εἰσιν αὐτοῦ, καὶ τοῖς ὀφθαλμοῖς οὐ δύναται
ἀνορᾶν,[270] καὶ αἱ τρίχες αἱ ἐν τῇ κεφαλῇ ὀρθαὶ ἵσταν-
ται, καὶ πυρετὸς ὀξὺς ἐπιλαμβάνει.

Τούτῳ χρὴ προσφέρειν τὰ αὐτὰ ἃ καὶ τοῖσι πρόσ-
θεν· οἱ δὲ πολλοὶ ἀποθνήσκουσιν ἐν τῇσιν ἕνδεκα
ἡμέρῃσι· παῦροι δ' ἐκφυγγάνουσιν.

30. Σπληνὸς ἡ πρώτη·[271] γίνεται δὲ διὰ θερμασίην
τοῦ ἡλίου, χολῆς κινηθείσης, ὅταν ἑλκύσῃ ὁ σπλὴν
ἐφ' ἑωυτὸν χολήν. πάσχει τάδε· πυρετὸς ὀξὺς ἐπιγίνεται κατ' ἀρχάς· προϊούσης δὲ τῆς νούσου ἀφίησι,
πλὴν κατ' αὐτὸν τὸν σπλῆνα· ταύτῃ δὲ αἰεὶ θέρμη
ἔχει. καὶ ὀδύνη ἄλλοτε καὶ ἄλλοτε ἐμπίπτει ὀξέη καὶ
ἐς τὴν λαπάρην καὶ ἐς τὴν κοιλίην. τὰ δὲ σιτία κατ'
ἀρχὰς προσίεται μέν, διαχωρέει δὲ οὐ μάλα· προϊούσης δὲ τῆς νούσου, ἥ τε χροιὴ ὠχρὴ γίνεται, καὶ
ὀδύνη ἰσχυρὴ ἐπιπίπτει, καὶ αἱ κληῖδες λεπτύνονται,
καὶ τὰ σιτία οὐχ ὁμοίως[272] προσίεται ὥσπερ κατ'
ἀρχάς, καὶ ἀπὸ ὀλίγου πίμπλαται. ὁ δὲ σπλὴν μέγας
ἄλλοτε αὐτῆς τῆς ἡμέρης, ἄλλοτ' ἐλάσσων γίνεται.

Τοῦτον, ὅταν οὕτως ἔχῃ, πῖσαι ἐλλέβορον, κάτω δ'
ὑποκαθῆραι Κνιδίῳ κόκκῳ· ἐς ἑσπέρην δὲ μετὰ τὴν
κάθαρσιν φακῆς τρυβλίον δοῦναι[273] ὀξυτέρης, καὶ
τευτλίων τρυβλίον λιπαρῶν, ἄλφιτα περιπάσας. τῇ δ'
ὑστεραίῃ καὶ τῇ τρίτῃ ἄρτον σμικρὸν δοῦναι· ὄψῳ δὲ
φακῇ χρήσθω καὶ οἰὸς κρέῃ[274] ἐν τῇ φακῇ ἐφθῷ

---

[270] Potter: ἂν ὁρᾶν Θ: ὁρῆν Μ.

self about, talks nonsense, and howls like a dog; his nails become red, he cannot look up with his eyes, the hairs of his head stand on end, and a sharp fever supervenes.

To this patient you must administer the same things as to the preceding ones; most die in eleven days; few escape.

30. First disease of the spleen: this disease arises from the heat of the sun, when bile is set in motion and the spleen draws it to itself. The patient suffers the following: sharp fevers set in at the outset but, as the disease advances, these abate, except at the site of the spleen itself; there heat is always present; sharp pains also attack from time to time, in both the flank and the cavity. At the beginning, the patient accepts his food, but does not pass it through very well; as the disease progresses, his colour becomes pale-yellow, violent pain besets him, his collarbones become lean, he no longer accepts his food as he did at the beginning, but now becomes full on a little; on one and the same day the spleen is at one time large, at another time smaller.

When the case is such, have the patient drink hellebore, and clean him downwards with Cnidian berry; towards evening after the cleaning, give a bowl of quite acid lentil-soup, and a bowl of beets boiled in grease, over which meal has been sprinkled. On the second and third days, give a small loaf of bread, and let the patient employ as main dish lentil-soup with boiled ground mutton in it;

---

271 Potter: τῆς πρώτης ΘΜ (sc. περὶ).

272 ὁμοίως ΘΜ: Galen seems to refer to this passage in his gloss on ἀμαλῶς (Kühn XIX. 77).

273 M adds ῥοφέειν.

274 Θ: ὑὸς κρέας M.

τετρυμένῳ· οἶνον δὲ πινέτω αὐστηρόν, μέλανα, ἀκρητέστερον κατ᾽ ὀλίγον, καὶ ἡσυχίην ἐχέτω ταύτας τὰς
ἡμέρας, πλὴν ἔνδον ὀλίγα περιπατείτω ἐν σκιῇ. τὸν δὲ
λοιπὸν χρόνον τάδε χρὴ προσφέρεσθαι· σίτων μὲν
ἄρτον αὐτοπυρίτην, ὄψον δ᾽ ἐχέτω κρέας κυνὸς μέζο-
νος ἢ οἰὸς ἢ αἰγὸς[275] τετρυμένον καὶ τάριχον Γα-
246 δειρικὸν ἢ σαπέρδην. καὶ τὰ ὀξέα καὶ τ᾽ ἁλμυρὰ
πάντα προσφερέσθω καὶ τὰ στρυφνά· πινέτω δὲ οἶνον
Κῷον ὑπόστρυφνον ὡς μελάντατον. τῶν δὲ γλυκέων
καὶ λιπαρῶν καὶ κνισωδέων ἀπεχέσθω, καὶ σίλφιον
μὴ προσφερέσθω,[276] μηδὲ κρέας ὕειον,[277] μηδὲ κε-
στρέα μηδὲ ταρίχηρον μήτε νεαρόν, μήτε ἔγχελυν,
μήτε λάχανον μηδὲν[278] ἄνευ ὄξους. τρωξίμων δέ ῥαφα-
νῖδι χρήσθω καὶ σελίνῳ καὶ[279] ἐς ὄξος βάπτων· καὶ
οἴνῳ ἄλφιτα φυρῶν ἐσθίειν, καὶ τοῦ οἴνου ἐπιρυφείτω
ἄκρητον· ἢν δὲ βούληται, καὶ τὸν ἄρτον ἐς τὸν οἶνον
ἐνθρυπτόμενον ἐσθιέτω θερμόν· διδόναι δὲ καὶ ἰχθύων
σκορπίον, κόκκυγα, κωβιόν, δράκοντα, καλλιώνυμον,
τούτους ἐφθοὺς καὶ ψυχρούς.

Διδόναι δὲ καὶ ἃ μέλλει τὸν σπλῆνα ἰσχναίνειν[280]
ἑκάστης ἡμέρης, ἀσφοδέλου τὸν καρπόν, ἢ κισσοῦ[281]
τὰ φύλλα, ἢ αἰγὸς κέρας, ἢ ἄγνου καρπόν, ἢ πήγανον,
ἢ διδυμαίου ῥίζαν· τούτων τρίβων ὅ τι ἂν βούλῃ,
διδόναι ἐν οἴνου κοτύλῃ αὐστηροῦ νῆστει. ἢν δὲ δυνατὸς ᾖ, ἀνάγκα[282] αὐτὸν πρίειν ξύλα τριήκοντα ἡμέρας·
καὶ ἀπ᾽ ἄκρων τῶν ὤμων παλαιέτω, καὶ περιπατείτω
δι᾽ ἡμέρης, καὶ ὄψιος εὑδέτω, πρώϊος δ᾽ ἐγειρέσθω, καὶ
εὐωχείσθω τὰ προειρημένα. ἢν δὲ μὴ ὑπὸ τούτων

let him drink dark dry wine, quite unmixed with water, a little at a time, and rest on these days except for walking a little inside in the shade. From then on, the following must be administered: of cereals let the patient have whole-wheat bread, as main dish ground meat of an adult dog, sheep or goat, and Cadiz salt-fish or saperdes; administer all the acid and salty foods, and also the sour ones. Let the patient drink very dark sour Coan wine. Have him abstain from foods that are sweet, fat, and that steam like roasted meat, and do not administer silphium, pork, grey mullet either salted or fresh, eel, or any vegetable without vinegar. Of raw vegetables let the patient employ radishes and celery, dipping them in vinegar; also let him eat meal mixed with wine, and afterwards drink pure wine; if he wishes, let him eat bread warm, crumbled into wine; of fish give scorpion fish, piper, goby, weever, and star-gazer, these boiled and cold.

Give medications to bring down the swelling of the spleen daily: asphodel seeds, ivy leaves, goat's horn, chaste-tree seed, rue, or orchis root: grind whichever of these you wish, and give it in a cotyle of wine to the fasting patient. If he is able, he must saw wood for thirty days; let him also wrestle with the tips of his shoulders, take walks all through the day, go to sleep late, be awakened early, and be well fed on the things mentioned above. If the per-

---

275 M: ὑὸς Θ.   276 M adds μηδὲ σκόροδα.
277 Θ: χοίρειον M.   278 M adds ἐφθὸν.
279 καὶ om. M.
280 Θ: λεπτύνειν M.
281 Θ: ἰξοῦ M.
282 Θ: ἀναγκάζειν M.

ῥαΐζῃ, καῦσαι τὸν σπλῆνα μύκησι δέκα ἐσχάρας μεγάλας,[283] ὅταν μέγιστος ᾖ ὁ σπλὴν καὶ ἐξηρμένος μάλιστα· ἢν γὰρ τύχῃς καύσας,[284] ὑγιᾶ ποιήσεις[285] ἐν τάχει.

Ἡ δὲ νοῦσος δεῖται θεραπείης· χαλεπὴ γὰρ καὶ χρόνιος, ἢν μὴ παραχρῆμα μελετηθῇ.

31. Ἄλλη σπληνός· γίνεται μὲν ἀπὸ τῶν αὐτῶν ὧν καὶ ἡ πρόσθεν· πάσχει δὲ ὑπὸ τούτου τοῦ νοσήματος τάδε. ἡ γα|στὴρ φυσᾶται μεγάλη,[286] καὶ ὁ σπλὴν οἰδέει καὶ σκληρός ἐστι, καὶ ὀδύναι ὀξεῖαι ἐμπίπτουσιν ἐς τὸν σπλῆνα. ἡ δὲ χροιὴ τρέπεται[287] —μέλας, ἔπωχρος, σιδιοειδής—καὶ ἐκ τοῦ ὠτὸς κακὸν ὄζει, καὶ τὰ οὖλα ἀφίσταται ἀπὸ τῶν ὀδόντων καὶ κακὸν ὄζει· καὶ ἐκ τῶν κνημέων ἕλκεα ἐκρήγνυνται, οἷά περ ἐπινυκτίδες· τὰ δὲ[288] γυῖα λεπτύνεται, καὶ ἡ κόπρος οὐ διαχωρέει.

Τοῦτον, ὅταν οὕτως ἔχῃ, μελετᾶν τοῖσιν αὐτοῖσι καὶ φαρμάκοισι καὶ ἐδέσμασι καὶ ποτῷ καὶ ταλαιπωρίῃσι καὶ τοῖσιν ἄλλοισι πᾶσι· καὶ κλύζειν, ὅταν μὴ θέλῃ ἡ κόπρος διαχωρέειν, τοῖσδε· μέλιτος ἡμικοτύλιον καὶ λίτρου Αἰγυπτίου ὅσον ἀστράγαλον οἱός· ταῦτα τρίψας διεῖναι ἀπὸ σεύτλων ἑφθῶν ὕδατι τέσσερσι κοτύλῃσιν· εἶτα τούτῳ κλύζειν. ἢ δὲ μὴ ὑπὸ τούτων ἡ νοῦσος καθίστηται, καῦσαι αὐτὸν τὸν σπλῆνα ὡς τὸν πρόσθεν· καὶ ἢν τύχῃς καύσας, ὑγιᾶ ποιήσεις.

283 μεγάλας om. Θ.

son does not improve with these measures, cauterize his spleen with fungi, making ten large eschars when the spleen is largest and most raised; for, if you succeed in your cautery, you will quickly make him well.

The disease requires attention, for it lasts a long time and is severe if not cared for at once.

31. Another disease of the spleen: this one arises from the same things as the preceding one, but in it the patient suffers the following: his belly is puffed up large, his spleen swells and is hard, and sharp pains occupy it. His colour is altered, becoming dark and yellowish like pomegranate-peel, he smells foully from the ear, and the gums separate from his teeth and smell foully; ulcers very like epinyctides break out on his legs below the knees. The limbs become lean, and stools do not pass off.

When the case is such, treat the patient with the same medications, foods, drinks, exercises and all the rest. In addition, when stools refuse to pass off, apply the following enema: a half-cotyle of honey, and Egyptian soda to the amount of a sheep's vertebra; grind these, and mix them in four cotylai of juice boiled from beets; then administer as a enema. If the disease does not go away with these measures, cauterize the spleen itself, as in the preceding case; if your cautery succeeds, you will bring about recovery.

---

284 M adds ὡς δεῖ.
285 M adds πλὴν οὐκ.
286 Θ: μετὰ δὲ M.
287 M adds γίνεται.
288 M: τότε Θ.

32. Ἄλλη σπληνός· γίνεται μὲν τοῦ ἔτεος θέρεος[289] ὥρῃ μάλιστα· ἡ δὲ νοῦσος γίνεται ἀπὸ αἵματος, ὅταν ὁ σπλὴν ἐμπλησθῇ[290] αἵματος, ἐκρήγνυται ἐς τὴν κοιλίην. καὶ ὀδύναι ὀξέαι ἐς τὸν σπλῆνα ἐμπίπτουσι καὶ ἐς τὸν τιτθὸν καὶ ἐς τὴν κληῗδα καὶ ἐς τὸν ὦμον καὶ ὑπὸ τὴν ὠμοπλάτην. ἡ δὲ χροιὴ τούτου[291] μο- λυβδοειδής, καὶ ἀμυχὰς ἐν τῇσι κνήμῃσι λαμβάνει, καὶ ἕλκεα μεγάλα γίνεται ἐξ αὐτῶν· καὶ τὰ κάτω ὑποχωρέοντα ἅμα τῇ κόπρῳ[292] αἱματώδεα καὶ οἰνώ- δεα[293] ὑπέρχεται. ἡ δὲ γαστὴρ σκληρή, καὶ ὁ σπλὴν ὥσπερ λίθος. οὗτος τῶν πρόσθεν θανατωδέστερος, καὶ ἐξ αὐτοῦ παῦροι ἐκφυγγάνουσι.

Τοῦτον, ὅταν οὕτως ἔχῃ, τοῖς αὐτοῖς οἷσι καὶ τοὺς πρόσθεν ἰᾶσθαι,[294] πλὴν ἄνω φάρμακον μὴ δῷς· κάτω 250 δὲ τοῦ Κνιδίου κόκκου δοῦ|ναι. τῆς δὲ ὑστεραίης ὀνείου ἢ[295] ἱππείου γάλακτος δοῦναι ἑφθοῦ ὀκτὼ κοτύ- λας, μέλι παραχέας· εἰ δὲ μή, αἰγείῳ ἢ βοείῳ ἑφθῷ δύο χοεῦσι μέλι παραχέων παρὰ τὴν ἑτέρην κύλικα, ἐναλλὰξ δὲ πίνειν·[296] ἐς ἑσπέρην δὲ διδόναι μετὰ τὴν κάθαρσιν ταῦτά ἃ καὶ τῷ πρόσθεν μετὰ τὸ φάρμακον· καὶ ἢν δοκέῃ, τοῦ αἵματος ἀφελεῖν ἀπὸ τοῦ ἀγκῶνος τοῦ ἀριστεροῦ τῆς ἐντὸς φλεβός. τὰς δὲ λοιπὰς τῶν ἡμερέων διδόναι νήστει ἑκάστης ἡμέρης βοείου γά- λακτος τέσσερας κοτύλας, τρίτον μέρος ἅλμης παρα- μίσγων. σιτίοισι δὲ καὶ ποτοῖσι καὶ τοῖσιν ἄλλοισι τοῖς αὐτοῖσι θεραπεύειν οἷσι καὶ τοὺς πρόσθεν. ἀπ-

[289] Θ: ἔαρος M.   [290] M adds μάλιστα.

32. Another disease of the spleen: this one occurs mainly in summer; it arises from the blood, when the spleen becomes filled with blood, and it breaks out into the cavity. Sharp pains befall the spleen, nipple, collar-bone, shoulder, and the region beneath the shoulder-blade. This patient's colour is leaden, and on his legs below the knees he has scratches from which large ulcers develop; what is evacuated with the stools passes off bloody and wine-coloured. The belly is hard, and the spleen is like a stone. Such a spleen is more often mortal than the preceding ones, and few patients survive it.

When the case is such, treat this patient with the same things administered to the ones above, except do not give any medication to act upwards; to act downwards give Cnidian berry. On the next day, give eight cotylai of boiled ass's or mare's milk with honey; if not that, then add honey to one of two cups containing two choes of goat's or cow's milk, and have the patient drink from these alternately; towards evening, after the cleaning, give the same things you gave to the preceding patient after his medication; if it seems advisable, draw blood from the inner vessel of the bend of the left arm. On the days that follow, give daily to the fasting patient four cotylai of cow's milk to which one third part of brine has been added. Treat with the same foods, drinks, and other measures used for the patients

---

291 Θ: τοῦ γυίου M.   292 Mercurialis (*Hippocratis Coi Opera quae extant*, Venice, 1588): τὸ πρωὶ Θ: τῷ πρώτῳ M.

293 Θ: ἰώδεα M.   294 οἷσι . . . ἰᾶσθαι Θ: θεράπευε οἷσι καὶ τὸν ἔμπροσθεν M.

295 τῆς δὲ . . . ἢ Θ: τῇ ὑστεραίῃ οἴνου M.

296 M adds χρή.

ἔχεσθαι δὲ λαγνείης καὶ οἰνοφλυγίης καὶ τοῦτον καὶ
τὸν πρόσθεν. καὶ ἢν σοι δοκέῃ, καῦσαι ὅταν παχύτατος ᾖ καὶ μέγιστος ὁ σπλήν· καὶ ἢν τύχῃς καύσας
κατὰ καιρόν,[297] ὑγιᾶ ποιήσεις· ἢν δὲ μὴ ὑπὸ ταύτης
τῆς ἰήσιος ὑγιὴς γένηται, φθειρόμενος χρόνῳ θνή-
σκει· ἡ γὰρ νοῦσος χαλεπή.

33. Ἄλλη σπληνός· προσπίπτει ἡ νοῦσος μάλιστα
ἦρος·[298] ὅταν φλέγμα ἀναλάβῃ ὁ σπλὴν ἐς ἑωυτόν,
μέγας παραχρῆμα γίνεται καὶ σκληρός· εἶτα αὖτις
καθίσταται· καὶ ὅταν μὲν ἐξηρμένος ᾖ, ὀδύναι ὀξέαι
ἐμπίπτουσιν· ὅταν δὲ λαπαρὸς ᾖ, ἀνώδυνός ἐστι. καὶ
ὅταν χρόνος ἐγγένηται τῇ νούσῳ, ἀμαυροτέρη ἡ νοῦσος, καὶ χρόνῳ[299] ἐπανίσταται, καὶ ταχέως καθίσταται. οὗτος, ὅταν οὕτως ἔχῃ, κατ' ἀρχὰς τοῦ νοσήματος
τὰ σιτία οὐ δύναται προσίεσθαι, καὶ καταλεπτύνεται
ταχέως, καὶ ἀκρασίην ἔχει πολλὴν τὸ σῶμα. τὸ δὲ
νόσημα, ἢν μὴ παραχρῆμα θεραπευθῇ, αὐτόματον
δὲ[300] καταστῇ, διαλιπὸν πέντε μῆνας ἐξαῦτις[301] ἐπαν-
ίσταται· διαλαμβάνει[302] δὲ μάλιστα τὸν χειμῶνα.

252   Τοῦτον ἢν λάβῃς κατ' ἀρχάς, καῦσον δέκα | ἐσχά-
ρας ἐς τὸν σπλῆνα, καὶ εὐθὺς ὑγιᾶ ποιήσεις. ἢν δὲ μὴ
καύσῃς, τοῖς αὐτοῖσι φαρμάκοισιν ἰᾶσθαι οἷσι καὶ
τοὺς πρόσθεν, καὶ ἐδέσμασι καὶ ποτοῖσι καὶ ταλαι-
πωρίησιν· οὕτω γὰρ ἂν τάχιστα ὑγιέα ποιήσειας.
τούτου ἡ χροιὴ γίνεται ἔκλευκος, ἄρτι ὕπωχρος, ἄρτι[303] αὐχμηρή.

[297] κ. κ. Θ: τοῦ καιροῦ Μ.

above. Both this patient and the preceding one must abstain from venery and drunkenness. If you think it advisable, cauterize when the spleen is thickest and largest. If you succeed in cauterizing at the opportune moment, you will bring about recovery; but, if the patient does not recover with this treatment, he wastes away, and in time dies; for the disease is severe.

33. Another disease of the spleen: this one attacks mainly in spring. When the spleen takes up phlegm into itself, it immediately becomes large and hard; then it goes down again. When it is raised, sharp pains are present, but when the swelling is gone, it is free of pain. As time passes, the disease becomes less pronounced, and after a while the spleen comes to swell and subside again quickly. When the case is such, at the beginning of the disease the patient cannot accept his food; he rapidly becomes lean, and he suffers great weakness in his body. The disease, if not treated at once, remits spontaneously, but, after a break of five months, recurs; the intermission occurs especially in winter.

If you take on this patient at the beginning, burn ten eschars over his spleen, and you will quickly make him well. If, however, you do not cauterize, treat with the same medications, foods, drinks and exercises used for the patients above; for with these you will most quickly bring about recovery. The patient's skin becomes quite white, sometimes slightly yellowish, sometimes parched.

---

298 μ. ἦ. Θ: καὶ αὕτη ἔαρος μ. M.
299 χρόνῳ om. M.   300 Θ: ἢ αὐτόματον M.
301 Θ: ἢ ἐξ αὖτις M.   302 Θ: -λιμπάνει M.
303 Θ: καὶ M.

34. Ἄλλη σπληνός· γίνεται μὲν μετοπώρου μάλιστα ἀπὸ χολῆς μελαίνης· γίνεται δὲ ἀπὸ λαχανοφαγίης τρωξίμων πολλῶν καὶ ἀπὸ ὑδροποσίης. πάσχει οὖν τάδε· ὅταν τὸ νόσημα λάβῃ, ἀλγέει τὸν σπλῆνα σφόδρα, καὶ ῥῖγος καὶ πυρετὸς ἐπιλαμβάνει, καὶ ἀσιτίη ἔχει αὐτόν, τό τε γυῖον συμπίπτει ταχέως. ὁ δὲ σπλὴν μέγας μὲν οὐ πάνυ γίνεται, σκληρὸς δέ, καὶ προσπίπτει πρὸς[304] τὰ σπλάγχνα καὶ μύζει προσκείμενος.

Τοῦτον, ὅταν οὕτως ἔχῃ, τοῖσιν αὐτοῖσιν ἰᾶσθαι καὶ φαρμάκοισι καὶ βρωτοῖσι καὶ ποτοῖσι καὶ ταλαιπωρίῃσιν, ὡς τοὺς πρόσθεν, καὶ ἤν σοι παράσχῃ, καῦσαι τὸν αὐτὸν τρόπον ὅνπερ καὶ τοὺς ἄλλους. αὕτη ἡ νοῦσος τοῖς πολλοῖσιν, ἢν παραχρῆμα μελετηθῇ, ἐξέρχεται ἐν τάχει.

35. Ἴκτεροι τέσσερες· ὅδε μὲν τοῦ θέρεος μάλιστα ἐπιλαμβάνει χολῆς κινηθείσης· ἵσταται δὲ[305] ἡ χολὴ ὑπὸ τῷ δέρματι καὶ ἐν τῇ κεφαλῇ, ὥστε εὐθέως ἀλλοτροπέει[306] τὸ σῶμα καὶ γίνεται ὠχρὸν οἷόν περ σίδιον· καὶ οἱ[307] ὀφθαλμοὶ ὠχροί, καὶ ἐν τῇ κεφαλῇ ὑπὸ τὰς τρίχας οἷον χνοῦς ὕπεστι, καὶ ῥῖγος καὶ πυρετὸς ἐπιλαμβάνει· καὶ οὐρέει ὠχρὸν τὸ οὖρον, καὶ ὑφίσταται ὑπ᾿ αὐτῷ[308] παχὺ ὕπωχρον. καὶ ἔωθεν, ἔστ᾿ ἂν νῆστις ᾖ, ἐς τὴν καρδίην καὶ τὰ σπλάγχνα μύζει, καὶ ὅταν τις αὐτὸν προσφθέγγηται ἢ[309] ἐρωτᾷ, ἀσᾶταί τε καὶ λυπεῖται, καὶ οὐκ ἀνέχεται ἀκροώμενος. ὁ δὲ
254 ἀπόπατος προϊὼν ὠχρὸς[310] | καὶ κάκοδμος. οὗτος, ὅταν

164

34. Another disease of the spleen: this one occurs mainly in autumn from dark bile; it arises from eating many raw vegetables and from drinking water. The patient suffers the following: when the disease sets in, he has great pain in his spleen, he is subject to chills and fever, he loses his appetite, and his body rapidly becomes emaciated. The spleen does not become very large, but it is hard, and it falls against the inward parts, and rumbles as it lies there.

When the case is such, treat this patient with the same medications, foods, drinks, and exercises used for the preceding ones and, if you are able, cauterize him as you did the others. This disease passes off quickly in most patients, if they are treated immediately.

35. Four jaundices: the first one occurs mainly in summer when bile is set in motion; for the bile comes to rest under the skin and in the head, so that the body rapidly changes colour and becomes yellow like pomegranate-peel. The eyes are pale-yellow, a kind of incrustation is laid down on the scalp under the hair, and chills and fever set in; the urine passed is pale-yellow, and a thick very pale-yellow sediment precipitates from it. In the early morning, before the patient has eaten, there is rumbling in his cardia and inward parts, and, when anyone addresses him or asks him a question, he is vexed and grieved, and cannot bear to listen. The stools pass off yellow and foul-smelling. When

---

304 Θ: παρα- παρὰ M.   305 Θ: οὖν M.
306 Θ: ἀλλοχροέει M.   307 οἱ om. Θ.
308 M: -ῶν Θ.   309 ἢ om. Θ.
310 Θ: ὠχρόλευκος M.

οὕτως ἔχῃ, θνήσκει μάλιστα ἐν τεσσερεσκαίδεκα ἡμέ-
ρῃσιν· ἢν δὲ ταύτας ἐκφύγῃ, ὑγιαίνεται.

Μελετᾶν δὲ αὐτὸν τόνδε χρὴ τὸν τρόπον[311] ὅταν ὁ
πυρετὸς ἀνῇ λούειν αὐτὸν[312] πολλῷ καὶ θερμῷ· πίνειν
δὲ διδόναι μελίκρητον, καὶ ῥυφήμασι διαχρήσθω πτισάνης χυλῷ, μέλι παραχέων—καὶ μὴ συνεψεῖν τὸ
μέλι—ἕως ἂν αἱ τεσσερεσκαίδεκα ἡμέραι παρέλθω-
σιν· αὗται γὰρ κρίνουσι[313] θανασίμους ἢ οὔ. μετὰ δὲ
ταῦτα λούσθω δὶς τῆς ἡμέρης, καὶ εὐωχέειν τὸ πρῶτον
ὀλίγοις ἅσσα ἂν μάλιστα προσίηται, ἔπειτα μέντοι
καὶ πλείω διδόναι· πινέτω δὲ καὶ οἶνον λευκὸν ὡς
πλεῖστον δι᾽ ἡμέρης. καὶ ἢν σοι δοκέῃ προϊόντος τοῦ
χρόνου ὠχρὸς εἶναι καὶ ἀσθενής, ἔμετον αὐτὸν[314] κελεύειν ποιέεσθαι ἀπὸ τῶν σιτίων, ὡς ἐν τῇσι νούσοισι
τῇσιν ἄνω ἔχει.

Καὶ ἢν μὲν ὑπὸ τούτων παύηται, ἅλις· εἰ δὲ μή,
ἐλλέβορον πῖσαι. μετὰ δὲ τὴν κάθαρσιν ἑσπέρης[315]
φαγέτω μᾶζαν μαλθακὴν ἢ ἄρτου τὸ ἐντός· ὄψον δ᾽
ἐχέτω[316] νεοσσὸν ἀλεκτορίδος δίεφθον ἐζωμευμένον
κρομμύῳ καὶ κοριάννῳ καὶ τυρῷ καὶ ἁλὶ καὶ σησάμῳ
καὶ σταφίδι λευκῇ· οἶνον δὲ πινέτω λευκόν, αὐστηρόν,
ὡς παλαιότατον. τὴν δὲ κάτω κοιλίην ὑποκαθῆραι, τῇ
δ᾽ ὑστεραίῃ, χυλῷ ἀπὸ ἐρεβίνθων λευκῶν· μέλι δὲ χρὴ
τῷ χυλῷ παραχέαι, ἐκπιέτω δὲ δύο χόεας[317] τοῦ χυλοῦ.

Τῇ δὲ τρίτῃ ἀρξάμενος, ἑκάστης ἡμέρης πινέτω
ὕδωρ ἀπὸ τῶνδε ἑψήσας· μορῶν[318] ῥίζας λεπτὰς πλή-

311 αὐτὸν . . . τρόπον Θ: χρὴ αὐτὸν ὧδε Μ.

the case is such, the patient usually dies in fourteen days; if he survives that long, he recovers.

You must treat this patient as follows: when his fever remits, wash him in copious hot water, give him melicrat to drink, and have him take barley-water gruel with honey—do not boil the honey in the barley-water—until the fourteen days have passed, for this is the critical period that decides whether or not patients will die. From then on, let the patient bathe twice a day, and feed him well, first on small amounts of whatever he is most willing to accept, and then, of course, give him more; let him also drink white wine in large amounts, all through the day. If, after a time has passed, you find he is pale-yellow and weak, order him to provoke vomiting by means of foods, as was done in the diseases above.

If, with these measures, the disease goes away, fine; if not, have the patient drink hellebore. After the cleaning, in the evening let him eat soft barley-cake or the inner part of a loaf of bread, have as main dish chicken well-boiled into a soup with onion, coriander, cheese, salt, sesame and white raisins, and drink a very old dry white wine. On the following day, clean the lower cavity downwards with juice from white chick-peas; you must add honey to the juice, and let the patient drink two choes of it.

Beginning on the third day, let the patient each day drink water boiled from the following: boil a pinch (the

---

312 Θ: χρὴ M.
314 αὐτὸν om. M.
316 Θ: ἐσθιέτω M.
317 Θ: κοτύλας M.
318 Θ: μαράθου M.

313 M adds ἦν.
315 Θ: ἐς ἑσπέρην M.

θος ὅσον τοῖσι τρισὶ δακτύλοισι περιλαβεῖν, δραχ-
μίδα· ταύτας ἑψεῖν ἐν τρισὶ[319] χοεῦσιν ὕδατος, ἑψεῖν δὲ
ἕως ἂν λειφθῇ τὸ ἥμισυ· ἑψεῖν δὲ[320] καὶ ἐρεβίνθων
λευκῶν χοίνικα ἐν δυσὶ χοεῦσι, καὶ τούτου λειπέτω[321]
τὸ ἥμισυ· ταῦτα ὅταν λειφθῇ, διηθήσας ἐξαιθριάσαι,
καὶ[322] ἀμφότερα μίξας, τούτῳ τῷ ὕδατι καὶ τὸν οἶνον
256 κιρνὰς πινέτω· | καὶ αὐτὸ τὸ ὕδωρ ψιλόν, ἢν βούληται,
πινέτω ὡς πλεῖστον, καὶ ἄλλο ὕδωρ μὴ πινέτω. ἢν δὲ
τοῦτο μὲν τὸ ποτὸν μὴ προσίηται, τόδε αὐτῷ σκευάσας διδόναι· λευκῶν ἰσχάδων χοίνικα ἑψήσας ἐν
δυσὶ χοεῦσι διηθῆσαι καὶ ἐξαιθριάσαι· ἔπειτα τοῦτο
τὸ ὕδωρ πινέτω, ἤν τε αὐτὸ ψιλὸν βούληται, ἤν τε σὺν
τῷ οἴνῳ κιρνάς· πινέτω δὲ μὴ πολὺ μηδὲ ἀθρόον, ἀλλὰ
κατὰ ἡμικοτύλιον, ὅπως ἂν μὴ διάρροια ἐπιγένηται·[323]
καὶ διαλιπὼν χρόνον ὀλίγον πινέτω. ἀγαθὰ δὲ καὶ
τάδε διδόναι πίνειν νήστει ἑκάστης ἡμέρης· τρίβοντα
λεῖα καὶ διέντα κοτύλην οἴνου παλαιοῦ λευκοῦ, σελίνου καρπόν, σικύου σπέρμα, μαράθου καρπόν, Αἰθιο-
πικὸν κύμινον, ἀδίαντον, κόριν τὴν ποίην, ἀσταφίδα
τὴν λευκήν· ταῦτα πίνων καὶ ἐκεῖνα, τάχιστα ὑγιὴς
ἔσται.

36. Ἄλλος ἴκτερος· ἐπιλαμβάνει χειμῶνος ὥρῃ[324] ἐκ
μέθης καὶ ῥίγεος· ἄρχεται δὲ πρῶτον μὲν τὸ ῥῖγος
ἐπιλαμβάνειν, εἶτα ὁ πυρετός· τὸ δὲ ὑγρὸν ἐν τῷ
σώματι τὸ ὑπὸ τῷ δέρματι πήγνυται ἅμα τῷ αἵματι·[325]
τοῖσδε δὲ ἀποδηλοῖ ὡς οὕτως ἔχει· πελιδνόν ἐστιν

319 Θ: δύο Μ.

amount you take with three fingers) of peeled roots of mulberry trees in three choes of water until half is left; also boil
a choenix of white chick-peas in two choes of water until
half is left; when these amounts remain, strain, expose
them to the open air, mix them together, and let the patient
mix his wine with this water, and drink it; also let him drink
just the water alone, if he wishes, in copious amounts, but
no other water. If he will not accept this beverage, prepare
the following one, and give it to him: boil a choenix of dried
white figs in two choes of water, strain, and expose to the
open air; then let the patient drink this water, either just by
itself, or mixed with his wine, whichever he prefers. Let
him drink neither a large amount, nor too quickly, but a
half-cotyle at a time, in order that diarrhoea does not come
on; then, leaving a short space of time, let him drink again.
The following is also good to give to the fasting patient to
drink each day: grind fine and soak in a cotyle of old white
wine: celery seed, cucumber seed, fennel seed, Ethiopian
cummin, maiden-hair, hypericum herb, and white raisins.
If the patient drinks this and the beverages mentioned
above, he will recover very quickly.

36. Another jaundice: this one comes on in winter from
drunkenness and chills. At the beginning, first chills set in,
then fever. The subcutaneous moisture in the patient's
body congeals with the blood, as is clear from the follow-

---

320 ἐψεῖν δὲ om. M.
321 Θ: δὲ πιέτω M.
322 καὶ om. Θ.
323 M adds αὐτῷ.
324 Θ: -ην M.
325 Θ: σώμ- M.

αὐτοῦ τὸ σῶμα καὶ ὑπόσκληρον, καὶ αἱ φλέβες διὰ
τοῦ σώματος τέτανται ὠχραὶ καὶ[326] μείζονες ἢ πρόσθεν καὶ παχύτεραι· τέτανται δὲ καὶ[327] ὑπομελάντεραι[328]
ἄλλαι φλέβες. καὶ ἢν τάμῃς τινὰ αὐτῶν, ῥεύσεται τὸ
αἷμα ὠχρόν, ἢν ὠχραὶ ἔωσιν αἱ φλέβες· ἢν δὲ μέλαιναι ὦσι, μέλαν τὸ αἷμα ῥεύσεται. καὶ τὸ ἱμάτιον πρὸς
τῷ χρωτὶ κείμενον οὐκ ἀνέχεται ὑπὸ τοῦ κνησμοῦ·
258 οὗτος πρόθυμος | περιφοιτᾶν,[329] ἀλλ᾽ ὑπὸ τῆς ἀσθενεί-
ης αὐτοῦ τὰ σκέλεα ὑποφέρεται· καὶ διψῇ σφόδρα.
αὕτη ἡ νοῦσος ἧσσον τῆς προτέρης θανασίμη· προέρχεται δὲ πλείω χρόνον, ἢν μὴ ἐν τῇσιν ἑπτὰ ἡμέρῃσιν
ὑγιὴς γένηται. ἢν δὲ ἡ νοῦσος ἀπομηκύνηται καὶ
γένηται ὄγδοος ἢ ἔνατος μείς, πίπτει[330] ἐς κλίνην,[331]
καὶ ἡ νοῦσος καὶ ἡ ἀλγηδὼν πιέζει μᾶλλον, καὶ
ἀνίστασθαι οὐ δύναται, καὶ οἱ πολλοὶ ἐν τούτῳ τῷ
χρόνῳ παραχρῆμα διαφθείρονται.

Τοῦτον, ὅταν οὕτως ἔχῃ, κατ᾽ ἀρχὰς μὲν τῇ νούσῳ[332] ἢν παραγένῃ, ὅταν αἱ ἑπτὰ ἡμέραι παρέλθωσιν,
ἐλλέβορον πῖσαι, τὴν δὲ κάτω κοιλίην χυλῷ ὑποκαθῆραι, ὡς ἐν τῇ πρόσθεν· καὶ τἆλλα τὰ αὐτὰ διδόναι
φάρμακα. διδόναι δὲ καὶ κανθαρίδας, ἄνευ τῶν πτερῶν
καὶ τῆς κεφαλῆς, τέσσερας τρίβων καὶ διεὶς οἴνου[333]
ἡμικοτυλίῳ λευκοῦ,[334] ἤδη καὶ μέλι παραχέαι ὀλίγον,
εἶτα οὕτως διδόναι πιεῖν· τοῦτο πινέτω δὶς ἢ τρὶς τῆς
ἡμέρης. προϊούσης δὲ τῆς νούσου λουτροῖσι καὶ πυρι-
ήμασι θεραπεύειν. ἐσθιέτω δὲ ἄσσα ἂν προσίηται·

326 M adds εἰσὶ.   327 δὲ καὶ Θ: τε Μ.

170

ing: the body is livid and somewhat hard, and the vessels through it are stretched, pale-yellow, larger than they were before, and wider; other vessels are stretched and more darkish; if you incise one of the vessels, pale-yellow blood will flow out if the vessel was pale-yellow; if it was dark, dark blood will flow out. The patient will not tolerate a blanket lying against his skin, for the itching; he is eager to walk about but, because of his weakness, his legs collapse under him; he is very thirsty. This disease is less often mortal than the preceding one but, unless the patient recovers in seven days, it goes on for a longer time. If the disease becomes prolonged, and the eighth or ninth month arrives, the patient falls into bed, the disease and the pain press him more intensely, and he is unable to get up; many suddenly perish at this time.

When the case is such, if you attend the disease at its beginning, after seven days have passed have the patient drink hellebore, and clean out his lower cavity with juice as in the preceding diseases; give the rest of the same medications. Give blister-beetles, too, with their wings and heads removed: grind four, dissolve in a half-cotyle of white wine, immediately add a little honey, and give thus to drink; let the patient drink this potion two or three times a day. As the disease progresses, treat with baths and vapour-baths. Let the patient eat whatever he will accept, and

---

328 M adds καί.
329 Θ: πρὸς τὸ φοιτῆν M.
330 μ. π. Θ: συμπίπτει M.
331 Θ: νοῦσον M.
332 Θ: τῆς νούσου M.
333 Θ: -ῳ M.   334 Potter: -ῷ ΘM.

οἶνον δὲ πινέτω λευκόν, αὐστηρόν· καὶ τἆλλα τὰ αὐτὰ προσφέρειν ἃ καὶ τῷ πρόσθεν. αὕτη ἡ νοῦσος χρονίη καὶ χαλεπή, ἢν κατ᾽ ἀρχὰς μὴ μελετηθῇ.

37. Ἄλλος ἴκτερος· ἐπιδήμιος καλέεται, διότι πᾶσαν ὥρην ἐπιλαμβάνει· γίνεται δὲ ἀπὸ πλησμονῆς μάλιστα καὶ μέθης ἐπειδὰν ῥιγώσῃ. εὐθὺς οὖν τὸ σῶμα ἀλλοιοτροπέει καὶ γίνεται ὠχρόν, καὶ οἱ ὀφθαλμοὶ σφόδρα ὠχροί, καὶ ὑπὸ τὰς τρίχας καὶ ὑπὸ τοὺς ὄνυχας ἡ νοῦσος προέρχεται.[335] καὶ ῥῖγος καὶ πυρετὸς βληχρὸς ἔχει, καὶ ἀσθενέει[336] τὸ σῶμα, καὶ ἐν τῇ κεφαλῇ[337] ὀδύνη ἔχει, καὶ οὐρέει ὠχρὸν καὶ παχύ. οὗτος ὁ ἴκτερος ἧσσον θανατώδης τῶν πρόσθεν, καὶ ἐξέρχεται μελετώμενος ἐν τάχει.

Τοῦτον, ὅταν οὕτως ἔχῃ, σχάζειν αὐτοῦ τοὺς ἀγκῶνας καὶ ἀφαιρέειν τοῦ αἵματος, ἔπειτα πυριάσας πῖσαι ἐλατήριον· κάτω δὲ αὖτις ὑποκαθῆραι τῇ τρίτῃ ὀνείῳ γάλακτι· τὰ δ᾽ ἄλλα ῥυφήματα καὶ ποτὰ καὶ ἐδέσματα ταὐτὰ διδόναι ἃ καὶ τῷ πρόσθεν, καὶ καθαίρειν τὴν κεφαλὴν αὐτοῦ θαμινά. καὶ ἀπὸ χαραδριοῦ[338] πίνειν, καὶ λούειν αὐτὸν πολλῷ καὶ θερμῷ, καὶ ξύων τοῦ χαραδριοῦ[339] τὴν σάρκα ἐν οἴνῳ λευκῷ διδόναι· καὶ ἐς τὰ ἄλλα πάντα αὐτῷ χρήσθω, καὶ ἐν τάχει ὑγιὴς ἔσται.

38. Ἄλλος ἴκτερος· γίνεται μὲν ἀπὸ φλέγματος, τῆς δὲ ὥρης χειμῶνος μάλιστα ἐπιλαμβάνει. καὶ ἡ χροιὴ αὐτοῦ λευκή, καὶ τὰ στήθη αὐτοῦ πλήρη γίνεται φλέγματος· καὶ ἀποπτύει τὸ σίελον πολύ,[340] καὶ ὅταν ἀποχρέμψηται, λυγμὸς αὐτῷ ἐμπίπτει.[341] καὶ οὐρέει

drink dry white wine; otherwise administer the same things as to the preceding patient. This disease lasts a long time and is severe if it is not cared for at the beginning.

37. Another jaundice: this one is called "common", because it occurs in every season; it arises mainly from fullness or drunkenness, when a person has a chill. At once, then, the body changes colour and becomes yellow, the eyes intensely so, and the disease invades beneath the hair and nails. Chills and mild fever are present, the body is weak, there is pain in the head, and the patient passes thick pale-yellow urine. This jaundice is less mortal than the preceding ones and, if treated promptly, it passes off.

When the case is such, incise the bends of the patient's arms, and draw blood; then apply a vapour-bath, and have him drink squirting-cucumber juice; clean him downwards again on the third day with ass's milk; otherwise give the same gruels, drinks and foods as to the preceding patient, and clean out the head frequently. Have the patient drink soup made from plover, and wash him in copious hot water; also shred plover fresh into white wine, and give this. For the rest, let the patient follow the same regimen; he will soon recover.

38. Another jaundice: this one arises from phlegm and occurs mainly in winter. The patient's colour becomes white, and his chest fills up with phlegm; he expectorates copious sputum and, as he is coughing it up, he suffers from an attack of hiccups; he passes thick white urine, and

---

335 Θ: ἐπ- M.   336 Later mss: -νείη ΘM.
337 Θ adds ἤ.   338 Θ: χαλαρίου M.
339 ξ. τ. χ. Mack: χυωνίου χαραδριοῦ Θ: ξύων M.
340 πολύ om. M.   341 Θ: γίνεται M.

λευκὸν καὶ παχύ, καὶ ἐπ' αὐτῷ ἐφίσταται[342] οἷον ἄλευρον. οὗτος ὁ ἴκτερος ἥκιστα θανάσιμος καὶ ἐν τάχει
ὑγιαίνεται.

Τοῦτον, ὅταν οὕτως ἔχῃ, πῖσαι Κνίδιον κόκκον, καὶ
μετὰ τὴν κάθαρσιν πτισάνης χυλῷ μέλι παραχέας,
δοῦναι ἐκρυφεῖν τέσσερας κοτύλας. τὴν δὲ ἄνω κοιλίην ἐμετοποιεύμενος καθαρὴν[343] παρεχέτω ὡς ἐν τοῖσι πρόσθεν· οὕτως γὰρ ἂν ῥᾷστα τὸ φλέγμα ἀνάγοι
ἀπὸ[344] τοῦ πλεύμονος καὶ τῶν ἀρτηρίων. καὶ ἀναγαρ-
γαρισμὸν δὲ αὐτῷ σκευάζειν θαμινά. οὗτος ἐνίοτε καὶ
πυρεταίνει πυρετῷ βληχρῷ, καὶ φρίκη λεπτὴ ἐπιγίνεται. καὶ τἆλλα μελετᾶν τοῖσιν αὐτοῖσιν οἷσι καὶ
τοὺς πρώτους[345] ἰκτέρους, καὶ φαρμάκοισι καὶ πυρίῃσι
καὶ λουτροῖσι καὶ ἐδέσμασι καὶ ποτοῖσι καὶ ῥυφή-
μασιν· οὕτω γὰρ ἂν τάχιστα[346] ὑγιέα ποιήσειας.

39. Τῖφος· τὸ νόσημα ἐπιλαμβάνει θέρεος ὥρῃ,
ὅταν ὁ κύων τὸ ἄστρον ἐπιτέλλῃ, χολῆς κινηθείσης
κατὰ τὸ σῶμα. εὐθέως οὖν πυρετοὶ ἔχουσιν ἰσχυροὶ
καὶ καῦμα ὀξύ, καὶ ὑπὸ τοῦ βάρεος ἀσθενείη καὶ
ἀκρησίη τῶν μελέων,[347] καὶ ἐκ τῶν χειρῶν ἄχρειος[348]
γίνεται μάλιστα. καὶ ἡ γαστὴρ ταράσσεται, καὶ τὰ
ὑποχωρεῦντα δυσώδεα, καὶ ἰσχυρὸς στρόφος γίνεται.
ταῦτά τε πάσχει, καὶ ἤν τις ἀνιστῇ αὐτόν, οὐ δύναται
ὀρθοῦσθαι· οὐδὲ τοῖσιν ὀφθαλμοῖσιν ἀνορᾶν δύναται
ὑπὸ τοῦ καύματος· καὶ ἤν τις αὐτὸν ἐρωτᾷ τι,[349] ὑπὸ
τοῦ πόνου ἀκούων οὐ δύναται ὑποκρίνεσθαι. ὅταν δὲ

262

---

342 Θ: ὑπ' α. ὑφ- Μ.   343 Θ: -αρσιν Μ.

material like meal rises to its surface. This jaundice is not very often mortal, and the patient recovers quickly.

When the case is such, have the patient drink Cnidian berry; after the cleaning, give him four cotylai of barley-water gruel with honey, to drink off. Let him clean out his upper cavity by provoking vomiting as in the preceding cases; for thus will he most easily bring up the phlegm from his lung and bronchial tubes. Prepare frequent gargles for him. This patient is sometimes mildly feverish and subject to light shivering. For the rest, treat with the same things used in the former jaundices: medications, vapour-baths, baths, foods, drinks and gruels; in this way you will very quickly make the patient well.

39. Typhus: this disease comes on in summer, when the Dog Star rises,[9] because of bile being set in motion through the body. At once, then, sharp burning heat and powerful fevers set in; there are weakness and loss of control in the limbs because of heaviness, and the patient becomes especially helpless in his arms. The belly is set in motion, the stools are foul-smelling, and there is violent colic. The patient also suffers the following: if anyone makes him stand up, he is not able to stand straight; nor can he look up with his eyes, on account of the burning heat; if anyone asks him a question, although he hears, on account of his distress he cannot reply. When the patient is

[9] High summer.

---

344 Θ: ἀπαγάγοι M.   345 Θ: πρόσθεν M.
346 Θ: μάλιστα M.   347 Θ: σκελέων M.
348 ἐκ . . . ἄχρειος Θ: τῶν χ. ἀχρεῖα γὰρ M.
349 τι om. M.

μέλλῃ ἀποθανεῖσθαι ὀξύ[350] τε ὁρᾷ καὶ φθέγγεται θαρσαλέως, καὶ πιεῖν καὶ φαγεῖν αἰτέει· καὶ ἢν δῷ τις καὶ φάγῃ, ἐν τάχει ἀφῆκε[351] τὴν ψυχὴν ἢν μὴ ἀπεμέσῃ. τούτῳ ἡ νοῦσος κρίνεται ἐν ἑπτὰ ἡμέρῃσιν ἢ τεσσερεσκαίδεκα· πολλοὶ δὲ διαφεύγουσι καὶ ἐς τὰς τέσσερας καὶ εἴκοσιν· ἢν οὖν ταύτας ἐκφύγῃ, ὑγιής ἐστιν, ἐν γὰρ ταύτῃσι τῇσιν ἡμέρῃσι διαδηλοῖ εἰ θανάσιμος ἢ οὔ.

Τοῦτον, ὅταν οὕτως ἔχῃ, ὧδε ἰᾶσθαι· ἐν τῇσι πρώτῃσι τῶν ἡμερέων λουτροῦ μὲν ἀπέχειν, ἀλείφειν δὲ οἴνῳ καὶ ἐλαίῳ χλιήνας ἐς κοίτην. καὶ σιτίων ἀπέχειν, ῥυφήματα δὲ λεπτὰ ἀποψύχων διδόναι· πινέτω δὲ οἶνον μέλανα, αὐστηρόν, ἢν ξυμφέρῃ· ἢν δὲ μή,[352] λευκόν, αὐστηρόν, ὑδαρέστερον. ἢν δὲ ἡ δίψα πιέζῃ ἰσχυρή, τοῦ ὕδατος ἀθρόον διδοὺς πιεῖν, κέλευε ἐξεμέειν· ταῦτα δὶς ἢ τρὶς ἐφεξῆς ποιῆσαι. καὶ ὅταν τὸ καῦμα ἔχῃ, ῥάκεα βάπτων ὕδατι ψυχρῷ προστιθέναι, ᾗ ἂν[353] μάλιστα δοκῇ[354] καίεσθαι· ἢν δὲ φρίξῃ τὸ σῶμα, ἀνιέναι τὰ ψύγματα. οὗτος ὅταν μάλιστα πονέῃ, ὑπὸ τῆς ὀδύνης κινδυνεύσει τότε ἀφεῖναι τὴν
264 ψυχήν· ἀλλὰ διδόναι[355] αὐτῷ τῆς ὀδύνης τὰ | φάρμακα ταῦτα ἃ καὶ τῷ ὑπὸ τῆς πλευρίτιδος ἐχομένῳ. ὅταν δὲ ἐξαναστῇ, ἀνακομίζειν σίτῳ καὶ ποτῷ καὶ λουτροῖσιν ὡς τάχιστα· ἡ γὰρ νοῦσος χαλεπή, καὶ παῦροι ἐκφυγγάνουσιν.

40. Ἄλλος τῖφος· ἐπιλαμβάνει μὲν τὸ νόσημα πᾶσαν ὥρην· γίνεται δὲ δι' ὑγρασίην τοῦ σώματος, ὅταν τὰ σιτία ὑγρὰ ἐόντα καὶ τὸ ποτὸν πολὺ αἱ σάρκες

on the verge of death, he sees keenly, talks confidently, and demands something to drink and eat; if anyone gives it to him, and he eats it, he soon gives up the ghost, unless he vomits. The disease reaches its crisis in this patient in seven or fourteen days, but many patients survive for twenty-four days; now, if they live beyond those, they have recovered, for that is the critical period within which a patient shows whether or not he will die.

When the case is such, treat the patient as follows: on the first days do not allow him to bathe, but anoint him at bedtime with warm wine and oil. Keep him away from food, but give thin chilled gruels; let him drink dry dark wine, if it benefits him; if this wine does not benefit him, let him drink very dilute dry white. If a violent thirst presses the patient, give him water to drink frequently, and order him to vomit it up; do this two or three times in a row. When burning heat is present, dip rags in cold water and apply them wherever the burning seems to be most intense; if the body shivers, though, remove the cold compresses. When this patient is suffering most severely, he will be in danger of dying from the pain; in order to prevent this, give him the same medications against pain as to a patient with pleurisy. Once the patient has got up, strengthen him very quickly with food, drink and baths; for the disease is severe, and few escape it.

40. Another typhus: this disease occurs in every season; it arises from moistness of the body, when the tissues soak

---

350 Θ: ὀξύτερά M. 351 Θ: μεθῇ καὶ M.
352 M adds ἀλλὰ. 353 ᾖ ἂν Cornarius: ἐπὴν Θ: ἢν M.
354 Θ: φῇ M. 355 M adds χρὴ.

ἀναπίωσι καὶ θαλεραὶ³⁵⁶ γένωνται· ἀπὸ τούτων μάλιστα τὸ νόσημα γίνεται. ἄρχεται οὖν πυρετὸς τριταῖος
ἢ τεταρταῖος κατ᾽ ἀρχὰς γινόμενος· καὶ πόνος ἰσχυρὸς ἐν τῇ κεφαλῇ ἐνέστηκεν, ἐνίοτε δὲ καὶ ἐν τῷ
σώματι καὶ διαλείπει.³⁵⁷ καὶ ἐμέει σίαλα, καὶ ἐρεύγεται
πυκινά, καὶ τὰς χώρας τῶν ὀφθαλμῶν ἀλγέει· καὶ τὸ
πρόσωπον ἀφύει, καὶ ἐς τοὺς πόδας οἴδημα κατέρχε-
ται· ἐνίοτε δὲ καὶ τὸ σῶμα ὅλον ἐποιδέει. καὶ ἐς τὰ
στήθεα καὶ ἐς τὸ μετάφρενον ἡ ὀδύνη ἔχει· ἐνίοτε δὲ³⁵⁸
καὶ ἡ γαστὴρ τετάρακται. καὶ τοῖσιν ὀφθαλμοῖσιν
ἐξορᾷ ἰσχυρῶς· καὶ τὸ σίαλον ἀποπτύει πολὺ καὶ
ἀφρῶδες, καὶ ἐν τῇ φάρυγγι δοκέει τι³⁵⁹ ἐνέχεσθαι
καὶ κέρχνειν αὐτήν, πολλάκις δὲ καὶ φλεγμαίνει ὁ
φάρυγξ. τοῦτον ὅταν οὕτως³⁶⁰ ὁ πόνος πιέζῃ, ἔστι δ᾽
ὅτε καὶ ὀρθοπνοίη ἰσχυρὴ ἐπιπίπτει, καὶ πολλάκις
ἐξαπίνης ὑπὸ τοῦ πόνου τὴν ψυχὴν ἀφῆκεν ἐν ἑπτὰ
ἡμέρῃσιν ἢ τεσσερεσκαίδεκα· πολλοὶ δὲ διαφεύγου-
σι³⁶¹ καὶ ἐς τὰς εἰκοσιτέσσαρας. πολλάκις δὲ ἐξαπίνης
ἡ νοῦσος ἀφῆκε,³⁶² καὶ δοκέει ὑγιὴς εἶναι· ἀλλὰ φυλάσσεσθαι χρή, ἕως ἂν τέσσερες καὶ εἴκοσιν ἡμέραι
παρέλθωσιν· ἢν δὲ ταύτας φύγῃ, οὐ μάλα θνήσκει.

Τοῦτον, ὅταν οὕτως ἔχῃ, ἐν τῇσι πρώτῃσι τῶν
266 ἡμερέων | ῥύφημα διδόναι ἄλευρον³⁶³ κάθεφθον λε-
πτόν, μέλι παραχέων· πίνειν δὲ διδόναι μέλανα οἶνον
κατ᾽ ὀλίγον, ὅπως ἂν βούληται κεκρημένον· σιτία δὲ
μὴ προσφερέσθω,³⁶⁴ πρὶν ἂν αἱ³⁶⁵ ἡμέραι παρέλθωσι·
πουλύποδας δὲ ἐφθοὺς διδόναι ἐν οἴνῳ ἐσθίειν, καὶ τὸν
ζωμὸν ῥυφάνειν· καὶ ῥαφανίδας τρωγέτω³⁶⁶ πολλάς.

up foods that are moist and drink that is excessive, and
become stout; it is from these things that the condition
usually arises. Fever begins, then, as a tertian or a quartan;
intense pain establishes itself in the head, and is sometimes also present intermittently in the body. The patient
vomits sputa, belches frequently, and has pain in the sock-
ets of his eyes; his face becomes white, and his feet swell
up; sometimes his whole body swells up, too. Pain occupies
the chest and back; sometimes the belly is set in motion.
The patient protrudes his eyes greatly; he coughs up copious frothy sputum, something seems to be caught in his
throat and to make it hoarse, and often his throat swells up,
too. When pain is pressing the patient like this, sometimes
severe orthopnoea comes on, and often, under the strain, a
patient has suddenly given up the ghost in seven or fourteen days; but many survive for twenty-four. Often the disease suddenly resolves, and the patient seems to have recovered; still, he must take care until twenty-four days
have passed; if he escapes those, death is rare.

When the case is such, in the first days give the patient
as gruel thin boiled-down meal with honey; to drink give
him dark wine, a little at a time, mixed however he wishes;
let him not take cereals until the critical days have passed,
but give him polyp boiled in wine to eat, and the sauce to
drink; let him eat many radishes. Also, roast cress seed,

---

356 Θ: πλαδαραὶ M.   357 καὶ δ. Θ: διαλείπων M.
358 ἡ ὀδύνη . . . δὲ Θ: ἐνίοτε δὲ καὶ ἡ ὀδύνη ἔχει M.
359 τι om. M.   360 Θ adds ἔχῃ.
361 Θ: -φέρουσι M.   362 Θ: ἀν- M.
363 Θ: ἄλητον M.   364 Θ: -φέρειν M.
365 αἱ om. Θ.   366 Θ: -ειν M.

καὶ καρδάμου καρπὸν φώξας, ἀλέσαι καὶ σῆσαι λε-
πτά· ἔπειτα ἐπ᾽[367] οἶνον ἐπιβαλὼν μέλανα στρυφνὸν
καὶ ἄλφιτα λεπτὰ ὀλίγα, διδόναι πίνειν ἔωθεν. λουτροῦ δὲ ἀπεχέσθω μέχρι ἂν αἱ ἡμέραι παρέλθωσιν·
οἴνῳ δὲ καὶ ἐλαίῳ χλιήνας ἀλείφειν ἐς κοίτην, καὶ
ἐκμάσσειν. καὶ γλυκυσίδης καρποῦ δέκα κόκκους
ἑψῶν ἐν οἴνῳ μέλανι, διδόναι πίνειν· καὶ γογγυλίδας
διεφθους ποιέων ῥυφανέτω[368] ἀρτύσας τυρῷ ἀνάλτῳ
καὶ μήκωνι καὶ ἐλαίῳ καὶ ἁλὶ καὶ σιλφίῳ καὶ ὄξει. ἢν
δὲ βούλῃ πῖσαι φάρμακον, τοῦ Κνιδίου κόκκου πῖσαι·
καὶ μετὰ τὴν κάθαρσιν ἀλεύρου ἐφθοῦ καὶ λιπαροῦ
δοῦναι δύο τρυβλία ἐκρυφεῖν· οἶνον δὲ πινέτω τὸν
αὐτόν. τούτων[369] τῶν φαρμάκων καὶ ῥυφημάτων καὶ
ποτῶν ὅ τι ἂν διδῷς ὀνήσεις, ἤν τε κατὰ ἓν ἤν τε καὶ[370]
πλείω προσφέρῃς, καὶ τάχιστα ὑγιᾶ ποιήσεις. ἡ δὲ
νοῦσος χαλεπή, καὶ παῦροι ἐκφυγγάνουσι.

41. Ἄλλος τῖφος· γίνεται μὲν διὰ τάδε· ὅταν ἡ χολὴ
σαπεῖσα μιγῇ τῷ αἵματι ἀνὰ τὰς φλέβας καὶ τὰ
ἄρθρα, καὶ[371] ὅταν στῇ, οἴδημα ἀνίσταται[372] μάλιστα
μὲν ἐν τοῖσιν ἄρθροισι καὶ[373] καταστηρίζει, ἐνίοτε
δὲ[374] καὶ ἐς τὸ ἄλλο σῶμα. καὶ ὀδύνας παρέχει ὀξείας,
καὶ οἱ πολλοὶ ἐκ ταύτης τῆς νούσου χωλοὶ γίνονται
ὅταν ἀποληφθεῖσα ἐν τοῖσιν ἄρθροισιν ἡ χολὴ
πωρωθῇ· ἡ δ᾽ ὀδύνη διαλείπουσα ἐπιλαμβάνει καὶ διὰ
τριῶν ἡμερέων καὶ διὰ τεσσάρων.

268   Τοῦτον, ὅταν οὕτως ἔχῃ, ὧδε μελετᾶν· ὅταν μὲν ἡ
ὀδύνη ἔχῃ ἐν τῷ σώματι, χλιάσματα προστιθέναι

---

[367] ἐπ᾽ Θ: πινέτω Μ.   [368] Μ adds τοῦ χυμοῦ.

grind and sift it fine, sprinkle it over sour dark wine along with a little fine meal, and give to drink early in the morning. Let the patient abstain from the bath until the critical days have passed, but anoint him with warm wine and oil at bedtime, and wipe this off. Boil ten grains of peony seed in dark wine, and give to drink; boil turnips well, and have the patient drink their juice, after seasoning it with unsalted cheese, poppy, oil, salt, silphium and vinegar. If you wish to have the patient drink a medication, let him drink Cnidian berry; after the cleaning, give him two bowls of rich boiled meal to drink off; let him drink the same wine. Whichever of these medications, gruels, and drinks you give, you will do good, whether you administer one alone or more, and you will very quickly make the patient well. Still, the disease is severe, and few escape it.

41. Another typhus: this one arises from the following: when bile that has become putrid mixes with the blood in the vessels and joints, and when this stands, swelling comes up and becomes established, mainly in the joints, but sometimes also in the rest of the body. This produces sharp pains, and most patients become lame from the disease, inasmuch as bile is cut off in their joints, and congeals. Intermittent pains are present every third or fourth day.

When the case is such, treat the patient as follows: when pains are present in the body, anoint with oil, and ap-

---

369 Littré: -ῳ Θ: καὶ M.   370 Θ: κατὰ M.

371 καὶ om. M.

372 Θ: τὸ αἷμα ἐνίσταται M.

373 καὶ om. M.

374 Θ: δ. ἐ. M.

ἐλαίῳ ὑπαλείψας. ὅταν δὲ ἀνῇ, δοῦναι αὐτῷ ἐλλέβορον
πυριάσας πρόσθεν ἅπαν τὸ σῶμα. τῇ δ᾽ ὑστεραίῃ
ὀρὸν αἰγὸς ἑψήσας, δοῦναι πιεῖν δύο χοέας, μέλι
παραχέας παρὰ τὸν ἕτερον χοέα, παρὰ δὲ τὸν ἕτερον
ἅλας παραβάλλων· εἶτα κύλικα παρὰ κύλικα ἐναλ-
λάσσων[375] πινέτω ἕως ἂν ἐκπίῃ ἅπαν. ἐς ἑσπέρην δὲ
μετὰ τὴν κάθαρσιν φακῆς τρυβλίον ῥυφείτω, καὶ
σεύτλων τρυβλίον[376] λιπαρῶν ἄλφιτα παραπάσας ἐκφαγέτω,[377] καὶ νεοσσοῦ ἀλεκτορίδος κρέας ἢ πελειάδος ἢ τρυγόνος ἢ οἰὸς ἢ ὑὸς πίονος.[378] τὸν δὲ ἐλλέβορον δι᾽ ἕκτης ἡμέρας διδόναι.

Καὶ ἤν που τῶν ἄρθρων ἀποιδίσκηται καὶ μὴ
θέλῃ[379] καθίστασθαι, σικύην προσβάλλων ἀφαιρέειν
τοῦ αἵματος· καὶ κέντρῳ ἀκίδος τριγώνου[380] ἐς τὰ
γούνατα κεντέειν,[381] ἢν ἐν τοῖσι γούνασιν ἐνῇ τὸ οἴδημα, τῶν δὲ ἄλλων ἄρθρων μηδὲν κεντρώσῃς.

Τὰς δὲ μεταξὺ τῶν ἡμερέων σιτίον προσφερέσθω
ἄρτον μὲν ὡς ὀπτότατον, μᾶζαν δὲ ψαιστὴν ὡς μάλι-
στα· ὄψον δὲ ἐχέτω μάλιστα μὲν ὄρνιθα ὀπτὴν ἄναλ-
τον· εἰ δὲ μή, καὶ ἐφθήν, ἐζωμευμένην, ἄνευ τυροῦ καὶ
σησάμου καὶ ἁλός. ἰχθύσι δὲ χρήσθω τοῖσι σαρκω-
δεστάτοισιν, ὀπτῶν[382] τὸν αὐτὸν τρόπον τοῖσι κρέασιν, ἢ[383] ἕψων, ὀριγάνου πάσας, ἐλαίῳ[384] ὑποχρίσας·
οἶνον δὲ πινέτω λευκόν, ἢν ξυμφέρῃ· εἰ δὲ μή, μέλανα.
καὶ περιόδοισι ταλαιπωρείτω δι᾽ ἡμέρης καὶ μετὰ
δεῖπνον καὶ ὄρθριος. ὀρὸν δὲ καὶ[385] γάλα τὴν ὥρην αἰεὶ

---

375 Θ: παρ- Μ.   376 τρυβλίον om. Θ.

182

ply fomentations; when the pains go away, first apply a vapour-bath to the whole body and then give hellebore. On the next day, boil goat's whey, and give two choes to drink, pouring honey into one chous, and sprinkling salt over the other one; let the patient drink these alternately, cup for cup, until he has drunk everything off. Towards evening after the cleaning, let him drink a bowl of lentil-soup, and eat a bowl of beets boiled in grease and with meal sprinkled over them; also let him eat flesh of chicken, pigeon, turtle-dove, sheep, or fat swine; give the hellebore every sixth day.

If the patient swells up in the joints somewhere, and the swelling does not want to go down, apply a cupping instrument, and draw blood; if the swelling is in the knees, pierce into the knees with the point of a triangular needle, but do not pierce any of the other joints.

On the days between medications, as cereal let the patient take very well-done bread, and especially ground barley-cake; as main dish let him best have broiled fowl without salt; if not broiled, then boiled and made into a soup without cheese, sesame or salt. Of fish let him have the fleshiest, broiling them in the same way as the meats; or boil them, sprinkle on marjoram, and coat them with oil. Let the patient drink white wine, if it benefits him; if not, then dark wine. Have him exercise by walking every day, both after dinner and early in the morning. Let him always

---

377 Θ: -πιέτω M.   378 Θ: πῖον M.   379 μὴ θ. M: **σος δὴ Θ.   380 καὶ κ. ἀ. τ. Potter: ἠκετρωακίδος τριγώνου Θ: κατακεντῶν ἀκίδι τριγώνῳ M.   381 κεντέειν om. M.   382 M: ὀπτοῖσι Θ.   383 ἢ om. Θ.   384 Θ: παστὰ ἐλαίου M.   385 καὶ om. M.

ΠΕΡΙ ΤΩΝ ΕΝΤΟΣ ΠΑΘΩΝ

πινέτω· ἢν δὲ δοκέῃ, καὶ ὄνειον διδόναι ἀφεψήσας.

270   Καὶ ἢν ὑγιὴς γένηται, ἐν φυλακῇ αὐτὸν | ἔχειν τοῦ
ψύχεος καὶ τοῦ πνίγεος, καὶ τῶν σιτίων μὴ λίην
πιμπλάσθω· κίνδυνος γὰρ αὖθις ὑποτροπάσαι τὴν
νοῦσον. αὕτη ἡ νοῦσος οὕτω θεραπευομένη, ἐν ἓξ
μησὶν ὑγιὴς γίνεται· οὗτοι γὰρ κρίνουσιν εἰ θανάσιμος ἢ οὔ, ἢν παραχρῆμα θεραπεύηται· ἡ γὰρ νοῦσος
χαλεπή, καὶ τοῖσι πλείστοισι συναποθνήσκει.

42. Ἄλλος τῖφος· γίνεται μὲν τὸ νόσημα ὀπώρης
μάλιστα, ὅταν πλησθῇ παντοίης ὀπώρης. πολλοῖσι δὲ
καὶ ἀπὸ τῶνδε ἐγένετο τὸ νόσημα φαγοῦσι[386] πλακοῦντα καὶ σησαμῆν[387] καὶ τῶν ἄλλων τῶν μελιτοέν-
των· τὸ γὰρ μέλι τὸ ἑφθὸν καυματῶδές ἐστι καὶ προσ-
πλάσσεται ἐπὶ τὴν κοιλίην.[388] ἔπειτα ὅταν καθεψηθῇ
ἐν τῇ κοιλίῃ διαχεῖται,[389] καὶ ἐξαπίνης ἡ γαστὴρ
αἴρεται καὶ πίμπραται, καὶ δοκέει διαρραγήσεσθαι,
καὶ[390] ἐξαπίνης διάρροια ἐγένετο· καὶ ὅταν ἅπαξ ἄρξηται χωρέειν, πολλὰς ἡμέρας καθαίρεται. καὶ πολλοὶ
μετὰ ταύτην τὴν κάθαρσιν ὑγιέες ἐγένοντο.

Ὅταν οὖν παύσηται αὐτόματος καθαιρόμενος, φακῶν χυλοῦ ἀναγκάσαι αὐτὸν ἐκπιεῖν τρία ἡμίχοα,
ἅλας παραβάλλων. μετὰ δὲ τὴν κάθαρσιν τὴν ἐκ τοῦ
χυλοῦ, ἐς ἑσπέρην φακῆς τρυβλίον ῥυφείτω ψυχρῆς
ἀνάλτου, σίλφιον δ᾽ ἐπιξύσθω[391] πολύ· καὶ σεύτλου
τρυβλίον ἀνηδύντου λιπαροῦ, ἄλφιτα παραπάσας φα-
γέτω·[392] οἶνον δὲ πινέτω μέλανα αὐστηρὸν κατ᾽ ὀλί-

386 Θ: -όντες Μ.   387 Θ: -ος καὶ -ῆς Μ.

184

drink whey and milk in season; if it seems advisable, also give boiled-down ass's milk.

If a patient recovers, guard him from both cold and stifling heat, and let him not overfill himself with food; for there is a danger that the disease will recur. Treated thus, the disease goes away in six months; for that is the critical period that decides whether or not the patient will die, if he is treated at once. The disease is severe, and stays with most patients until they die.

42. Another typhus: this disease arises mainly in late summer, when a person stuffs himself on all sorts of the season's fruits; in many, it has also arisen from eating flat-cake, sesame-seed cake, and some of the others sweetened with honey, for boiled honey is burning-hot, and adheres to the cavity. When these foods have been boiled-down in the cavity, they are dispersed, and all at once the belly becomes raised and distended, and seems about to burst; diarrhoea suddenly comes on, and when this once begins to flow, the person is cleaned for many days; after the cleaning, many patients have recovered.

Now, when the spontaneous cleaning has come to an end, compel the person to drink off three half-choes of lentil-juice with salt. After the cleaning brought on by the juice, towards evening let him drink a bowl of cold lentil-soup without salt, but with much silphium grated over it; also, have him eat a bowl of unseasoned beets boiled in grease and with meal sprinkled over them, and drink dry dark wine, a little at a time. From then on, let the patient

---

388 ἐπὶ τὴν κ. Θ: τῇ κοιλίῃ M.　　389 Θ: ἀναζέεται M.
390 M adds ἔπειτα.　　391 Littré: ἐπεξύσθω Θ: ἐπεξέσθω
M.　　392 φαγέτω om. M.

*γον. τὸν δὲ λοιπὸν χρόνον ταῦτά τε ποιείτω, καὶ σιτία
προσφερέσθω ἄρτον ἔξοπτον, μᾶζαν δὲ ψαιστὴν ὡς
μάλιστα· τὰς δὲ πρώτας τῶν ἡμερέων ἄλευρον ῥυφανέτω[393] κάθεφθον ἀποψύχων, μέλι παραχέων, οἶνον δὲ*

272 *πινέτω μέλανα στρυφνόν· ἕως ἂν καταστῇ ἡ νοῦσος,
ταῦτα προσφερέσθω.*

Ὁ δὲ ὑπὸ τῆς ὀπώρης ληφθεὶς τῇ νούσῳ φῦσαν
παρέχει καὶ στρόφον καὶ ὀδύνην· ἡ ὀπώρη[394] καὶ τὰ
σιτία οὐκ ἐθέλει διαχωρέειν· καὶ ἡ γαστὴρ μεγάλη καὶ
σκληρὴ αὐτοῦ ἐστι, καὶ ῥῖγος καὶ πυρετὸς ἔχει. τούτῳ
ἢν μὲν αὐτομάτη ταραχθῇ ἡ κοιλίη, ἐν εἴκοσι δὲ
ἡμέρῃσι τὸ ἐλάχιστον καθαίρεται, καὶ ὅταν παύσηται
καθαιρόμενος, ὑγιὴς παραχρῆμά ἐστιν. ἢν δὲ μὴ αὐτομάτη ταραχθῇ,[395] καθαίρειν αὐτὸν τῷ τοῦ ἱππόφεω
ὀπῷ ἢ τῷ Κνιδίῳ κόκκῳ. ἐς ἑσπέρην δὲ ταὐτὰ διδόναι
ἃ καὶ τῷ αὐτομάτῳ καθαιρομένῳ· τῇ δ᾽ ὑστεραίῃ, ἢν
μὲν πυρετὸς ἔχῃ, ἡσυχίην ἐχέτω, διδόναι δὲ αὐτῷ
πίνειν τοῦ αὐτοῦ οἴνου ὡς ἐν ψυχροτάτῳ ὕδατι. ἢν δὲ
μὴ ἔχῃ πυρετός, διαιτάσθω δίαιταν μὴ ὑγρὴν ἀλλὰ
ἰσχυροτέρην, καὶ περιπατείτω πρὸς τὰ σιτία τεκμαιρόμενος. ὑπὸ τούτου νοσήματος πολλοῖς ἤδη ὕδερος
ἐγένετο· καὶ ἢν δοκέῃ σοι κλύζειν, τοῖσιν αὐτοῖσιν[396]
οἷσι καὶ τὸν ὑδερῶντα. ἢν δὲ βούλῃ, τοισίδε κλύζειν·
ἐς μελικρήτου[397] κοτύλην ποίην[398] θαψίης ἐγχύσαι,
εἶτα οὕτως ἐγκλύσαι. οὗτος οὕτω θεραπευόμενος
τάχιστα ὑγιὴς ἔσται.

continue to do the same, and as cereal take well-baked bread and especially ground barley-cake; on the first days, let him take boiled-down meal, cooled and with honey, and drink sour dark wine; until the disease subsides, these are the things he should take.

The patient that has the disease as the result of eating late summer fruits has flatulence, colic and pain; the fruits and foods do not want to pass off, the patient's belly is large and hard, and chills and fever are present. If the cavity is set in motion spontaneously, the person is cleaned out in twenty days, at the soonest, and, when the cleaning stops, he is at once well. If the cavity is not set in motion spontaneously, clean it out with hippopheos juice or Cnidian berry; towards evening, give the same things as to a patient cleaned spontaneously. On the following day, if fever is present, keep the patient quiet, and give him some of the same wine to drink, in very cold water; if fever is not present, let him employ a regimen that is not moist but quite strong, and take walks, determining their distance according to his foods. (In many cases, dropsy has developed out of this disease.) If it seems advisable to you to apply an enema, use the same kind as for a patient with dropsy. If you wish, make the following one: into a cotyle of melicrat shred thapsia herb; then inject. If he is treated in this way, the patient will very quickly recover.

---

393 Θ: φαγέτω M.
394 ἡ ὀπώρη om. M.
395 α. τ. Θ: καθαρθῇ M.
396 M adds κλύσον.
397 M: -ον Θ.
398 A later ms., Cornarius: ποσὶ Θ: ποιεῖν M.

43. Ἄλλος τῖφος· γίνεται μὲν τὸ νόσημα, ὅταν τὸ ὑγρὸν τὸ ἐν τῷ σώματι συμπαγῇ[399] καὶ ἀναξηρανθῇ μᾶλλον τοῦ καιροῦ. γίνεται δὲ τὴν ἰδέην, ὅταν τῷ νοσήματι ἔχηται, δίυγρος, ὕπωχρος,[400] διαφανής, κύστει πλήρει ἐοικὼς[401] οὔρου, οὐκ οἰδέει δέ,[402] ἀλλὰ λεπτὸς καὶ σκελιφρός ἐστι καὶ ἀσθενής· μάλιστα δὲ τοῦ σώματος λεπτύνεται τὰς κληῖδας· καὶ τὸ πρόσωπον ἰσχυρῶς κάτισχνος, καὶ οἱ ὀφθαλμοὶ ἔγκοιλοι σφόδρα. ταῦτα μὲν ὑπὸ τοῦ νοσήματος πάσχει. ἔστι δ᾽ ὅτε ἡ[403] χροιὴ τοῦ σώματος μέλαινα,[404] τάδε δὲ αἴτιά ἐστιν· ἐς τὰ φλέβια καὶ ὑπὸ[405] τὸ δέρμα ὅταν χολὴ

274 μέλαινα ὑπέλθῃ, καὶ ἐπὶ τούτοισιν ὅταν θέρμη ἐπιγένηται, ἀνάγκη οὖν ὑπὸ τοῦ θερμοῦ συγκαίεσθαι καὶ ἀναξηραίνεσθαι τὰ φλέβια, ὥστε[406] τὸ αἷμα μὴ χωρέειν[407] κατὰ τὰ φλέβια. τάδε οὖν πάσχει οὗτος πρὸς ἐκείνοις·[408] λεπτὸς γίνεται ἰσχυρῶς,[409] καὶ τοῖσιν ὀφθαλμοῖσιν ἀραιὰ καρδαμύσσει, καὶ τὰς μυίας ἀπὸ τοῦ ἱματίου θηρεύει, καὶ βορὸς τῶν σιτίων μᾶλλόν[410] ἐστιν ἢ ὑγιαίνων, καὶ λύχνου ἀπεσβεσμένου τῇ ὀδμῇ ἥδεται, καὶ ἐξονειρώσσει θαμινά· πολλάκις δὲ καὶ βαδίζοντι αὐτῷ προέρχεται ἡ γονή.

Τοῦτον, ὅταν οὕτως ἔχῃ, καθαίρειν τὴν κοιλίην τὴν μὲν ἄνω[411] ἐλλεβόρῳ, τὴν δὲ κάτω ὀπῷ σκαμωνίης. μετὰ δὲ τὴν κάθαρσιν ταὐτὰ διδόναι ἃ καὶ τοῖσιν ἄλλοισι· καὶ ὀρὸν καὶ γάλα βοὸς ἢ αἰγὸς διδόναι τὴν ὥρην· διδόναι δὲ καὶ ὄνειον γάλα[412] ἐς ὑποκάθαρσιν.

---

399 Θ: σαπῇ Μ.   400 Θ: ἔπ- Μ.

43. Another typhus: this disease arises when the moisture in the body congeals and dries up more than it should. When a person is attacked by it, he takes on a watery yellowish translucent appearance, like a bladder filled with urine; however, he does not swell up, but is lean, parched and weak; the part of his body in which he becomes leanest are the collar-bones; his face is very emaciated, and his eyes are very sunken; this is what a person generally suffers in the disease. Sometimes the body also becomes dark in colour, and for the following reason: when dark bile finds its way into the small vessels and under the skin, and when heat follows in these parts, the vessels necessarily become over-heated and dried up, so that the blood cannot move through them. Such a patient suffers the following, in addition: he becomes very lean, he seldom blinks his eyes, he chases flies away from his blanket, he has a greater hunger than when he was healthy, he takes pleasure in the smell of the extinguished lamp, and he has frequent nocturnal emissions; often semen even passes when he is walking.

When the case is such, clean out the upper cavity with hellebore, and the lower one with scammony juice. After the cleaning, give the same things as to the other patients; in season give cow's or goat's whey and milk, and also ass's milk, to clean downwards. If treated in such a way, this pa-

---

401 π. ἐ. Θ: ἔοικε Μ.   402 Θ: πλὴν οὐκ οἰδέει Μ.
403 ἔστι . . . ἡ Θ: ἡ δὲ Μ.   404 Μ adds ἦ.
405 Θ: ἐς Μ.   406 Μ: ὅταν Θ.   407 Littré: μὴ χωρέη
Θ: ἐκχωρέειν Μ.   408 πρὸς ἐκείνοις om. Μ.
409 ἰ. Θ: καὶ κάτισχνος καὶ ἰσχυρός Μ.
410 μᾶλλόν om. Μ.   411 τὴν μὲν ἄνω Θ: τῷ μέλανι Μ.
412 Μ adds ἀφεψῶν.

οὗτος οὕτω μελετώμενος μάλιστα ἐν δυσὶν ἔτεσιν
ὑγιὴς γίνεται. σιτία δὲ ἃ βούλεται προσφερέσθω·
εὐωχείσθω δ' ὡς μάλιστα, καὶ περιπατείτω πρὸς τὰ
σιτία τεκμαιρόμενος. αὕτη ἡ νοῦσος[413] λαμβάνει
πρεσβύτερον εἰκοσαετοῦς· ὅταν δὲ λάβῃ, ἢν μὴ κατ'
ἀρχὰς τοῦ νοσήματος τούτου μελετηθῇ, οὐκ ἐκλείπει,
πρὶν ἂν[414] εἴκοσιν ἔτεα ἄλλα[415] παρέλθῃ, ἀλλὰ προσ-
ίσχει· ἔπειτα ἐνίοισι μελετωμένη ἐξέρχεται. ἡ δὲ νοῦσος χαλεπή.

44. Εἰλεοὶ δὲ τάδε νοσήματα καλέονται. γίνονται[416]
δὲ ἀπὸ τῶνδε μάλιστα· ἢν τοῦ χειμῶνος θερμῇ τῇ
διαίτῃ καὶ ὑγρῇ χρῆται καὶ μηδὲ περιόδοισι ταλαι-
πωρέῃ πρὸς τὰ σιτία τεκμαιρόμενος, πιμπλάμενος
δὲ εὕδῃ, εἶτα ἐξαπίνης ἀναγκασθῇ βαδίσαι μακρὴν
ὁδόν, ψύχεος ἐόντος, εἶτα ῥιγώσῃ ὑπὸ τὰ ὀστέα, τάδε
οὖν πάσχει. φῦσα ἐγγίνεται ἐν τῷ σώματι παντί, καὶ
ἡ χροιὴ αὐτοῦ γίνεται μολυβδοειδής, καὶ ῥιγοῖ αἰεί,
276 ὥστε οἱ θερμοῦ καταχεομένου[417] οὐ | δοκέει θερμὸν
εἶναι. τὸ δὲ σῶμα λουομένου αὐτοῦ λεπίζεται ὑπὸ τοῦ
θερμοῦ, μάλιστα δὲ ἡ ὄρχεα·[418] καὶ ἢν τῷ δακτύλῳ τοῦ
σώματός που πιέζῃς,[419] ἐνθλάσεις, ἐκεῖ ἐκμάσσεταί[420]
σοι ὥσπερ ἐν σταιτί·[421] μάλιστα δ' ἐν τοῖσι ποσὶν
ἐνθλᾶται. τὰ δὲ σκέλεα βαρέα αὐτοῦ, καὶ ἢν περι-
φοιτέῃ, τρέμει· καὶ ἢν πρὸς αἶπος βαδίζῃ, πνευστιᾷ
σφόδρα. καὶ αἱ ὠλέναι δοκέουσιν ἀποκρέμασθαι, καὶ
τὴν κεφαλὴν ἀλγέει, καὶ αἱ ὀφρύες δοκέουσιν ἀποκρέ-

---

413 Θ adds οὐ.   414 πρὶν ἂν Θ: εἰ μὴ Μ.

tient usually recovers in two years. Let him eat whichever
foods he prefers and in as large amounts as possible, and
take walks, determining their distance according to his
foods. This disease attacks persons over twenty years of
age; when it does, if they are not cared for at the beginning,
it continues and does not go away before another twenty
years have passed; then, in some, if treated, it passes off.
The disease is severe.

44. The following diseases are called ileuses. They
arise, in most cases, from the following: if, in winter, a
person employs a hot, moist diet and does not exercise by
taking walks in accordance with his food but goes to bed
in a full state, and is then suddenly obliged to walk a long
distance when it is cold, and then gets chilled to the bone,
he suffers the following: tympanites develop in his whole
body, his colour becomes leaden, and he has a perpetual
chill, so that even when hot water is poured over him
he does not think it is hot. When he is bathed, his body
is excoriated by the hot water, especially his scrotum. If
you press his body at any point with your finger, indenting
it, it receives your impression there as if in dough; it is
most indented in the feet. The patient's legs are heavy; if
he walks about, he trembles; if he walks up a grade, he
pants violently. His forearms seem to hang down, he has a
headache, his eyebrows seem to overhang, thirst is present

---

415 ἄλλα om. M.   416 Θ: -εται M.
417 οἱ θ. κ. Θ: θερμὸν καὶ καταχεομένῳ M.
418 Θ: ὄσχη M.   419 π. π. Θ: ὑποπιέζῃς M.
420 ἐ. ἐ. Potter: καὶ ἐκμανεῖταί Θ: καὶ σημαινέεταί M.
421 ἐν σ. M: Θ has a blank space of suitable length.
422 καὶ τὴν . . . ἀποκρέμασθαι om. M.

μασθαι,[422] καὶ δίψα ἔχει τὰς νύκτας, τὰ δὲ σιτία ὠμὰ
διαχωρέει, ἄσσ᾽ ἂν φάγῃ.

Τοῦτον, ὅταν οὕτως ἔχῃ, πυριήσας αὐτὸν τοῦ κνεώρου δοῦναι πιεῖν ἢ τοῦ ἱππόφεω[423] ἢ τοῦ Κνιδίου
κόκκου. μετὰ δὲ τὴν κάθαρσιν ταῦτα διδόναι ἃ καὶ
τοῖσι πρόσθεν. τῇ δ᾽ ὑστεραίῃ ὀνείου γάλακτος ἑφθοῦ
χοέα δοῦναι δὶς ἐκπιεῖν ἅλας παραβάλλων. ἐς ἑσπέρην δὲ δειπνείτω ἄρτον, ὄψον δ᾽ ἐχέτω οἰὸς κρέα ἑφθὰ
καὶ πουλύποδας ἑφθοὺς ἐν οἴνῳ μέλανι ἡψημένους·[424]
καὶ τὸν ζωμὸν ῥυφείτω. καὶ φακὴν ἐχέτω ὧδε ἐσκευα-
σμένην· κοτύλην φακῶν ἑψήσας, τρίψας λείην, ἔπειτα ἄλευρον μίξας[425] καὶ σίλφιον ξέσας[426] ἅλας ἐμ-
βαλεῖν, καὶ ὄξος ἐπιχέαι, καὶ σκόροδα συνεψεῖν χρή·
ἐπὶ ταῦτα[427] ὕδωρ ἐπιχέας ζέσαι δὶς ἢ τρίς, καὶ τορυ-
νᾶν ἅμα. ἔπειτα ἀφελὼν ὀψάσθω·[428] ἔστω δὲ μὴ λίην
παχέη· ἐμβεβλήσθω δὲ καὶ γλήχων ἑψομένη τῆς εὐ-
ωδίης εἵνεκα. τὰς δὲ μεταξὺ τῶν ἡμερέων ἐμέτους
ποιείσθω δι᾽ ἕκτης ἡμέρης· πυριᾶσθαι δὲ χρὴ ἄλλοτε
καὶ ἄλλοτε πρὸ τοῦ ἐμέτου καὶ τῆς φαρμακοπωσίης.
καὶ διὰ τρίτης ἡμέρης λούσθω, ἢν συμφέρῃ· εἰ δὲ μή,
ἀλειφέσθω, καὶ περιπατείτω ἢν δυνατὸς ᾖ, πρὸς τὰ
σιτία τεκμαιρόμενος. καὶ ἀκτῆς φύλλα καὶ κονύζης
τῆς αἰεὶ ἁπαλῆς ἑψῶν διδόναι πίνειν.[429]

278   Οὕτω γὰρ | μελετώμενος ῥᾷστ᾽ ἂν διάγοι, καὶ ἡ
νοῦσος ἐκλείποι ἂν ἐνιαυσίη. πολλοῖσι δὲ ἤδη ὑγιέσι
γενομένοισι διὰ δύο ἐτέων πάλιν[430] ἡ νοῦσος ὑπετρό-

---

[423] M adds ὁποῦ.   [424] ἡψημένους om. M.

during the nights, and his food passes off undigested, just as he ate it.

When the case is such, apply vapour-baths to the patient, and give him spurge-flax, hippopheos, or Cnidian berry to drink. After the cleaning, give the same things as to the preceding patients. On the following day, twice give a chous of boiled ass's milk with salt, for the patient to drink off, and towards evening have him eat for dinner bread and, as main dish, boiled mutton and polyp boiled in dark wine, and drink the sauce too; also let him have lentil-soup prepared thus: boil a cotyle of lentils and mash them fine; then mix in meal, shred in silphium, and add salt; add vinegar, and boil in garlic; over this pour water, bring to a boil two or three times, and stir together; then let the patient take this and eat it as a main dish; it should not be too thick; also add boiled penny-royal, for the aroma. On the days between medications, let the patient induce vomiting every sixth day. You must apply vapour-baths, now and then, before the patient vomits and before he drinks his medication. Every third day, let him bathe, if it helps; if not, let him be anointed; also let him take walks, if he is able, determining their distance in accordance with his food. Boil leaves of the elder tree and of the fleabane that is always tender, and give these to the patient to drink.

Treated in this way, the patient will fare best, and the disease will go away in a year. In many who had already recovered, the disease has recurred after two years; if it does

---

425 Θ: παραμῖξαι M.   426 Θ: ἐπιξέσαι, καὶ M.

427 Θ: ἔπειτα M.

428 Θ: ἐψήσθω M: *frigefacito* Cornarius, implying ἐψύχθω.

429 Θ: ἐσθίειν M.   430 πάλιν om. M.

πασεν· ἀλλὰ χρή, ἢν ὑποτροπάσῃ, τὴν αὐτὴν ἴησιν
ἰᾶσθαι. ἢν δὲ τὸ τρίτον ὑποτροπάσῃ, οἴδημα μὲν οὐκ
ἐγγίνεται, λεπτὸς δὲ γίνεται καὶ κάτισχνος· ἄρχεται
δὲ λεπτύνεσθαι ἀπὸ τοῦ προσώπου, καὶ τὴν χροιὴν
ἔκλευκος γίνεται μᾶλλον ἢ τὸ πρόσθεν. τούτῳ ἐνίοτε
ὕδερος ἐγγίνεται ἐν τῇ κοιλίῃ· ἢν οὖν ἐγγένηται,
τάμνειν μὲν οὐ χρή· ἀποθανεῖται γάρ· θεραπεύειν δὲ
τοῖς αὐτοῖς οἷσι καὶ τὸν ἀπὸ τοῦ σπληνὸς ὑδεριῶντα.
τοῦτον μάλιστα μὲν κατ' ἀρχὰς βούλεσθαι λαβὼν
ἰᾶσθαι,[431] ταχὺ γὰρ ὑγιᾶ ποιήσεις. ἡ δὲ νοῦσος δεῖται
μελέτης· χαλεπὴ γάρ.

45. Ἄλλος εἰλεός· ἐπιλαμβάνει μὲν μάλιστα θέρεος
ὥρῃ[432] ἐν ἑλώδεσι χωρίοισι,[433] μάλιστα δ' ἐπιλαμβάνει
ἀφ' ὑδροποσίης· πολλοὶ δὲ ἤδη καὶ πρὸς τὸν ἥλιον
εὐνασθέντες[434] τὸ νόσημα ἔλαβον, τὴν κεφαλὴν
ἀλγήσαντες. τὰ δ' ἄλλα παραπλήσια τῷ πρόσθεν
πάσχουσι, πλὴν τῆς χροιῆς, οὗτος γὰρ ὠχρὸς γίνεται
οἷόν περ σίδιον· καὶ οἱ ὀφθαλμοὶ ἔστιν ὅτε ἰκτέρου
πίμπλανται.

Τοῦτον ὅταν οὕτως ἔχῃ, θεραπεύειν τοῖς αὐτοῖς
οἷσι καὶ τὸν πρόσθεν· διδόναι δὲ καὶ ἀπὸ ἐρεβίνθων
λευκῶν ἑψῶν τὸ ὕδωρ πίνειν, καὶ ἐν τῷ οἴνῳ κιρνὰς
διδόναι· καὶ τὴν κεφαλὴν αὐτοῦ καθαίρειν τῷ τετραγώνῳ. οὗτος ἧσσον τοῦ πρόσθεν θανατώδης· καλέεται
δὲ εἰλεὸς ἰκτεριώδης.

280   46. Ἄλλος ὅδε[435] εἰλεός· τὰ μὲν ἄλλα πάσχει πλῆθος ταὐτὰ τοῖσι πρόσθεν, ἄρχεται δὲ μετοπώρου γίνεσθαι τὸ νόσημα, τάδε δ' ἐν τῷ νοσήματι τούτῳ[436]

recur, you must employ the same treatment. If it recurs a third time, swelling is not present, but the patient becomes lean and emaciated; he begins by losing flesh from his face, and his colour becomes very white, even whiter than before; sometimes dropsy develops in the cavity; if it does, you must not make an incision; for, if you do, the patient will die; instead, treat with the measures used in a dropsy arising from the spleen. Much prefer to take this patient for treatment at the beginning of the disease, for then you will quickly make him well. The disease requires care, for it is severe.

45. Another ileus: this one occurs mainly in summer, in swampy areas; in most cases, it comes on as the result of drinking water, but many persons have also taken the disease from sleeping exposed to the sun when they had a headache. In this disease the patient suffers the same things as in the preceding one in everything else except his colour; for he becomes yellow like pomegranate-peel, and his eyes, too, are sometimes filled with jaundice.

When the case is such, treat this patient with the same things used for the preceding one; in addition, give water boiled from white chick-peas to drink, this also mixed with wine; clean out the head with square-berry. This disease is less often mortal than the preceding one. It is called icteric ileus.

46. Another ileus: generally this patient suffers the same things as the preceding ones, except that the disease begins in late autumn, and includes the following things in

---

431 λ. ἰ. Θ: ἦν μή, καὶ οὕτω M.   432 Θ: ὥρην M.
433 M adds οὗτος.   434 Θ: εἰληθέντες M.
435 ὅδε om. M.   436 τούτῳ om. M.

195

προσγίνεται· ἐκ τοῦ στόματος κακὸν ὄζει, καὶ ἀπὸ τῶν
ὀδόντων τὰ οὖλα ἀφίσταται, καὶ ἐκ τῶν ῥινῶν αἷμα
ῥεῖ. ἐνίοτε δὲ καὶ ἐκ τῶν σκελέων ἕλκεα ἐκφυεῖ,[437] καὶ
τὰ μὲν ὑγιαίνεται τὰ δ' ἄλλα παραγίνεται,[438] καὶ ἡ
χροιὴ μέλαινα, καὶ λεπτόδερμος· περιφοιτᾶν δὲ καὶ
ταλαιπωρέειν πρόθυμος.

Τοῦτον, ὅταν οὕτως ἔχῃ, τοῖσιν αὐτοῖσιν[439] ἰᾶσθαι
οἷσι καὶ τοὺς πρόσθεν· καὶ κλύζειν τοῖσδε· σικύου τοῦ
ἀγρίου πέντε φύλλα τρῖψαι λεῖα, καὶ παραμῖξαι μέλιτος ἡμικοτύλιον, καὶ ἁλῶν δραξάμενος τῇ μιῇ χειρί,
καὶ ἐλαίου ἡμικοτύλιον, καὶ ἀπὸ σεύτλων ἑφθῶν τοῦ
χυλοῦ τέσσαρας κοτύλας· διδόναι δὲ ἐς ὑποκάθαρσιν
ὀνείου γάλακτος ἑφθοῦ ὀκτὼ κοτύλας μέλι παραχέας.
πινέτω δὲ καὶ βόειον τὴν ὥρην, τεσσεράκοντα πέντε[440]
ἡμέρας· πινέτω δὲ [καὶ][441] ἕωθεν τοῦ βοείου γάλακτος
δύο κοτύλας, τρίτον μελικρήτου παραμίσγων τὰς
μεταξὺ τῶν ἡμερέων. αὕτη ἡ νοῦσος δεῖται πολλῆς
ἰήσιος· εἰ δὲ μή, οὐκ ἐξέρχεται, ἀλλὰ συναποθνήσκει·
καλέεται δὲ εἰλεὸς αἱματίτης.

47. Παχέα δὲ τάδε τὰ νοσήματα καλέεται· τάδε δὲ
ἀπὸ τῶνδε μάλιστα γίνεται. ὅταν φλέγμα καὶ χολὴ
μιχθῇ κατὰ τὸ σῶμα, συρρεῖ ἐς τὴν κοιλίην, καὶ ὅταν
ἁλισθῇ,[442] τὴν κοιλίην ἀείρει,[443] καὶ ἄνω τε καὶ κάτω
ἔρχεται ὥσπερ κῦμα. καὶ ῥῖγος καὶ πυρετὸς ἐπι-

---

[437] Θ: ἐκφλυνδάνει Μ.   [438] Θ: προσ- Μ.
[439] τ. α. Θ: τούτοισι Μ.   [440] πέντε om. Μ.
[441] Del. Potter, following Calvus.
[442] Μ adds ἐς.   [443] Potter: ἀεὶ ῥεῖ Θ: ἀείρεται Μ.

addition: the patient smells foully from the mouth, the gums separate from his teeth, and blood flows from his nostrils. Sometimes also ulcers break out on his legs—and while some heal, others develop—his colour is dark, and his skin is thin; the patient is eager to walk about and to exert himself.

When the case is such, treat this patient with the same things used for the preceding ones. Administer an enema composed of the following: grind fine five leaves of squirting-cucumber, and add a half-cotyle of honey, a handful of salt, a half-cotyle of oil, and four cotylai of juice boiled from beets. Also give eight cotylai of boiled ass's milk with honey, to clean downwards. In season let the patient drink cow's milk for forty-five days: let him drink[10] two cotylai of cow's milk with one third-cotyle of melicrat early in the morning on the days between medications. This disease requires much attention; otherwise, it does not go away, but clings to the person until he dies. It is called sanguinous ileus.

47. The following diseases are called the "thick ones"; they arise mainly from the following: when phlegm and bile mix with one another through the body, they are inclined to collect in the cavity, and, when they collect there, they raise the cavity up, and fluctuate both upwards and

---

[10] Calvus' translation (*per anni tempus idoneum et lac bubulum quadraginta dies potato, mane duo acetabula cotulasue cum tertia mulsi parte . . .*) makes better sense than Littré's (*Dans la saison le malade boira du lait de vache pendant quarante jours. Il boira aussi, le matin, deux cotyles de lait de vache, avec mélange d'un tiers de mélicrat . . .*).

λαμβάνει, καὶ ἐν τῇ κεφαλῇ ὀδύνη καθέστηκε· καὶ
ὅταν πρὸς τὰ σπλάγχνα ἡ ὀδύνη καταστῇ, πνῖγμα
282 παρέχει· | καὶ εὐθὺς[444] ἐμέει λάμπην ὀξέην, ἐνίοτε
ἁλμυρήν, καὶ ὅταν ἀπεμέσῃ, πικρὸν τὸ στόμα δοκέει
αὐτῷ εἶναι. ἐν δὲ τῇσι πλευρῇσιν ἐρυθήματα
κατάκειται·[445] ἅτε γὰρ τοῦ φλέγματος ἐν τῇ κοιλίῃ
ἐνόντος, τὸ δ᾽ αἷμα ὑπὸ τῆς θερμασίης ἀλισθὲν
προσέπεσε πρὸς τὰς πλευρὰς καὶ ἐρυθήματα παρέχει
ἐν τῇσι πλευρῇσι· καὶ δηγμὸς[446] καὶ θερμασίη ἔχει
μάλιστα τὰς πλευράς. καὶ τὸ μετάφρενον ἔγκυρτον
αὐτοῦ γίνεται· καὶ ὅταν πονέῃ μάλιστα, οὐκ ἀνέχεται
ψαυομένου[447] τοῦ σώματος· ἀλγέει γὰρ ὥσπερ ἕλκος.
καὶ αἱ σάρκες πάλλονται ὑπὸ τῆς ἀλγηδόνος, καὶ οἱ
ὄρχιες ἕλκονται,[448] καὶ ἐς τὴν καθέδρην καὶ ἐς τὴν
κύστιν θέρμη[449] καὶ ὀδύνη ἐμπίπτει. καὶ οὐρέει παχὺ
οἷόν περ ὕδρωπα, καὶ αἱ τρίχες ἐκ τῆς κεφαλῆς
ἐκρέουσι, καὶ τὰ σκέλεα καὶ οἱ πόδες αἰεὶ ψυχροί, καὶ
ὀδύνη πιέζει μάλιστα τὰς πλευρὰς καὶ τὸ μετάφρενον
καὶ τὸν τράχηλον· πρὸς δὲ τῷ δέρματί οἱ δοκεῖ τι
προσέρπειν. ἡ δὲ νοῦσος τότε μὲν πιέζει, τότε δὲ
ἀνίησι· προϊοῦσα δὲ ἡ νοῦσος[450] συνεχέστερον πιέζει·
καὶ τῆς κεφαλῆς τὸ δέρμα παχὺ καὶ ἐρυθρὸν γίνεται.
οὗτος μέχρι μὲν ἐξ ἐτέων τοιαῦτα πάσχων διατελέει·
ἔπειτα ἱδρώς τε πολὺς καταχεῖται καὶ κά- κοσμος.
πολλάκις δὲ καὶ ἐν τῷ ὕπνῳ τὸ λάγνευμα ὕφαιμον
προέρχεται ὑποπέλιδνον. τοῦτο τὸ νόσημα γίνεται διὰ

[444] Θ: ἐνίοτε Μ.   [445] Θ: ἐρυθήματι κατέχεται Μ.

198

downwards. Chills and fever come on, and pain establishes itself in the head; when pain enters the inward parts, it produces choking; at once the patient vomits up a sharp or sometimes salty scum, and after he has vomited his mouth has a pungent taste. In the sides, red patches appear; for, inasmuch as phlegm is present in the cavity, blood attracted by the heat of the phlegm falls against the sides and produces red patches in them; itching and heat are also present, particularly in the sides. The patient's back becomes crooked, and, when his suffering is most severe, he will not tolerate his body being touched, for he feels pain as if in an ulcer. His muscles quiver from pain, his testicles become ulcerated,[11] and heat and pain attack the posteriors and bladder. The patient passes thick urine similar to dropsy fluid, hair falls out of his head, his legs and feet are perpetually cold, and pain presses most intensely in his sides, back and neck; he imagines that something is crawling over his skin. At one time the disease presses more intensely, at another it relents; however, as it progresses, it tends to press more and more continuously. The skin of the head becomes thick and red. Such a patient goes on suffering in this way until the sixth year; then copious foul-smelling sweat pours down. Also, in his sleep he often passes somewhat livid semen charged with blood. This

[11] Variant text: "are drawn up."

---

446 καὶ δηγμὸς om. M.
447 Θ: -ος M.
448 Θ: -ονται M.
449 M adds τις.
450 Θ: -σης δὲ τῆς -ου M.

θερμασίην τοῦ ἡλίου καὶ ὑδροπωσίην.[451]

Τοῦτον ὅταν οὕτως ἔχῃ, τοῦ κνεώρου διδοὺς ὑποκαθαίρειν ἢ τοῦ Κνιδίου κόκκου ἢ τοῦ ἱππόφεω. δίδου δὲ
καὶ γάλα ὄνειον, ἑψήσας πίνειν ὀκτὼ κοτύλας, μέλι
παραχέας· τῇ δ' ὑστεραίῃ μετὰ τὴν κάθαρσιν ταὐτὰ
προσφέρειν ἃ καὶ τοῖσιν ἄλλοισι. τὰς δὲ πρώτας |

284   ἡμέρας εὐωχεέσθω τὰ αὐτὰ ἃ[452] καὶ ὃς[453] ὑπὸ τοῦ
ὑδέρου ἑάλωκε· πονείτω καὶ[454] περιόδοισιν, ἢν δυνατὸς
ᾖ· ἢν δ' ἀδύνατος ᾖ ὑπὸ τῶν πυρετῶν καὶ ἐσθίειν μὴ
δύνηται τὰ σιτία, χρήσθω ῥυφήματι φακῇ· ποτῷ δὲ
οἴνῳ μέλανι[455] αὐστηροτάτῳ. αὕτη ἡ νοῦσος ἐπιλαμβάνει μάλιστα μετοπώρου καὶ[456] ὀπώρης ἐούσης. οὗτος ἢν μὴ[457] ἰηθῇ ἐν τοῖσιν ἓξ ἔτεσι, προσίσχει ἡ
νοῦσος[458] καὶ ἄχρι ἐτέων[459] δέκα· πολλοῖσι δὲ καὶ
συναποθνήσκει, ἢν μὴ παραχρῆμα μελετηθῇ.

48. Ἄλλο[460] παχύ· γίνεται μὲν ἀπὸ χολῆς, ὅταν
χολὴ[461] ἐπὶ τὸ ἧπαρ ἐπιρρυῇ καὶ ἐς τὴν κεφαλὴν
καταστῇ. τάδε οὖν πάσχει· τὸ ἧπαρ οἰδέει, καὶ ἀνα-
πτύσσεται πρὸς τὰς φρένας ὑπὸ τοῦ οἰδήματος, καὶ
εὐθὺς ἐς τὴν κεφαλὴν ὀδύνη ἐμπίπτει, μάλιστα δὲ ἐς
τοὺς κροτάφους· καὶ τοῖσιν ὠσὶν οὐκ ὀξὺ ἀκούει,
πολλάκις δὲ καὶ τοῖσιν ὀφθαλμοῖσιν οὐχ ὁρᾷ· καὶ
φρίκη καὶ πυρετὸς ἐπιλαμβάνει. ταῦτα μὲν κατ' ἀρχὰς
τοῦ νοσήματος αὐτῷ[462] γίνεται· γίνεται δὲ[463] διαλιμ-
πάνοντα, τότε μὲν σφόδρα, τότε δὲ ἧσσον· ὅσῳ δ' ἂν ὁ

[451] τοῦτο . . . ὑδροπωσίην om. Θ.
[452] τὰς . . . ἃ om. Θ.   [453] Θ: ὅστις Μ.

disease arises because of the heat of the sun, and from
drinking water.

When the case is such, clean the patient downwards by
giving spurge-flax, Cnidian berry or hippopheos. Also, give
him eight cotylai of boiled ass's milk with honey to drink;
on the day after the cleaning, give the same things as were
given to the other patients. On the first days, let the patient
be well fed on the same foods as a person with dropsy, and
exert himself by taking walks, if he is able; if, on account of
his fevers, he is not able to take walks, and if he cannot eat
regular foods, let him employ lentil-soup as gruel, and
drink dry dark wine. This disease has its accesses mainly in
fall and later summer. If the patient is not treated during
the six years, the disease continues until the tenth year; in
many, it even clings to them until they die, if it has not been
treated immediately.

48. Another "thick" disease: this one arises from bile,
when bile collects in the liver, and also settles in the head.
The patient suffers the following: his liver swells up and, by
its swelling, expands against the diaphragm; pain immediately attacks his head, especially the temples; he does not
hear clearly, and often he cannot see, either; shivering and
fever set in. These things affect the patient at the beginning of the disease; they occur intermittently, sometimes
more intensely, sometimes less so. The longer the disease

454 Θ: ἑάλω· καὶ πονεέτω Μ.   455 Μ adds ὡς.
456 καὶ om. Θ.   457 Θ: μὲν Μ.
458 π. ἤ ν. Θ: εἰ δὲ μή, προσέχει Μ.   459 Θ: τῶν Μ.
460 Θ: Τάδε παχέα γίνεται τῶν νοσημάτων ἀπὸ χολῆς.
ἀλλὰ Μ.   461 χολὴ om. Μ.   462 Μ: αὐτοῦ τοῦ
νοσήματος Θ.   463 γίνεται δὲ om. Μ.

χρόνος τῇ νούσῳ προΐῃ, ὅ τε πόνος πλείων ἐν τῷ
σώματι. καὶ αἱ κόραι σκίδνανται τῶν ὀφθαλμῶν, καὶ
σκιαυγέει, καὶ ἢν προσφέρῃς τὸν δάκτυλον πρὸς τοὺς
ὀφθαλμούς, οὐκ αἰσθήσεται διὰ τὸ μὴ ὁρᾶν· τούτῳ δ'
ἂν γνοίης ὅτι οὐχ ὁρᾷ, οὐ γὰρ[464] καρδαμύσσει προσ-
φερομένου τοῦ δακτύλου. καὶ τὰς κροκύδας ἀφαιρέει
τοῦ ἱματίου, ἤν περ ἴδῃ, δοκέων φθεῖρας εἶναι. καὶ
ὅταν τὸ ἧπαρ μᾶλλον ἀναπτυγῇ πρὸς τὰς φρένας,
παραφρονέει· καὶ προφαίνεσθαί οἱ δοκέει πρὸ τῶν
286 ὀφθαλμῶν ἑρπετὰ | καὶ ἄλλα παντοδαπὰ θηρία καὶ[465]
ὁπλῖται μαχόμενοι, καὶ αὐτὸς ἐν αὐτοῖσι δοκέει μάχε-
σθαι· τοιαῦτα λέγει ὡς ὁρῶν καὶ ἐπέρχεται, καὶ ἀπει-
λεῖ, ἢν μή τις αὐτὸν ἐᾷ ἔξω ἐξιέναι· καὶ ἢν ἀναστῇ, οὐ
δύναται αἴρειν τὰ σκέλεα, ἀλλὰ πίπτει. οἱ δὲ πόδες
αἰεὶ ψυχροί· καὶ ὅταν καθεύδῃ, ἀναΐσσει ἐκ τοῦ
ὕπνου[466] ὅταν ἐνύπνια ἴδῃ φοβερά. τῷδε δὲ γινώσκο-
μεν, ὅτι ἀπὸ ἐνυπνίων ἀΐσσει καὶ φοβεῖται· ὅταν
ἔννοος γένηται, ἀφηγεῖται τὰ ἐνύπνια τοιαῦτα ὁρᾶν[467]
ὁποῖα καὶ τῷ σώματι ἐποίει καὶ τῇ γλώσσῃ ἔλεγε.
ταῦτα μὲν οὕτω πάσχει. ἔστι δ' ὅτε καὶ κεῖται ἄφωνος
ὅλην τὴν ἡμέρην καὶ τὴν νύκτα ἀναπνέων ἀθρόον
πολὺ τὸ πνεῦμα. ὅταν δὲ παύσηται παραφρονέων,
εὐθὺς [παραχρῆμα][468] ἔννοος γίνεται, καὶ ἢν ἐρωτᾷ τις
αὐτόν, ὀρθῶς[469] ἀποκρίνεται, καὶ γινώσκει πάντα τὰ
λεγόμενα. εἶτα αὖθις ὀλίγον ὕστερον ἐν τοῖς αὐτοῖσιν
ἄλγεσι κεῖται. αὕτη ἡ νοῦσος προσπίπτει μάλιστα ἐν

464 γὰρ om. Θ.   465 καὶ om. Θ.

goes on, the more pain there is in the body. The pupils of
the eyes are dilated, the patient sees dimly, and if you bring
your finger up to his eyes, he does not perceive it, because
he cannot see; this is how you can tell that he does not see:
he does not blink when the finger is brought near. He re-
moves pieces of wool from his blanket, if he does see them,
believing they are lice. When his liver expands even more
against the diaphragm, the patient becomes deranged;
there seem to appear before his eyes reptiles and every
other sort of beasts, and fighting soldiers, and he imagines
himself to be fighting among them; he speaks out as if he is
seeing such things, and he attacks and threatens, if someone will not allow him to go outside; if he does stand up,
though, he cannot lift his legs, but falls. His feet are perpetually cold; when he goes to bed, he starts up out of his
sleep on seeing fearful dreams. We know that his starting
up and fear are due to dreams, from the following: when he
comes to his senses, he reports having had dreams that correspond to the way he moved his body and spoke with his
tongue. These things he suffers as described. Sometimes,
he may also lie speechless the whole day and night, taking
frequent deep breaths. When his derangement ceases, he
immediately regains his senses, and, if someone questions
him, he answers correctly and understands everything that
is said. But then, a little later, he labours again under
the same distress. This disease usually attacks abroad, if a

---

466 M adds καὶ φοβέεται.
467 τοιαῦτα ὁρᾶν om. M.
468 Del. Vander Linden.
469 Θ: εὐθὺς M.

ἀλλοδημίη καὶ ἤν που ἐρήμην ὁδὸν βαδίσῃ καὶ φόβος
αὐτὸν λάβῃ· λαμβάνει δὲ καὶ ἄλλως.

Τοῦτον οὖν,[470] ὅταν οὕτως ἔχῃ, πῖσαι τοῦ μέλανος
ἐλλεβόρου πέντε ὀβολοὺς στήσας, διδόναι δ᾽ ἐν[471]
γλυκεῖ οἴνῳ· ἢ κλύζειν αὐτὸν τοῖσδε· λίτρου Αἰγυπτίου ὅσον ἀστράγαλον οἷός, τοῦτο τρῖψαι λεῖον, καὶ
παραμῖξαι μέλιτος ὡς[472] καλλίστου ἡμικοτύλιον δ᾽
ἐψήσας, ἐν θυΐῃ,[473] καὶ ἡμικοτύλιον ἐλαίου καὶ ἀπὸ
σεύτλων ἑφθῶν ὕδατος τέσσαρας κοτύλας ἐξαιθρι-
άσας· ἢν δὲ βούλῃ, ἀντὶ σεύτλου ὄνειον γάλα ἑψήσας
παραμῖξαι. ταῦτα τρίψας[474] κλύζειν, ἤν τε ὁ πυρετὸς
ἐπῇ ἤν τε μή. ῥυφήμασι δὲ χρήσθω πτισάνῃ καθέφθῳ, μέλι παραχέων· πινέτω δὲ μέλι καὶ ὕδωρ καὶ
ὄξος συγκεράσας, ἕως ἂν κριθῇ ἡ νοῦσος· κρίνεται δ᾽
288   ἐν τεσ|σεράκοντα[475] ἡμέρῃσι τὸ μακρότατον, εἰ θανά-
σιμος ἢ οὔ. πολλοῖσι δὲ ἤδη τοῦ νοσήματος πεπαυ-
μένου πάλιν ἡ νοῦσος ὑπετρόπασεν· ἢν οὖν ὑποτροπάσῃ, κίνδυνος αὐτὸν διαφθαρῆναι· κρίνεται δ᾽ ἡ
νοῦσος ἐν ἑπτὰ ἡμέρῃσιν, ἢν ὑποτροπάσῃ, εἰ[476] θανάσιμος ἢ οὔ. ἢν δὲ ταύτας ἐκφύγῃ, οὐ μάλα θνήσκει,
ἀλλὰ τοῖσι πολλοῖσι μελεδαινομένη ἐξέρχεται. ὅταν
δὲ παύσηται ἡ νοῦσος, διαίτῃ χρήσθω, ἡσύχως προσ-
άγων ὁπόσον ἂν ἡ κοιλίη προσδέξηται ὡς[477] μὴ
συγκαυθῇ, μήτε διάρροια ἐπιγένηται· ἀμφότερα γὰρ
δοκέει κινδυνώδεα εἶναι. καὶ λούσθω δὲ ἑκάστης ἡμέρης, καὶ περιπατείτω ὀλίγα μετὰ σιτία· καὶ ἐσθῆτα

____

470 οὖν om. M.   471 ἐν om. Θ.

person is travelling a lonely road somewhere, and fear seizes him, although it does also occur under other circumstances.

When the case is such, then, give the patient five obols weight of black hellebore to drink in sweet wine. Or administer an enema as follows: Egyptian soda to the amount of a sheep's vertebra, grind this fine; mix it together in a mortar with a half-cotyle of the finest honey you have, boiled, a half-cotyle of oil, and four cotylai of juice boiled from beets and exposed to the open air; if you wish, instead of the beet juice add boiled ass's milk; mash these together, and administer as an enema, whether fever is present or not. As gruel let the patient take boiled-down barley-water with honey; let him drink a mixture of honey, water and vinegar until the disease reaches its crisis, which is in forty days at the longest, these deciding whether or not the patient will die. In many, the disease has recurred after it had gone away; now, if it does recur, there is a danger that the patient will perish. If it recurs, the disease has its crisis in seven days, which decide whether or not the patient will die. If he escapes these, he is not likely to die, but in most cases, if treated, the disease goes away. When the disease is over, let the patient follow the regimen of cautiously increasing his intake as much as his cavity is able to accept without becoming burnt up or without diarrhoea coming on, for these are both held to be dangerous. Let him bathe each day, and walk about a little after his meals; let him

---

472 ὡς om. M.   473 Θ: θυμίη M.
474 Θ: μίξας M.   475 Θ: δεκατέσσαρσι M.
476 ἢν . . . εἰ Θ: ἢ M.
477 Potter: -έχηται ὡς Θ: -δέξηται καὶ M.

κούφην ἐχέτω καὶ μαλθακήν· καὶ γαλακτοποτείτω τὴν
ὥρην καὶ ὀροποτείτω πέντε καὶ τεσσεράκοντα ἡμέρας.
ταῦτα ἢν ποιέῃ, τάχιστα ὑγιὴς ἔσται. ἡ δὲ νοῦσος
χαλεπὴ καὶ δεῖται μελέτης πολλῆς.

49. Ἄλλο παχύ· γίνεται μὲν ἀπὸ φλέγματος σα-
πέντος· τῷδε <δὲ>[478] δῆλον γίνεται, ὅτι σαπρόν ἐστιν·
ἐρεύγεται ἀπ᾽ αὐτοῦ τὴν ὀδμὴν ἔχον, οἷόν περ ῥαφανῖδας φαγόντος. τὸ δὲ νόσημα ἄρχεται ἀπὸ τῶν σκελέων γινόμενον, εἶτα ἀνέρχεται ἐκ τῶν σκελέων ἐς τὴν
κοιλίην, καὶ ὅταν ἐν τῇ κοιλίῃ εἴῃ,[479] αὖτις ἀνέρχεται
πρὸς τὰ σπλάγχνα, καὶ ὅταν στῇ πρὸς τοῖσι σπλάγ-
χνοισι, μύζει καὶ ἔμετον ἄγει, ἅμα καὶ λάπην ὀξέην
ὑπόσαπρον· καὶ ὅταν ἀπεμέσῃ, οὐκ ἔχει ἑωυτόν. ἔπειτα ἀπορίαι πρὸς τοῖσι σπλάγχνοισιν, ἐνίοτε δὲ καὶ ἐς
τὴν κεφαλὴν ἐξαπίνης ὀδύνη στηρίζει ὀξέη, ὥστε οὔτε
290 τοῖσιν ὀφθαλμοῖσιν | ἀνορᾶν οὔτε τοῖσιν ὠσὶν ἀκούειν
δύναται ἀπὸ τοῦ βάρεος. ἱδρώς τε πολὺς καταχεῖται
καὶ κάκοδμος, μάλιστα μὲν ἢν ἡ ὀδύνη ἔχῃ, καταχεῖται δὲ καὶ ὅταν[480] ἡ ὀδύνη λωφᾷ, καὶ τῆς νυκτὸς[481]
μάλιστα. ἡ δὲ χροιὴ αὐτοῦ μάλιστα[482] ἰκτερώδης γίνεται. αὕτη ἡ νοῦσος τῆς προτέρης ἧσσον θανατώδης.

Τοῦτον, ὅταν οὕτως ἔχῃ, καθαίρειν τὴν κοιλίην,
κάτω μὲν τῷ ἱππόφεῳ, ἄνω δὲ ἐλλεβόρῳ· καὶ τὴν
κεφαλὴν κάθαιρε τῷ τετραγώνῳ. καὶ ὅταν ὑπὸ τοῦ
ἐλλεβόρου κεκαθαρμένος ᾖ, τῇ ὑστεραίῃ ὀνείῳ γάλακτι ἐφθῷ ὑποκαθῆραι, τῇ δὲ τρίτῃ αἰγείῳ ἐφθῷ, καὶ

[478] Cornarius: τῷδε Θ: τόδε Μ.

wear clothes that are light and soft, and drink milk and whey in season for forty-five days. If he does these things, he will very quickly become well. The disease is severe and requires much care.

49. Another "thick" disease: this one arises from putrefied phlegm; the following shows that the phlegm is putrid: the patient's belches have an odour, from the phlegm, like those of a person that has eaten radishes. This disease begins from the legs, then migrates from the legs into the cavity, and, when it has arrived in the cavity, it migrates again to the inward parts; when it has settled in the inward parts, it rumbles and provokes vomiting, and with it a sharp somewhat putrid scum; after he vomits, the patient is no longer himself. There is distress in the inward parts, and sometimes sharp pain also suddenly fixes itself in the head, so that the patient can neither look up with his eyes, nor hear with his ears, from its gravity. Sweat pours down copious and foul-smelling, especially if the pain is present, though also when it abates, and particularly at night. The patient's colour becomes very jaundiced. This disease is less mortal than the preceding one.

When the case is such, clean the patient's cavity downwards with hippopheos, and upwards with hellebore; clean out his head with square-berry. When he has been cleaned with the hellebore, the next day clean him downwards with boiled ass's milk, the third day with boiled goat's milk, and

---

479 Θ: στῇ M.
480 Θ: ἢν M.
481 λωφᾷ . . . νυκτὸς Θ: τε καὶ λωφᾷ τῆς νούσου M.
482 μάλιστα om. M.

τῇ τετάρτῃ καὶ τῇ πέμπτῃ· ἄλλας δ'[483] εἴκοσιν ὠμὸν
βόειον ἢ αἴγειον γάλα διδόναι, τρίτον μέρος μελικρήτου παραχέων, πινέτω δὲ τοῦ γάλακτος χοέα. μετὰ δὲ
τὴν κάθαρσιν[484] τῶν φαρμάκων ταῦτα προσφερέσθω
ὅσα καὶ ὁ[485] ὑπὸ τοῦ ὑδέρου ἐχόμενος·[486] τὸν δὲ λοιπὸν
χρόνον, τὸ γάλα πίνειν,[487] δειπνείτω δὲ ἄρτον ἔξοπον,
ὄψον δ' ἐχέτω σκορπίον ἢ καλλιώνυμον ἢ κόκκυγα ἢ
ῥίνης τέμαχος ἐφθὸν ἐν ἀρτύμασι· κρέα δὲ οἰὸς ἢ
ἀλεκτρυόνος νεοσσοῦ ταῦτα ἐφθά· οἶνον δὲ πινέτω
λευκόν, ἢν ξυμφέρῃ· εἰ δὲ μή, ἄλλον μέλανα αὐστηρόν. εἶτα[488] περιπατείτω ὀλίγα[489] μετὰ τὸ δεῖπνον,
φυλασσόμενος ὅπως ἂν[490] μὴ ῥιγώσῃ. τούτῳ ἢν ξυμ-
φέρῃ, τὰ σιτία διδόναι, ἢν δὲ μὴ συμφέρῃ τὰ σιτία,[491]
ῥύφημα διδόναι πτισάνης ἢ κέγχρου. ταύτῃ τῇ νούσῳ
ἢν τριήκοντα ἡμέραι παρέλθωσιν, ὑγιαίνεται ὤνθρω-
πος· αὗται γὰρ κρίνουσιν εἰ θανάσιμος ἢ οὔ. ἡ δὲ
νοῦσος χαλεπή.

292   50. Ἄλλο παχύ· γίνεται τὸ νόσημα ἀπὸ φλέγματος
λευκοῦ· ξυνίσταται δὲ ἐν τῇ κοιλίῃ, ὅταν πυρετοὶ
πολυχρόνιοι κατάσχωσι τὸ σῶμα. ἄρχεται δὲ τὸ νόσημα ἐκ τοῦ προσώπου γινόμενον, καὶ οἴδημα ἐγγί-
νεται· εἶτα κατέρχεται ἐς τὴν κοιλίην, καὶ ὅταν ἐν τῇ
κοιλίῃ στῇ, ἀείρει μεγάλην τὴν γαστέρα· καὶ τὸ σῶμα
κοπιᾷ ὡς ὑπὸ ταλαιπωρίης· καὶ ἐν τῇ κοιλίῃ βάρος
ἴσχει[492] καὶ πόνος ἰσχυρός· καὶ οἱ πόδες οἰδέουσι. καὶ
ὅταν ὕσῃ τῆς γῆς καὶ τῆς κονίας οὐκ ἀνέχεται ὀδμώ-
μενος· ἢν δ' ἑστηκὼς τύχῃ ἐν τῷ ὑσμένῳ[493] καὶ ὀδμή-
σηται[494] τῆς γῆς, ἐξαπίνης πίπτει.

also the fourth and fifth; for twenty more days give raw
cow's or goat's milk, adding one third part of melicrat; let
him drink a chous of this milk. After the cleaning, let the
patient receive the same medications as a patient with
dropsy; from then on, let him drink milk, have for dinner
well-baked bread and, as main dish, scorpion fish, star-
gazer, piper, or a slice of angel-fish boiled in seasonings, or
boiled meat of sheep or chicken; let him drink white wine,
if it benefits him; if not, then a different sort of dry dark
wine. Then let him go for short walks after dinner, taking
care not to have a chill. If it benefits this patient, give him
food, but if food does not benefit him, give barley-water or
millet gruel. In this disease, if thirty days pass, the person
recovers, for this is the critical period that decides whether
or not he will die. The disease is severe.

50. Another "thick" disease: this one arises from white
phlegm, which congeals in the cavity when longstanding
fevers beset the body. The disease begins with a swelling of
the face; then it moves down to the cavity and, on becoming established there, raises the belly up large; the body
suffers weariness, as if from exertion; in the cavity there is
heaviness and a violent ache; the feet swell. When it rains,
this patient cannot stand to smell the earth and the dust; if
he happens to be standing in the rain, and he smells the
earth, he immediately falls down.

---

483 δ' om. Θ.   484 M adds τὴν κάτω.   485 Potter:
καὶ ὅσα θ' Θ: ἃ καὶ ὃς M.   486 Θ: ἔχεται M.
487 Θ: ὁκόταν τὸ γάλα πίνῃ M.   488 M: εἰ δὲ μὴ Θ.
489 ὀλίγα om. M.   490 φ. ὅ. ἂν Θ: ὅκως M.
491 τὰ σιτία . . . τὰ σιτία om. Θ.
492 Θ: ἐς τὴν -ίην β. ἐνῇ M.   493 Θ: ὑετῷ M.

Αὕτη ἡ νοῦσος καὶ διαπαύουσα τῆς προτέρης πλείω χρόνον ἐπιλαμβάνει καὶ χρονιωτέρη ἀπαλλάσσεται. μελετᾶν δὲ χρὴ τοῖσιν αὐτοῖς οἷσι καὶ τὸν ὑδεριῶντα, πυρίῃσι καὶ φαρμάκοισι καὶ ἐδέσμασι καὶ ταλαιπωρίῃσιν. αὕτη ἡ νοῦσος προσίσχει μάλιστα ἓξ ἔτεα, εἶτα ἐξέρχεται μελετωμένη ἐν χρόνῳ, ἢν μὴ κατ' ἀρχὰς ἰηθῇ· ἡ γὰρ νοῦσος χαλεπὴ καὶ μελέτης δεῖται πολλῆς.

51. Ἰσχιὰς δὲ ἀπὸ τῶνδε μάλιστα γίνεται τοῖσι πολλοῖσιν· ἢν εἰληθῇ[495] ἐν ἡλίῳ πολὺν χρόνον καὶ τὰ ἰσχία διαθερμανθῇ καὶ τὸ ὑγρὸν ἀναξηρανθῇ ὑπὸ τοῦ καύματος τὸ ἐνεὸν ἐν τοῖσιν ἄρθροισιν. ὡς δὲ ἀναξηραίνεται καὶ πήγνυται, τόδε μοι τεκμήριον· ὁ γὰρ νοσέων στρέφεσθαι καὶ κινέειν τὰ ἄρθρα οὐ δύναται ὑπὸ τῆς ἀλγηδόνος τῶν ἄρθρων καὶ τοῦ ξυμπεπηγέναι τοὺς σφονδύλους. ἀλγέει δὲ μάλιστα τὴν ὀσφὺν καὶ τοὺς σφονδύλους[496] τοὺς ἐκ πλαγίων τῶν ἰσχίων καὶ τὰ γούνατα. ἵσταται δὲ ἡ ὀδύνη πλεῖστον χρόνον ἐν 294 τοῖσι | βουβῶσιν ἅμα καὶ τοῖσιν ἰσχίοις ὀξέη καὶ καυματώδης· καὶ ἤν τις αὐτὸν ἀνιστῇ ἢ μετακινέῃ, οἰμώζει ὑπὸ τῆς ἀλγηδόνος ὅσον ἂν μέγιστον δύνηται. ἐνίοτε δὲ καὶ ὁ σπασμὸς ἐπιγίνεται καὶ ῥῖγος καὶ πυρετός. γίνεται δὲ τὸ νόσημα[497] ἀπὸ χολῆς· γίνεται δὲ καὶ ἀπὸ φλέγματος[498] καὶ ἀπὸ αἵματος· καὶ ὀδύναι παραπλήσιαι ἀπὸ πάντων τούτων[499] τῶν νοσημάτων· καὶ ῥῖγος καὶ πυρετὸς ἐνίοτε ἐπιλαμβάνει βληχρός.

Ἀλλὰ χρὴ ὧδε μελετᾶν τὸν ἀπὸ τοῦ ἡλίου νοσέοντα. ὑγραίνειν αὐτοῦ τὸ σῶμα τῇ πυριήσει καὶ ἀπὸ

This disease, even though intermittent, attacks for a longer time than the preceding one, and is relieved later. You must treat with the same things prescribed for a patient with dropsy: vapour-baths, medications, foods and exercises. The disease usually continues for six years; then, if cared for, it gradually goes away, even if at first it was not treated; the disease is severe, and requires much care.

51. Sciatica generally arises from the following, in the majority of cases: if a person is exposed to the sun for a long time, and his hip-joints become heated, and the moisture present in them is dried up by the burning heat. My proof that the moisture is dried up and congealed is this: the patient cannot turn and move his joints, because of the pain in them, and because the vertebrae have become fixed. He has pain especially in the loins, in the vertebrae that grow out of the oblique part of the hip-bone, and in the knees. Sharp burning pain persists longest in the groins, and also in the hip-joints; if someone stands the patient up, or shifts him, he cries out at the top of his voice from the pain. Also, sometimes a convulsion supervenes, or chills and fever. The disease can arise from bile, but also from phlegm and blood; the pains from all of these diseases are similar; sometimes mild chills and fever are present.

You must treat the patient whose illness is from the sun thus: moisten his body with a vapour-bath, and by means of

---

494 Θ: ὀδμηθῇ M.
495 Coray, Littré: ἔλθῃ ΘM.
496 ἀλγέει . . . σφονδύλους om. Θ.
497 τὸ ν. Θ: καὶ M.
498 γ. δὲ κ. ἀ. φ. Θ: καὶ φλέγματος· γίνεται δὲ M.
499 ἀπὸ π. τ. Θ: ὡς καὶ ὑπὸ πάντων M.

τῶν σιτίων καὶ ἀπὸ τῶν⁵⁰⁰ ποτῶν καὶ ἀπὸ τῶν ἄλλων
τῶν διδομένων· διδόναι δὲ χλιαρὰ καὶ ὑγρά, ταῦτα δὲ
πάντα ἐφθά. σίτῳ δὲ χρήσθω μάζῃ μαλθακῇ ἀτρίπτῳ·
οἶνον δὲ πινέτω λευκὸν ὑδαρέα. καὶ τῷ σώματι ἡσυχίην ἐχέτω· ἢν δὲ καὶ δυνατὸς⁵⁰¹ ᾖ ἀνίστασθαι,⁵⁰² ὀλίγα περιπατείτω καὶ μὴ ριγούτω· καὶ ἑκάστης ἡμέ-
ρης⁵⁰³ λούσθω μὴ πολλῷ. καὶ ὅταν σοι δοκέῃ καλῶς
ἔχειν καὶ ὑγρὸς εἶναι τὸ σῶμα, πυράσαι σφόδρα
βληχρῇ τῇ πυρίῃ· μᾶλλον γὰρ ἀνήσει καὶ ὑγρανεῖ τὸ
συμπεπηγὸς ἐκ τῶν ἄρθρων. εἶτα τῇ ὑστεραίῃ πῖσαι
τοῦ Κνιδίου κόκκου. ἢν δὲ μὴ ὠφελήσῃ αὐτόν,⁵⁰⁴ κλύσαι τοῖσδε χρὴ αὐτόν· τρίβειν κυμίνου ἡμικοτύλιον,
σικύην ἄτμητον τῶν σμικρῶν καὶ στρογγύλων συγκό-
ψας ἐν τῷ ὅλμῳ, καὶ σήσας ὡς λεπτότατα λίτρου
ἐρυθροῦ Αἰγυπτίου τεταρτημόριον μνᾶς, ὀπτήσας,
τρίψας λεῖον, ταῦτα μίξας⁵⁰⁵ ἐμβάλλειν ἐς χυτρίδα,
καὶ ἐπιχέαι ἐλαίου κοτύλην, μέλιτος ἡμικοτύλιον, οἴνου λευκοῦ γλυκέος⁵⁰⁶ κοτύλην, σεύτλου δύο κοτύ-
λας.⁵⁰⁷ ταῦτα ἕψειν, ἕως ἂν δοκέῃ σοι καλῶς ἔχειν τοῦ
πάχεος· εἶτα διηθήσας δι' ὀθονίου, παραμίξαι αὐτοῖσι
296 μέλιτος Ἀττικοῦ | κοτύλην, ἢν μὴ βούλῃ συνεψεῖν τὸ
μέλι· ἢν δὲ μὴ ἔχῃς Ἀττικόν, κοτύλην τοῦ καλλίστου
παραμίξας δ' ἐψῆσαι ἐν θυΐῃ· ἢν δὲ τὸ κλύσμα παχύτερον ᾖ, οἴνου τοῦ αὐτοῦ παραχεῖ πρὸς τὸ πάχος
τεκμαιρόμενος· τούτῳ κλύζειν. εἶτα ἐὰν χρὴ μέχρι
τριῶν ἡμερέων καθαίρεσθαι· ἢν δὲ πλέονας ἡμέρας

---

⁵⁰⁰ ἀπὸ τῶν om. M.   ⁵⁰¹ Θ: ἀδύνατος M.

foods, drinks and other things you give; give these warm and moist, all of them boiled. As cereal let him employ soft unkneaded barley-cake, and drink white wine well mixed with water. Let him rest his body; if he is able to stand up, let him take short walks, but avoid having a chill; let him bathe each day with very little water. When the patient seems to you to be in a good state, and moist of body, apply a very gentle vapour-bath; for that will be effective in moistening and resolving the congealed matter from his joints. Then on the next day have him drink Cnidian berry. If this does not help, you must clean him with the following enema: grind a half-cotyle of cummin, bray an uncut bottle-gourd of the small round kind in a mortar, sift the fourth part of a mina of red Egyptian soda as fine as possible, roast, grind fine, mix these together, and pour into a pot; add a cotyle of oil, a half-cotyle of honey, a cotyle of sweet white wine, and two cotylai of beets; boil these until you think they have the proper consistency; then strain through a linen cloth, and add a cotyle of Attic honey to them, if you do not wish to boil the honey together with them; if you do not have Attic honey, mix in a cotyle of the best kind you have, and boil in a mortar; if the fluid is too thick, pour in some of the same wine, judging according to the thickness; administer as an enema. Allow the patient to be cleaned out for three days; however, if the cleaning goes

---

502 M adds πλὴν.
503 ἑ. ἡ. after περιπατείτω in M.
504 αὐτόν om. M.
505 μίξας om. M.
506 γλυκέος om. M.
507 δ. κ. Θ: χυλοῦ M.

καθαίρηται, βοείου ἢ αἰγείου κοτύλας τρεῖς γάλακτος δοῦναι ἐκπιεῖν. εἶτα σεῦτλα λιπαρά, περιπάσαντα,[508] δοῦναι ἐσθίειν ἄναλτα. ἐκ ταύτης τῆς νούσου πολλοὶ ἤδη χωλοὶ ἐγένοντο.

Ἢν δὲ ἀπὸ χολῆς ἡ νοῦσος γένηται, πῖσαι αὐτὸν ἐλλέβορον κάτω ἢ ὀπὸν σκαμωνίης· μετὰ δὲ τὴν κάθαρσιν πτισάνης, μέλι παραχέας, δύο τρυβλία δοῦναι ἐκρυφεῖν. τῇ δ' ὑστεραίῃ ἢ τῇ τρίτῃ πυριάσας γάλακτι ὀνείῳ ἐφθῷ ὑποκαθῆραι· ἐς ἑσπέρην δὲ σεύτλων ἐφθῶν λιπαρῶν τρία[509] τρυβλία ἐκφαγέτω ἄλφιτα περιπάσας· οἶνον δὲ πινέτω λευκόν, ὑδαρέα, γλυκὺν[510] καὶ νῦν καὶ μετὰ τοῦ φαρμάκου τὴν δόσιν.[511]

Ἢν δὲ ἀπὸ φλέγματος νοσήσῃ, πῖσαι αὐτὸν τοῦ Κνιδίου κόκκου ἢ τοῦ ἱππόφεω πυριάσας· μετὰ δὲ τὴν κάθαρσιν ταὐτὰ προσφέρειν καὶ ῥυφήματα καὶ ποτὰ καὶ γαλακτοπωσίην. τὰς δὲ μεταξὺ τῶν ἡμερέων διαίτῃ χρήσθω ὡς κουφοτάτῃ. καὶ ἢν μὲν ὑπὸ τούτων ὠφελῆται, ἅλις· ἢν δὲ μή, καῦσαι αὐτόν, τὰ μὲν ὀστώδεα μύκησι, τὰ δὲ σαρκώδεα σιδηρίοισι πολλὰς ἐσχάρας καὶ βαθείας.

Ἢν δὲ ὑφ' αἵματος νοσέῃ, πυριήσας σικύην προσβάλλειν, καὶ φλεβοτομέειν τὰς ἐν τῇσιν ἰγνύῃσι φλέβας· ἢν δὲ δοκέῃ, καὶ τοῦ Κνιδίου κόκκου πῖσαι αὐτόν. διαίτῃ δὲ χρήσθω ὡς ξηροτάτῃ· οἶνον δὲ μᾶλλον μὲν μὴ πινέτω· ἢν δ' ἄρα καὶ πίνῃ, ὡς ἐλάχιστον καὶ ὑδαρέστατον. καὶ περιπατεῖν κελεύειν, ἢν δυνατὸς ᾖ,

[508] Θ: -παστα Μ.

on for longer, give him three cotylai of cow's or goat's milk to drink off. Then give him beets boiled in grease to eat, without salt, but generously sprinkled with meal. From this disease many persons have become lame.

If the disease has arisen from bile, have the patient drink hellebore to clean downwards, or scammony juice; after the cleaning, as gruel give him two bowls of barley-water with honey, to drink off. On the next day, or the day after that, administer a vapour-bath, and clean downwards with boiled ass's milk; towards evening, let the patient eat three bowls of beets boiled in grease and sprinkled with meal; let him drink a sweet white wine diluted with water, both now and after the medication has been given.

If the patient's illness is caused by phlegm, administer a vapour-bath, and have him drink Cnidian berry or hippopheos; after the cleaning, administer the same gruels, potions and milk-drinks. On the days between, let him follow as light a regimen as possible. If he is benefited by these things, fine; if not, cauterize him, the osseous parts with fungi, the fleshy ones with irons, making many deep eschars.

If the patient's illness is caused by blood, after a vapour-bath apply a cupping instrument and phlebotomize the vessels in the ham; if it seems advisable, also have him drink Cnidian berry. Let him follow a very dry regimen; wine he had better not drink, but if he does drink it, let it be very little and very dilute. Order him to take walks, if he

---

509 σεύτλων . . . τ. Θ: τεύτλου λιπαροῦ δύο M.
510 γλυκὺν om. Θ.
511 Θ: κάθαρσιν M.

298 ὡς πλεῖστα· συμφέρει δὲ καὶ τῷ | ἀπὸ τοῦ φλέγματος
νοσέοντι ταὐτὰ ποιέειν. ἢν δὲ μὴ δύνηται ἵστασθαι, ἐν
τῇ κλίνῃ χρὴ περιστρέφειν⁵¹² ὡς πυκνότατα—ἀφ᾿ οὗ
ἂν νοσέῃ, πάντας⁵¹³ ὁμοίως—ὅπως ἂν ἐντὸς μὴ ξυμφυῇ ὁ χόνδρος· ἢν δὲ ξυμφυῇ καὶ τὰ ἄρθρα συμπαγῇ,
πᾶσα ἀνάγκη χωλόν ἐστι γενέσθαι.⁵¹⁴ καὶ ἢν κλύσαι
βούλῃ τὸν ἀπὸ τοῦ αἵματος νοσέοντα, ὥστε αἷμα
ἀπάγειν καὶ φλέγμα ἀπὸ τῶν ἰσχίων, τοῖσδε χρὴ
κλύζειν· ἁλὸς δραχμίδα τρίψας, παραμῖξαι ἐλαίου
κοτύλην καὶ ἀπὸ κριθέων ὀπτῶν τρεῖς⁵¹⁵ κοτύλας, εἶτα
οὕτω κλύζειν τοῦτον. οὕτω μελετῶν τάχιστα ὑγιᾶ
ποιήσεις. ἡ δὲ νοῦσος⁵¹⁶ χρονίη.

52. Τέτανοι τρεῖς· ἢν μὲν ἐπὶ τρώματι γένηται,
πάσχει τάδε· αἱ γέννες πήγνυνται, καὶ τὸ στόμα
διαίρειν οὐ δύναται· καὶ οἱ ὀφθαλμοὶ δακρύουσι καὶ
ἕλκονται, καὶ τὸ μετάφρενον πέπηγε, καὶ τὰ σκέλεα οὐ
δύναται ξυγκάμπτειν, οὐδὲ τὰς χεῖρας καὶ⁵¹⁷ τὴν
ῥάχιν. ὅταν δὲ θανατώδης ᾖ, καὶ τὸ ποτὸν καὶ τὰ
βρώματα, ἃ πρότερον ἐβεβρώκει, ἀνὰ τὰς ῥῖνας ἀνέρχεται ἐνίοτε.

Τοῦτον, ὅταν οὕτως ἔχῃ, πυριᾶν, καὶ ἀλείψαντα
λιπαρῶς, πρὸς πῦρ ἔκαθεν⁵¹⁸ θάλπειν· καὶ χλιάσματα
προστιθέναι ὑπαλείψας τὸ σῶμα. καὶ⁵¹⁹ ἀψίνθιον ἢ
φύλλα δάφνης ἢ τοῦ ὑοσκυάμου τὸν καρπὸν τρίψας
καὶ λιβανωτόν, εἶτα οἴνῳ λευκῷ διεὶς ἐγχέαι εἰς
χυτρίδα καινήν· εἶτα ἐπιχέαι ἔλαιον ἴσον τῷ οἴνῳ, καὶ

---

⁵¹² Θ: -φέρειν Μ.   ⁵¹³ Θ: καὶ ἄπαντα Μ.

is able, and as many as possible; it also benefits the person ailing because of phlegm to do the same. If this patient cannot stand up, you must turn him over in bed very frequently—all patients alike, whatever the source of their sickness—in order that the cartilage within does not grow together; if it does grow together, and the joints become fixed, the patient will inevitably become lame. If you wish to administer an enema to the patient whose disease is caused by blood, in order to draw blood and phlegm out of his hip-joints, you must employ the following one: grind a handful of salt, mix together with it a cotyle of oil and three cotylai of juice from parched barley; then apply this to the patient. If you treat in this way, you will very quickly make the patient well. The disease lasts a long time.

52. Three tetanuses: if tetanus follows a wound, the patient suffers the following: his jaws are fixed, and he is unable to open his mouth; his eyes shed tears and look awry; his back becomes rigid; he cannot bend his legs, nor his arms and spine. When he is near death, sometimes both the drink and the food that he has taken earlier come up through his nostrils.

When the case is such, treat the patient with vapour-baths, anoint him generously with oil, and warm him in firelight from a distance; anoint his body, and apply fomentations. Grind wormwood, bay leaves, or henbane seed with frankincense; soak this in white wine, and pour it into a new pot; add an amount of oil equal to the wine, warm,

514 M adds τὸν ἄνθρωπον.
516 M adds χαλεπὴ καὶ.
518 M: πυραναφθὲν Θ.
519 M: ἢ Θ.

515 ὀ. τ. Θ: ἐφθῶν δύο M.
517 Θ: οὐδὲ M.

θερμήνας ἀλεῖψαι τὸ σῶμα θερμῷ πολλῷ[520] καὶ τὴν
κεφαλήν. εἶτα κατακλίνας †ἐν σκάφῃ[521] ἄνω τὴν σάρκα ποιῆσαι,[522] † καὶ ἀμφιέσαι ἱμάτια μαλθακὰ καὶ
300 καθαρά, ὅπως ἂν ἐξιδρώσῃ σφόδρα. καὶ μελίκρητον
χλιαρὸν δοῦναι, ἢν μὲν δύνηται, κατὰ τὸ στόμα, εἰ δὲ
μή, κατὰ τὰς ῥῖνας ἐγχεῖν· διδόναι δὲ καὶ οἶνον λευκὸν
κὸν ὡς ἥδιστον πίνειν καὶ πλεῖστον. ταῦτα χρὴ ποιέειν ἑκάστης ἡμέρης, ἕως ἂν ὑγιὴς γένηται. ἡ δὲ
νοῦσος χαλεπὴ καὶ δεῖται μελέτης παραχρῆμα.

53. Ὁ δὲ ὀπισθότονος τὰ μὲν ἄλλα πάσχει πλῆθος
τὰ αὐτά· γίνεται δὲ ὅταν τοὺς ἐν τῷ αὐχένι τένοντας
τοὺς ὄπισθεν νοσήσῃ· νοσέει δὲ ὑπὸ κυνάγχης ἢ
σταφυλῆς ἢ τῶν ἀμφιβραγχίων ἐμπύων γενομένων·
ἐνίοισι[523] δὲ καὶ ἀπὸ τῆς κεφαλῆς πυρετῶν ἐπιγενο-
μένων σπασμὸς ἐπιγίνεται· ἤδη δὲ καὶ ὑπὸ τρωμάτων.
οὗτος ἕλκεται ἐς τοὔπισθεν, καὶ ὑπὸ τῆς ὀδύνης τὸ
μετάφρενον καὶ τὰ στήθεα οἰμώζει· καὶ οὕτω[524] σπᾶται
σφοδρῶς, ὥστε μόγις αὐτὸν κατέχουσιν οἱ παρεόντες,
ὥστε[525] μὴ ἐκ τῆς κλίνης ἐκπίπτειν. οὗτος πολλάκις
μὲν τῆς ἡμέρης πονέει, πολλάκις δὲ κουφότερον δι-
άγει.

Τούτῳ τὰ αὐτὰ προσφέρειν, ἃ καὶ τῷ πρόσθεν. ἡ δὲ
νοῦσος προσίσχει ἡμέρας τεσσεράκοντα τὸ μακρό-
τατον· ἢν δὲ ταύτας διαφύγῃ, ὑγιαίνεται.

54. Ὅδε δὲ ὁ[526] τέτανος ἧσσον θανατώδης ἢ ὁ
πρόσθεν, γίνεται δὲ ἀπὸ τῶν αὐτῶν. πολλοὶ δὲ καὶ

[520] πολλῷ om. Θ.   [521] Potter: ἐν βάπτει Θ: ἐμβατη Μ.

and anoint the patient's body copiously with the warm fluid, and also his head. Then, laying him down in a basin, make his tissue . . . ,[12] and clothe him in clean soft blankets, in order that he will perspire profusely. Give him warm melicrat, if he is able to take it, through the mouth; if he is not able, pour it into his nostrils; also give him very sweet white wine to drink, in large quantities. You must do these things each day until the patient recovers. The disease is severe, and requires immediate attention.

53. The patient with opisthotonus suffers, on the whole, the same, but the disease arises when he is affected in the posterior tendons of the neck; his illness arises from angina, from staphylitis, or from a suppuration occurring in the parts about the tonsils; also, in some cases, such a convulsion originates from the head, when there are fevers; occasionally it also follows wounds. This patient is drawn backwards, and cries aloud from the pain in his back and chest; he is drawn so forcefully that the attendants can hardly prevent him from falling out of bed. In the course of one day, he often suffers severely, but at other times goes along more easily.

Administer to this patient the same things as to the preceding one. The disease continues for forty days, at the longest; if the patient survives these, he recovers.

54. The following tetanus is less often mortal than the preceding one, but it arises from the same things; many

___

[12] The sense of this passage is lost.

___

522 Θ: -αντα M. 523 Θ: ἐνίοτε M.
524 κ. ο. Θ: οὗτος M. 525 Θ: ὡς M.
526 Θ: Ὁ δὲ M.

πεσόντες ἐς τοὔπισθεν ἔλαβον τὸ νόσημα. πάσχει οὖν τάδε· σπᾶται ὁμοίως πᾶν[527] τὸ σῶμα· ἐνίοτε δὲ καὶ ὅπῃ ἂν τύχῃ τοῦ σώματος σπᾶται. καὶ περιφοιτᾷ μὲν τὸ πρῶτον· ἔπειτα προϊόντος τοῦ χρόνου ἐς τὴν κλίνην πίπτει· καὶ αὖτις ἀνῆκεν ὁ πόνος καὶ ὁ σπασμός. καὶ ἀναστὰς ἢν περιέλθῃ ὀλίγας ἡμέρας, ἔπειτα αὖτις ἐν

302 τοῖσιν αὐτοῖσιν ἄλγεσιν κεῖται. ταῦτα | πάσχει καὶ μεταβάλλει συχνὸν χρόνον. καὶ ἤν τι φάγῃ, οὐ διαχωρέει κάτω, εἰ μὴ μόγις, καὶ ταῦτα συγκεκαυμένα· ἀλλ' ἐν τοῖσι στήθεσιν ἔχει[528] τὸ σιτίον καὶ πνῖγμα παρέχει.

Τοῦτον τοῖς αὐτοῖσι θεραπεύειν, οἷσι καὶ τοὺς πρόσθεν, καὶ τάχιστα ὑγιᾶ ποιήσεις· κλύζειν πτισάνῃ ἐφθῇ λεπτῇ καὶ μέλιτι.

[527] ὁ. π. Θ: ὅλον Μ.
[528] Θ: ἄρχεται Μ.

people have also taken it by falling backward. Now the patient suffers the following: he has a convulsion equally through his whole body; sometimes, though, the convulsion is just in some random part of the body. At first, he walks about; then, after a time, he falls into bed; then the pain and convulsions remit again; then, if he gets up and goes about for a few days, he labours once more under the same pains. This is the nature of his symptoms and how they continually change. If the patient eats anything, it does not pass off below or, if it does, only a little, and that burnt up; more likely, he retains his food in his chest, and it provokes choking.

Treat this patient with the same things as the preceding ones, and you will very quickly make him well. As enema, employ thin boiled barley-water and honey.

REGIMEN IN ACUTE DISEASES
(APPENDIX)

INTRODUCTION[1]

This piece of text is a continuation of *Regimen in Acute Diseases* (Loeb *Hippocrates*, II. 57–125). The manuscript A introduces it as "the spurious addition to *Barley-Gruel*",[2] M and V as the "beginning of the spurious part".

Erotian lists *Barley-Gruel* (Περὶ πτισάνης)[3] among the therapeutic writings of Hippocrates and, from the words he includes in his glossary, it is clear that he accepts *the Appendix* as a part of the work.[4]

Galen has left us a long commentary on *Regimen in Acute Diseases* including the *Appendix*. In the introduction to the part of his commentary devoted to the *Appendix*, he discusses the various arguments that were advanced in his time for and against authenticity.[5]

Athenaeus quotes a word from:

---

[1] The transmission of the text of this treatise is discussed above, p. xi f.

[2] Galen too knew this title; see Kühn VII. 913.

[3] Nachmanson p. 9.

[4] See in particular (Nachmanson p. 81) Σ48 σποράδες νοῦσοι from *Regimen in Acute Diseases* 5, Σ49 σπατίλη from *Regimen in Acute Diseases* 28, and Σ50 σησαμοειδές from the *Appendix*, ch. 60.

[5] Kühn XV. 732–734 = CMC V 9,1 pp. 271 f.

Hippocrates in *Barley-Gruel*, which is half spurious, or according to some wholly. . . .[6]

Caelius Aurelianus makes reference to several passages from the *Appendix*, four times citing Hippocrates' book *Against the Cnidian Maxims*,[7] and once Hippocrates' *On Regimen*.[8] From this, it would seem that he did not distinguish between "genuine" and "spurious" parts *of Regimen in Acute Diseases*.

Modern discussion on the relationship of the two parts is summarized well by Joly in the introduction to his edition.[9]

About the *Appendix* as a treatise, there is little to say. Its chapters, although for the most part internally coherent, bear little relationship to one another, and, except for certain poorly defined centres of interest (e.g. in 1–18 nosology; in 19–29 semiology; in 43—50 dietetics), no ordering principle is discernible. It is not known how the treatise came into its present form.

The *Appendix* has been included in all the standard collected editions and translations of the Hippocratic Collection, and in the editions of *Regimen in Acute Diseases*.

[6] *Deipnosophistae* II 57c.

[7] *Acute Diseases III* 25 (Drabkin 314) refers to *Appendix* 9; *Acute Diseases III* 83 f. (Drabkin 352) refers to *Appendix* 37; *Chronic Diseases III* 139 f. (Drabkin 802) refers to *Appendix* 52 and 58; *Chronic Diseases IV* 77 (Drabkin 862–4) refers *to Appendix* 53.

[8] *Acute Diseases II* 154 (Drabkin 236) refers *to Appendix* 34.

[9] Pp. 11–14.

Besides the works mentioned by Jones (II. 61), I have also made reference to:

Τὸ περὶ διαίτης ὀξέων νοσημάτων, ἤτοι περὶ πτισσάνης. *De victu ratione in morbis acutis, siue De ptisana* . . . Ioanne Vassaeo interprete. Paris, 1531.

Robert Joly, *Hippocrate, Du régime des maladies aiguës, Appendice,* . . . , Budé VI(2), Paris, 1972. (= Joly)

The standard edition of Galen's commentary on *Regimen in Acute Diseases* is:

Georg Helmreich, *Galeni in Hippocratis De uictu acutorum commentaria IV*, CMG V 9,1, Leipzig and Berlin, 1914. pp. 115–366.

An English translation of the *Appendix* appeared in:

Francis Adams, *The Genuine Works of Hippocrates*, London, 1849, I. 313–36.

ΠΕΡΙ ΔΙΑΙΤΗΣ ΟΞΕΩΝ
(ΝΟΘΑ)

II 394
Littré
1. Καῦσος γίνεται, ὁπόταν ἀναξηρανθέντα τὰ φλέβια ἐν θερινῇ ὥρῃ ἐπισπάσηται δριμέας καὶ χολώδε-
ας ἰχῶρας ἐφ᾽[1] ἑωυτά· καὶ πυρετὸς πολὺς ἴσχει, τό τε
σῶμα ὡς ὑπὸ κόπου[2] ἐχόμενον κοπιᾷ καὶ ἀλγέει.
γίνεται δὲ ὡς ἐπὶ τὸ πολὺ καὶ ἐκ πορείης μακρῆς καὶ
δίψεος μακροῦ, ὁπόταν ἀναξηρανθέντα τὰ φλέβια
δριμέα καὶ θερμὰ ῥεύματα ἐπισπάσηται. γίνεται δ᾽ ἡ
γλῶσσα τρηχείη καὶ ξηρὴ καὶ μέλαινα, καὶ τὰ περὶ
τὴν νηδὺν δακνόμενος ἀλγέει, τά τε ὑποχωρήματα
ἔξυγρα καὶ ὠχρὰ γίνεται· καὶ δίψαι σφοδραὶ ἔνεισι,
396
καὶ ἀγρυπνίη, ἐνίοτε δὲ καὶ | παραλλάξιες φρενῶν.

Τῷ τοιῷδε πίνειν μὲν ὕδωρ τε καὶ μελίκρητον δίδου
ἐφθὸν ὑδαρές, ὁπόσον ἐθέλει. καὶ ἢν πικρὸν τὸ στόμα
γίνηται, ἐμέειν συμφέρει καὶ τὴν κοιλίην ὑποκλύσαι·
ἢν δὲ μὴ πρὸς ταῦτα λύηται, γάλακτι ὄνου ἀφεψήσας
κάθαιρε. ἁλμυρὸν δὲ μηδὲν μηδὲ δριμὺ προσφέρειν,
οὐ γὰρ ὑποίσει· ῥύφημα δέ, ἔστ᾽ ἂν ἔξω τῶν κρίσεων[3]
γίνηται, μὴ δίδου. καὶ ἢν αἷμα ἐκ τῆς ῥινὸς ῥυῇ,

---

[1] A: ἐς MV. [2] A: ὀστεοκόπου MV.

228

REGIMEN IN ACUTE
DISEASES
(APPENDIX)[1]

1. Ardent fever occurs when the small vessels are dried up in summer, and attract sharp bilious sera; there is great fever, and the body, as if from toil, suffers weariness and pain. This disease usually follows a long journey and prolonged thirst, when the vessels, being dried up, attract sharp hot fluxes. The tongue becomes rough, dry and dark, the patient has a gnawing pain about the belly, and his evacuations are watery and pale yellow; great thirst is present, sleeplessness, and sometimes also aberrations of the mind.

Give such a patient water and dilute boiled melicrat to drink, as much as he wants. If his mouth is bitter, it helps to have him vomit, and to clean out his cavity by means of an enema; if, with this, the cavity is not opened, clean it with boiled-down ass's milk. Administer nothing salty or sharp, for the patient will not tolerate it; do not give gruels until he is beyond his crises. If blood flows from the nose, the

---

[1] Literally ΝΟΘΑ means "spurious".

---

[3] A: κρισίμων MV.

λύεται τὸ πάθος, καὶ ἢν ἱδρῶτες ἐπιγένωνται[4] κριτικοὶ
γνήσιοι μετ' οὔρων παχέων, λευκῶν καὶ λείων ὑφιστα-
μένων, καὶ ἢν ἀπόστημά που γένηται. ἢν δ' ἄνευ
τούτων λυθῇ, ὑποστροφὴ πάλιν ἔσται τῆς[5] ἀρρωστίης
ἢ ἰσχίου[6] ἢ σκελέων ἄλγημα συμβήσεται. καὶ πτύσε-
ται παχέα, ἢν μέλλῃ ὑγιὴς ἔσεσθαι.

2. Καύσου γένος ἄλλο. κοιλίη ὑπάγουσα δίψεος
398 μεστή, γλῶσσα τρηχεία, ξηρή, ἁλυκώδης, οὔρων
ἀπόληψις, ἀγωνίη,[7] ἀκρωτήρια ἐψυγμένα. τῷ τοιούτῳ
ἢν μὴ αἷμα ἐκ ῥινὸς ῥυῇ ἢ ἀπόστημα περὶ τράχηλον
γένηται ἢ σκελέων ἄλγημα καὶ πτύσματα παχέα
πτύσῃ—ταῦτα δὲ συστάσης τῆς κοιλίης γίνεται—ἢ
ἰσχίων[8] ὀδύνη ἢ αἰδοίων[8] πελίωμα, οὐ κρίνεται· καὶ
ὄρχις ἐνταθεὶς κριτικόν. ῥυφήματα ἐπισπαστικὰ δίδου.

3. (2 L.) Τὰ ὀξέα πάθεα φλεβοτομήσεις, ἢν ἰσχυρὸν
φαίνηται τὸ νόσημα καὶ οἱ ἔχοντες ἀκμάζωσι τῇ
ἡλικίῃ καὶ ῥώμῃ. ἢν μὲν οὖν σύναγχος ᾖ, ἐκλεικτῷ
ἀνακάθαιρε, ἢν δὲ ἄλλο τι, τῷ πλευριτικῷ· ἢν δὲ
ἀσθενέστεροι φαίνωνται ἢ καὶ πλέον τοῦ αἵματος
ἀφέλῃς, κλυσμῷ κατὰ[9] κοιλίην χρῆσθαι διὰ τρίτης
ἡμέρης, ἕως ἂν ἐν ἀσφαλείῃ γένηται ὁ νοσέων, καὶ
λιμῷ εἰ χρήζοι.[10]

4 MV: -γεννῶνται A.
5 τῆς om. A.
6 A: -ων MV.
7 A: ἀγρυπνίη MV.
8 A: -ου MV.

affection is resolved; also if true critical sweats supervene, accompanied by thick white urines having a fine sediment, and also if an abscess forms anywhere. If it is resolved without these, there will be a recurrence of the ailment, or pain in the hip or legs will follow. If a patient is going to recover, he will also expectorate thick sputa.

2. Another kind of ardent fever: the cavity in downward motion, full of thirst; tongue rough, dry, salty; stoppage of urine; distress; extremities cold. In such a patient, unless blood flows from the nose, or an abscess forms about the neck, or there is pain in the legs, and he expectorates thick sputa—these things happen when the cavity is contracted—or unless pain arises in the hips, or there is lividness of the genital organs, the disease does not reach a crisis; a testicle being stretched tight is also an indication of crisis. Give gruels that attract.[2]

3. The acute affections you treat with phlebotomy, if the disease seems to be severe, and patients are at the height of their youth and strength. If it is a case of angina, clean upwards with a lozenge, if some other disease, employ the treatment used in pleurisy.[3] If patients appear too weak, or you have drawn much blood, use an enema for the cavity every other day until the patient reaches safety, and fasting if he should require it.

[2] I.e. that attract and eliminate the peccant material.
[3] Perhaps a reference to chapter 31 below.

---

[9] MV: κάτω A.
[10] Littré: λειμῶ εἰ χ. A: λιμοῦ χρήζοι MV: λιμῷ χρήζῃ Joly.

400   4. (3 L.) Φλεγμαίνοντα ὑποχόνδρια, τῇ[11] πνευμάτων
ἀπολήψει φρενῶν ἐντάσιες, πνευμάτων προστάσιες
ὀρθοπνοίης ξηρῆς—οἷσι μὴ πύον ὕπεστιν, ἀλλὰ ὑπὸ
πνευμάτων ἀπολήψιος τὰ πάθεα ταῦτα ὑπογίνεται[12]—
καὶ ἥπατος περιωδυνίαι καὶ σπληνὸς βάρεα καὶ ἄλλαι
φλεγμασίαι τε καὶ ὑπὲρ φρενῶν περιωδυνίαι καὶ συ-
στροφαὶ νοσημάτων[13] οὐ δύνανται λύεσθαι, ἤν τις
πρῶτον ἐπιχειρέῃ φαρμακεύειν· ἀλλὰ φλεβοτομίη τῶν
τοιῶνδε ἡγεμονικόν ἐστιν. ἔπειτα δὲ ἐπὶ κλυσμόν, ἢν
402   μὴ μέγα καὶ ἰσχυρὸν | τὸ νόσημα ᾖ· εἰ δὲ μὴ καὶ
ὕστερον φαρμακείης δεῖ· δεῖται δὲ ἀσφαλείης καὶ
μετριότητος μετὰ φλεβοτομίην φαρμακείη.

5. Ὁπόσοι δὲ τὰ φλεγμαίνοντα ἐν ἀρχῇ τῶν νούσων ἐπιχειροῦσιν λύειν φαρμακείῃ, τοῦ μὲν συντετα-
μένου καὶ φλεγμαίνοντος οὐδὲν ἀφαιρέουσιν—οὐ γὰρ
διαδιδοῖ ὠμὸν ἐὸν τὸ πάθος—τὰ δ' ἀντέχοντα τῷ
νοσήματι καὶ ὑγιεινὰ συντήκουσιν. ἀσθενέος δὲ τοῦ
σώματος γινομένου τὸ νόσημα ἐπικρατέει· ὅταν δὲ τὸ
νόσημα ἐπικρατήσῃ τοῦ σώματος, τὸ τοιῶνδε ἀνιήτως
ἔχει.

6. (4 L.) Τὸ δὲ ἄφωνον ἐξαίφνης γενέσθαι, φλεβῶν
404   ἀπολήψιες λυπέουσιν, ἢν ὑγιαίνοντι τόδε συμβῇ ἄνευ
προφάσιος ἢ ἄλλης αἰτίας ἰσχυρῆς. φλεβοτομέειν οὖν
τὸν βραχίονα τὸν δεξιὸν τὴν ἔσω φλέβα καὶ ἀφαιρέειν τοῦ αἵματος κατὰ τὴν ἕξιν καὶ τὴν ἡλικίην
διαλογιζόμενον τὸ πλεῖον καὶ τὸ ἔλασσον. συμπίπτει
δὲ τοῖσι πλείστοισι τοιάδε· ἐρυθήματα προσώπου, καὶ
ὀμμάτων στάσιες, καὶ διαστάσιες χειρῶν, τρισμοὶ

4. Swelling of the hypochondrium, tension of the diaphragm from the stoppage of air, blockage of air in dry orthopnoea—in patients without suppuration, but in whom these affections are due to a stoppage of air—severe pains in the liver, heaviness of the spleen and other phlegmasias, and severe pains and disease complexes above the diaphragm can all not be resolved if a person undertakes first to treat with medications; but phlebotomy is the master of such things. Then an enema, unless the disease is great and severe; if it is great and severe, the patient must employ a medication later as well; the use of medications after phlebotomy requires caution and moderation.

5. Those who undertake to resolve swellings at the beginning of diseases, by using medications, draw off nothing of what is stretched and swollen—for the affection does not go away as long as it is unmatured—but consume the healthy elements that are resisting the disease. The body weakens and the disease is victorious, and when the disease wins out over the body, such a thing is incurable.

6. Suddenly becoming speechless: stoppage of the vessels produces this evil, if it befalls a healthy person without any antecedent condition or other potent cause. Phlebotomize the inner vessel of the right arm and draw blood, reckoning whether more or less according to the patient's condition and age. The following occur in most cases: red patches on the face, fixation of the eyes, spreading of the

---

11 Potter: μὴ AMV: ἐπὶ Reinhold.
12 MV add μάλιστα δὲ.
13 A adds ἃ.

ὀδόντων, σφυγμοί, σιαγόνων συναγωγή, καὶ κατάψυ-
ξις ἀκρωτηρίων.

7. (5 L.) Πνευμάτων ἀπολήψιες ἀνὰ φλέβας. ὁπόταν
406 ἀλγήματα προσγένηται, μελαίνης χολῆς | καὶ δρι-
μέων ῥευμάτων ἐπιρρύσιες γίνονται· ἀλγέει[14] δὲ τὰ
ἐντὸς δακνόμενος· δηχθεῖσαι δὲ καὶ λίην ξηραὶ γενό-
μεναι αἱ φλέβες ἐντείνονταί τε καὶ φλεγμαίνουσαι
ἐπισπῶνται τὰ ἐπιρρέοντα· ὅθεν διαφθαρέντος τοῦ
αἵματος καὶ τῶν πνευμάτων οὐ δυναμένων ἐν αὐτῷ τὰς
κατὰ φύσιν ὁδοὺς βαδίζειν καταψύξιές τε γίνονται
ὑπὸ τῆς στάσιος καὶ σκοτώσιες καὶ ἀφωνίη καὶ καρη-
βαρίη ἢ καὶ σπασμοί, ἢν ἤδη ἐπὶ τὴν καρδίην ἢ τὸ
ἧπαρ ἢ ἐπὶ τὴν φλέβα ἔλθῃ· ὅθεν ἐπίληπτοι γίνονται
ἢ παραπλῆγες, ἢν ἐς τοὺς περιέχοντας τόπους ἐμπέσῃ
τὰ ῥεύματα καὶ ὑπὸ τῶν πνευμάτων οὐ δυναμένων
διεξιέναι καταξηρανθῇ.

408   8. Ἀλλὰ χρὴ τοὺς | τοιούτους προπυριῶντα φλεβο-
τομέειν ἐν ἀρχῇ εὐθέως μετεώρων ἐόντων πάντων τῶν
λυπεόντων ῥευμάτων·[15] εὐβοηθητότερα γάρ ἐστιν· καὶ
ἀναλαμβάνοντα καὶ τὰς κρίσιας ἐπιθεωρέοντα φαρμακεύειν, ἢν μὴ κουφίζηται, ἄνω· τὴν δὲ κάτω κοιλίην,
ἢν μὴ ὑποχωρέῃ κλυσμῷ, ὄνου γάλα ἑφθὸν δίδου, καὶ
πινέτω μὴ ἔλασσον δώδεκα κοτυλῶν· ἢν δὲ ῥώμη
περιέχῃ, πλείω [ἑκκαίδεκα].[16]

9. (6 L.) Σύναγχος δὲ γίνεται, ὁπόταν ἐκ τῆς κεφαλῆς ῥεῦμα πολὺ καὶ κολλῶδες ὥρην χειμερινὴν ἢ

14 MV: ἀλγεῖται A.

234

fingers, grinding of the teeth, throbbing, closing of the jaws, and coldness of the extremities.

7. Stoppage of air through the vessels: when pains come on, affluxes of dark bile and sharp fluids are occurring; the patient suffers gnawing pains in his inward parts. The vessels, being irritated and too dry, stretch tight, swell up, and attract the fluxes; from this, since the blood becomes disordered and the air is no longer able to follow its normal paths through the blood, chills occur as a result of stasis, along with darkening of the vision, loss of speech, heaviness of the head, and even convulsions, if the fluxes have already reached the heart, liver, or vessel[4]; from this, patients become epileptic or are paralysed, if the fluxes invade the surrounding parts and are dried up because air cannot pass through.

8. After first subjecting such patients to a vapour-bath, you must phlebotomize right at the onset, while all the injurious fluids are unsettled; for in this state the condition is more easily helped. Then restore the patient and, paying attention to his crises, employ a medication to clean him upwards, unless he has already been lightened. For the lower cavity, unless it has been evacuated by means of an enema, give boiled ass's milk; let the patient drink not less than twelve cotylai; if he is quite strong, even more.

9. Angina arises when, in winter or spring, a massive viscid flux occurs from the head into the jugular vessels,

---

[4] Presumably one of the large vessels such as the aorta or the vena cava.

---

[15] Kuehlewein: πνευμάτων A: πνευμάτων καὶ ῥευμάτων MV.   [16] Del. Kuehlewein.

ἐαρινὴν[17] ἐς τὰς σφαγίτιδας φλέβας ἐπιρρυῇ, καὶ τὸ
410 ῥεῦμα[18] πλέον | διὰ τὴν εὐρύτητα ἐπισπάσωνται. ὅταν
δὲ ψυχρόν τε ἐὸν καὶ κολλῶδες ἐμφράξῃ τοῦ τε πνεύματος τὰς διεξόδους καὶ τοῦ αἵματος ἀποφράσσον,
πήγνυσι τὰ σύνεγγυς τοῦ αἵματος καὶ ἀκίνητον καὶ
στάσιμον ποιέει φύσει ψυχρὸν ἐὸν καὶ ἐμφρακτικόν.
διὰ τοῦτο πνίγονται τῆς γλώσσης ἀποπελιουμένης
καὶ στρογγυλουμένης καὶ ἀνακαμπτομένης διὰ τὰς
φλέβας τὰς ὑπὸ τὴν γλῶσσαν· τῆς γὰρ ὑποτεινομέ-
νης[19] σταφυλῆς—οἱ δὲ κιονίδα καλεῦσιν—ἑκατέρωθεν
412 φλὲψ παχείη. ὁπόταν οὖν | πλήρεις αὗται ἐοῦσαι ἐς
τὴν γλῶσσαν ἐναποστηρίζωνται ἀραιὴν ἐοῦσαν καὶ
σπογγώδεα, διὰ τὴν ξηρασίην ὑπὸ βίης τὸ ἐκ τῶν
φλεβῶν δεχομένη ὑγρὸν ἐκ πλατείης μὲν στρογγύλη
γίνεται, ἐξ εὐχρόου δὲ πελιδνή, ἐκ μαλθακῆς δὲ σκληρή, ἐξ εὐκάμπτου δὲ ἄκαμπτος· ὥστε ταχέως ἀποπνί-
γεσθαι, ἢν μή τις ταχέως βοηθῇ φλεβοτομίην τε
ποιεύμενος ἀπὸ βραχιόνων καὶ τὰς ὑπὸ τὴν γλῶσσαν
φλέβας ὑποτάμνων καὶ φαρμακεύων τοῖσιν ἐκλεικτοῖ-
σι καὶ ἀναγαργαρίζων θερμοῖσι καὶ κεφαλὴν ξυρῶν.
καὶ κήρωμα[20] τραχήλῳ περιτιθέναι καὶ εἰρίοισι περιελίσσειν καὶ σπόγγοισι μαλθακοῖσιν ἐν ὕδατι θερμῷ
ἐκπιεζέοντα πυριᾶν. πίνειν δὲ ὕδωρ καὶ μελίκρητον μὴ
414 ψυχρά· | χυλὸν δὲ προσφέρειν, ὁπόταν ἐκ κρίσιος ἐν
ἀσφαλείῃ ἤδη ᾖ.

10. Ἄλλο εἶδος συνάγχου· ὅταν ἐν θερινῇ ἢ μετ-
οπωρινῇ ὥρῃ ἐκ κεφαλῆς θερμὸν τὸ ῥεῦμα καταρρυῇ
καὶ νιτρῶδες ᾖ,[21] ἅτε ὑπὸ τῆς ὥρης δριμὺ καὶ θερμὸν

and these, because of their wideness, attract the increased flow. When the flux, which is cold and viscid, stops up the passage-ways that carry air and blood by blocking them, being by nature cold and obstructive it coagulates whatever blood is nearby, making it fixed and still. Because of this, choking occurs, as the tongue becomes livid, globular, and rigid from the vessels underneath it; for on each side of the pendant uvula, which they call the little pillar, there is a wide vessel; when these vessels become filled, they fix themselves in the spongy rarified tongue, and it, being dry, accepts the moisture coming to it under force from the vessels: from flat, it becomes globular, from normal-coloured, livid, from soft, hard, and from flexible, rigid. Thus patients rapidly suffocate, unless someone quickly helps them by phlebotomizing from their arms, by cutting the vessels under their tongue, by administering medications in the form of lozenges, by having them gargle hot liquids, and by shaving their head. Also apply wax-salves around the neck, and wrap it with wool; foment the patient with soft sponges dipped in warm water and squeezed out. Have him drink warm water and melicrat. When he is safely past the crisis, administer barley-water.

10. Another form of angina: when, in summer or fall, a hot flux descends from the head, and it is like soda, being sharp and hot like this because of the season, it irritates,

---

17 ἢ ἐαρινὴν om. A.

18 A: πνεῦμα MV.

19 Reinhold: ἀποταμν- A: ὑποτεμν- MV.

20 MV add κεφαλῇ καὶ.

21 ᾗ om. MV.

γεγενημένον, δάκνει τοιόνδε ἐὸν καὶ ἑλκοῖ καὶ πνεύματος ἐμπίπλησι, καὶ ὀρθόπνοια παραγίνεται καὶ ξηρα-
σίη πολλή, καὶ τὰ ὁρώμενα²² ἰσχνὰ φαίνεται· καὶ τοὺς
ὄπισθεν τένοντας ἐν τῷ τραχήλῳ συντείνεται, καὶ
δοκέει οἱ τέτανος ἐντετάσθαι, καὶ ἡ φωνὴ ἀπέρρωγε
416 καὶ τὸ πνεῦμα σμικρὸν καὶ ἡ ἀντίσπασις τοῦ | πνεύ-
ματος πυκνὴ καὶ βιαίη παραγίνεται. οἱ τοιοίδε τὴν
ἀρτηρίην ἑλκοῦνται καὶ τὸν πνεύμονα πίμπρανται οὐ
δυνάμενοι τὸ ἔξωθεν πνεῦμα ἐπάγεσθαι. τοῖσι τοιούτοισι δὲ ἢν μὴ ἐς τὰ ἔξω μέρη τοῦ τραχήλου ἑκουσίη
ἀποφέρηται, δεινότερα καὶ ἀφυκτότερά ἐστι διὰ τὴν
ὥρην καὶ ὅτι ἀπὸ θερμῶν καὶ δριμέων.

11. (7 L.) Ἢν πυρετὸς λάβῃ, παλαιῆς κόπρου ὑπ-
418 εούσης νεο|βρῶτι ἐόντι, ἤν τε σὺν ὀδύνῃ πλευροῦ, ἤν
τε μή, ἡσυχίην ἄγειν μέχρι οὗ καταβῇ τὰ σιτία
πρῶτον ἐς τὴν κάτω κοιλίην. πόματι δὲ χρῆσθαι
ὀξυμέλιτι. ὁπόταν δὲ ἐς τὴν ὀσφὺν βάρος ἥκῃ, κάτω
κλύσαι κλυσμῷ ἢ καθᾶραι²³ φαρμάκῳ. ὅταν δὲ καθαρθῇ, διαιτᾶν ῥυφήματι πρῶτον καὶ πόματι μελικρήτῳ,
ἔπειτα σιτίοισι καὶ ἰχθύσιν ἐφθοῖσι καὶ οἴνῳ ὑδαρεῖ
ἐς νύκτα ὀλίγῳ, ἡμέρης δὲ ὑδαρὲς μελίκρητον. ὁπόταν
δὲ αἱ φῦσαι δυσώδεες ἔωσιν, οὕτως ἢ βαλάνῳ ἢ
κλυσμῷ· εἰ δὲ μή, ἐπισχεῖν ὀξύμελι πίνοντα, ἕως ἂν ἐς
τὴν κάτω κοιλίην καταβῇ, εἶθ' οὕτω κλυσμῷ ὑπ-
αγαγεῖν.

12. Ἢν δὲ λαπαρῷ ἐόντι καῦσος ἐπιγένηται, ἤν σοι
δοκῇ φαρμακεύειν ἐπιτηδείως ἔχειν, ἔσω τριῶν ἡμερέων μὴ φαρμακεύειν[, ἀλλ' ἢ τεταρταῖον].²⁴ ὁπόταν δὲ

produces ulceration, and inflates; orthopnoea and great dryness prevail; the visible part of the neck does not appear swollen; the posterior tendons of the neck are pulled tight, the person seems to be convulsed, his voice is broken off, his breath is slight, and the ebb and flow of air is frequent and laborious. Such patients become ulcerated in the bronchial tube, inflated in the lung, and are incapable of drawing in air from the outside. Unless this condition moves spontaneously into the outer parts of the neck, it is quite terrible and difficult to get rid of, because of the season, and because it arises from hot sharp substances.

11. If fever occurs in a person that has just eaten, while old fecal material was still present in his body, whether there is pain in the side or not, have him rest until the food first passes down into his lower cavity; as drink give oxymel. When a heaviness arrives in the loins, clean him downwards with an enema or medication. After the patient has been cleaned, first employ a regimen of gruel and, as drink, melicrat, then cereals and boiled fish, and towards night a little dilute wine; during the day give dilute melicrat. If the flatus is foul-smelling, employ a suppository or an enema; if not, have the patient continue to drink oxymel until the food descends into his lower cavity, and then evacuate it with an enema.

12. If, on the other hand, ardent fever attacks a person that is in a state of emptiness, and it seems suitable to you to administer a medication, do not do so for three days;

---

22 A: θεωρεύμενα MV.
23 καθᾶραι om. A.
24 MV: del. Reinhold: μὴ δὲ τ. A.

ΠΕΡΙ ΔΙΑΙΤΗΣ ΟΞΕΩΝ (ΝΟΘΑ)

420 φαρμακεύσῃς, | τοῖσι ῥυφήμασι χρέω, φυλάσσων
τοὺς παροξυσμοὺς τῶν πυρετῶν, ὅκως μηδέποτε
προσοίσεις[25] μελλόντων ἔσεσθαι τῶν πυρετῶν,[26] ἀλλὰ
ληγόντων ἢ παυσαμένων καὶ ὡς προσωτάτω[27] ἀπὸ τῆς
ἀρχῆς.

13. Ποδῶν δὲ ψυχρῶν ἐόντων μήτε ποτὸν μήτε
ῥύφημα μήτε ἄλλο μηδὲν δίδου τοιόνδε, ἀλλὰ μέγιστον ἡγέου τοῦτ' εἶναι, διαφυλάσσεσθαι, ἕως ἂν διά-
θερμοι γένωνται σφόδρα· εἶθ' οὕτω τὸ συμφέρον
πρόσφερε. ὡς γὰρ ἐπὶ τὸ πολὺ σημεῖόν ἐστι μέλλον-
τος παροξύνεσθαι τοῦ πυρετοῦ ψύξις ποδῶν. εἰ δ' ἐν
τοιούτῳ καιρῷ προσοίσεις, ἅπαντα τὰ μέγιστα ἐξ-
αμαρτήσεις· τὸ γὰρ νόσημα αὐξήσεις οὐ σμικρῶς.
ὅταν δὲ ὁ πυρετὸς λήγῃ, τοὐναντίον θερμότεροι οἱ
πόδες γίνονται τοῦ ἄλλου σώματος· αὔξεται μὲν γὰρ
ψύχων τοὺς πόδας, ἐξαπτόμενος ἀπὸ[28] τοῦ θώρηκος
καὶ ἐς τὴν κεφαλὴν ἀναπέμπων τὴν φλόγα. συνδεδρα-
μηκότος δὲ ἀλέος τοῦ θερμοῦ ἄπαντος ἄνω καὶ ἀνα-
422 θυμιωμένου ἐς | τὴν κεφαλήν, εἰκότως οἱ πόδες ψυχροὶ
γίνονται, ἄσαρκοι καὶ νευρώδεες φύσει ἐόντες. ἔτι δὲ
πολὺ ἀπέχοντες τῶν θερμοτάτων τόπων ψύχονται,
συναθροιζομένου τοῦ θερμοῦ ἐς τὸν θώρηκα· καὶ
πάλιν ἀνὰ λόγον λυομένου τοῦ πυρετοῦ καὶ κατακερ-
ματιζομένου ἐς τοὺς πόδας καταβαίνει· κατὰ δὲ τὸν
χρόνον τοῦτον ἡ κεφαλὴ καὶ ὁ θώρηξ κατέψυκται.

14. Τοῦ δ' εἵνεκα τότε οὐ προσαρτέον, ὅτι, ὅταν οἱ
πόδες ψυχροὶ ἔωσιν, θερμὴν ἀνάγκη τὴν ἄνω[29] κοιλίην
424 εἶναι καὶ πολλῆς ἄσης μεστὴν καὶ ὑπο|χόνδριον

240

after you have given the medication, employ gruels; be careful to avoid the paroxysms of fever, so as never to make administrations when fevers are incipient, but only when they are declining or stopped, and as far as possible from their onset.

13. When the feet are cold, do not give drink, gruel, or anything else of the sort, but consider it most important to pay heed until they become quite warm; only then apply the fitting treatment; for usually coldness of the feet indicates that the fever is about to grow virulent, and if you make an administration at that moment, you will be committing all the greatest mistakes, since you will increase the disease by no small measure. When the fever diminishes, the opposite happens, and the feet become hotter than the rest of the body. For fever increases by cooling the feet, by being kindled from the thorax, and by sending its flame up into the head; when the heat has all collected together into one mass above, and steamed up into the head, the feet naturally become cold, since they have but little flesh and are cord-like in structure. They are chilled even more by being far away from the hottest area, since the heat is collected in the thorax. Conversely, when a fever is resolved and gradually diminishes, heat descends into the feet; about that time, the head and thorax cool down.

14. For this reason food is not to be given at that time, for when the feet are cold, it necessarily follows that the upper cavity is hot and much filled with nausea, the hypo-

---

25 MV add ἐόντων μηδὲ.
26 τῶν πυρετῶν om. MV.
27 A: πορρ- MV.   28 A: ἐκ MV.
29 Ermerins after Galen: κάτω A: om. MV.

ΠΕΡΙ ΔΙΑΙΤΗΣ ΟΞΕΩΝ (ΝΟΘΑ)

ἐντεταμένον καὶ ῥιπτασμὸν τοῦ σώματος διὰ τὴν
ἔνδον ταραχὴν καὶ μετεωρισμὸν[30] καὶ ἀλγήματα· καὶ
ἕλκεται καὶ ἐμέειν ἐθέλει, καὶ ἢν πονηρὰ ἐμέῃ, ὀδυνᾶται. θέρμης δὲ καταβάσης ἐς τοὺς πόδας καὶ οὔρου
διελθόντος, κἢν μὴ ἱδρώσῃ, πάντα λωφᾷ. κατὰ τόνδε
οὖν τὸν καιρὸν[31] τὸ ῥύφημα διδόναι· τότε δὲ ὄλεθρος.

15. (8 L.) Οἷσι δὲ[32] διὰ τέλεος ἡ κοιλίη ἐν τοῖσι
πυρετοῖσιν ὑγρή, τούτοισι διαφερόντως τοὺς πόδας
θερμαίνων καὶ περιστέλλων κηρώμασι καὶ ταινιδίοισι
περιελίσσων πρόσεχε, ὡς μὴ ἔσονται ψυχρότεροι τοῦ
ἄλλου σώματος. θερμοῖσι δ' ἐοῦσι θέρμασμα μηδὲν
πρόσφερε, ἀλλὰ παρατήρει, ὅκως μὴ ψυχθήσονται.
πόματι δὲ χρῆσθαι ὡς ἐλαχίστῳ ψυχρῷ ὕδατι ἢ μελικρήτῳ.

426   16. Ὁπόσοισι δὲ[33] κοιλίη ὑγρὴ καὶ γνώμη | τετα-
ραγμένη, οἱ πολλοὶ τῶν τοιούτων τὰς κροκύδας ἀφαι-
ρέουσι καὶ τὰς ῥῖνας σκάλλουσι καὶ κατὰ βραχὺ μὲν
ἀποκρίνονται τὸ ἐρωτώμενον, αὐτοὶ δὲ ἀφ' ἑωυτῶν
οὐδὲν λέγουσιν κατηρτημένον· δοκέει οὖν μοι τὰ τοιάδε μελαγχολικὰ εἶναι. ἢν δὲ τοιῶνδε ἐόντων ἡ κοιλίη
ὑγρὴ ᾖ,[34] δοκέει μοι τὰ ῥυφήματα ψυχρότερα καὶ
παχύτερα προσφέρειν, καὶ τὰ πόματα στατικά, καὶ
οἰνωδέστερα ἢ καὶ[35] στυπτικώτερα.

17. Ὁπόσοισι δὲ τῶν πυρετῶν δινοί τε ἀπ' ἀρχῆς
καὶ σφυγμοὶ τῆς κεφαλῆς εἰσι καὶ οὖρον λεπτόν,
τούτοισι προσδέχεσθαι πρὸς τὰς κρίσιας παροξυνθη-
σόμενον τὸν πυρετόν· οὐ θαυμάσαιμι δ' ἂν οὐδ' εἰ
παραφρονήσειαν. οἷσι δὲ ἐν ἀρχῇ τὰ οὖρα νεφελοει-

242

chondrium stretched, and the body restless because of its internal upheaval, swelling, and pain; the patient retches and wants to vomit, and if he vomits up things that are injurious, he suffers pain. When, however, the heat descends into the patient's feet, and he passes urine, even if he does not sweat he recovers from everything. Take this opportunity to give gruel; before then it would be fatal.

15. In patients whose cavity remains moist all through their fevers, attend especially to the feet by warming them, by wrapping them in wax-salves, and by winding them with strips of linen so that they will not be colder than the rest of the body. If they are warm, do not employ any measures to warm them, but take care of them in order that they do not become cold. As drink, use a very little cold water or melicrat.

16. Cases in which the cavity is moist and the mind disturbed: most of these patients pluck off bits of wool, pick their noses, answer shortly whatever is asked, but of their own accord say nothing sensible; such things seem to me to be due to dark bile. If, in such patients, the cavity is moist, I hold it advisable to administer gruels that are very cold and thick, and drinks that tend to constipate, both the more vinous or even the more astringent.

17. In fevers accompanied from the beginning by vertigo, throbbing in the head, and thin urine, wait for the crises when the fever has its paroxysm; it would come as no surprise if such patients even lose their wits. In those whose urines are cloudy and thick at the beginning, clean

---

30 MV add γνώμης.   31 MV add δεῖ.
32 δὲ om. A.   33 MV add ἐν πυρετοῖς.
34 MV add καὶ συντήκῃ.   35 καὶ om. MV.

428 δέα καὶ παχέα, τοὺς | τοιούσδε ὑποκαθαίρειν, ἢν καὶ
τὰ ἄλλα συμφέρῃ. ὁπόσοισι δὲ ἐν ἀρχῇ τὰ οὖρα
λεπτά, μὴ φαρμάκευε τοὺς τοιούσδε, ἀλλ᾽ ἢν δοκέῃ,
κλύσαι. τοὺς τοιούτους συμφέρει οὕτω θεραπεύεσθαι·
τῷ σώματι ἡσυχίην ἄγοντας ἀλείφοντά τε καὶ περι-
στέλλοντα ὁμαλῶς· ποτῷ δὲ χρῆσθαι μελικρήτῳ ὑδαρεῖ καὶ ῥυφήματι χυλῷ πτισάνης ἐς ἑσπέρην. κοιλίην
δὲ ὕπαγε κατ᾽ ἀρχὰς κλυσμῷ· φάρμακα δὲ μὴ πρόσ-
430 αγει | τούτοισιν· ἢν γάρ τι κινήσῃς κατὰ κοιλίην, τὸ
οὖρον οὐ πεπαίνεται, ἀλλ᾽ ἄνιδρός τε καὶ ἄκριτος ὁ
πυρετὸς ἐπὶ πολὺν χρόνον ἔσται. τὰ δὲ ῥυφήματα, τῶν
κρισίων ὁπόταν ἐγγὺς ᾖ, μὴ δίδου ἢν θορυβῆται, ἢν
δὲ ἀνῇ καὶ ἐπιδιδῷ ἐπὶ τὸ βέλτιον. φυλάσσεσθαι δὲ
δεῖ καὶ τῶν ἄλλων πυρετῶν τὰς κρίσιας καὶ ἀφαιρεῖν
τὰ ῥυφήματα κατὰ τοῦτον τὸν καιρόν. μεμαθήκασι δὲ
432 μακροὶ οἱ πυρετοὶ οἱ τοιοίδε γίνεσθαι καὶ ἀποστήματα ἴσχειν, ἢν μὲν τὰ κάτω ψυχρὰ ᾖ, περὶ ὦτα καὶ
τράχηλον· ἢν δὲ μὴ ψυχρὰ ᾖ, ἄλλας ἴσχει[36] μετα-
βολάς. ῥεῖ δὲ καὶ αἷμα ἐκ ῥινῶν καὶ κοιλίη τοιούτοισιν
ἐκταράσσεται.

18. Ὁπόσοι δὲ ἐν πυρετοῖσιν[37] ἀσώδεές εἰσι καὶ
ὑποχόνδρια συντείνουσι, καὶ κεκλιμένοι οὐκ ἀνέχον-
ται ἐν τῷ αὐτῷ, καὶ τὰ ἄκρεα ψύχονται πάντα, πλεί-
στης ἐπιμελείας καὶ φυλακῆς δέονται. διάγειν δὲ τού-
τοισι προσφέροντα μηδὲν ἄλλο ἢ ὀξύμελι ὑδαρές, |
434 ἕως ἂν λήξῃ καὶ οὖρον πεπανθῇ. κατακλίνειν δὲ ἐς
ζοφερὰ οἰκήματα, καὶ κεκλίσθαι ὡς ἐπὶ μαλθακωτά-
τοισι στρώμασι πολὺν χρόνον ἐπὶ ταὐτὰ καρτερέοντα

downwards if the other signs are propitious. In those whose urines are thin at the beginning, do not employ medications but, if it seems advisable, an enema. It benefits such patients to treat them as follows: have them keep their bodies at rest, and anoint and wrap them carefully. As drink give dilute melicrat, and as gruel barley-water, towards evening. Evacuate the cavity at the beginning with an enema, but do not administer medications to these patients, for if you set anything in motion through the cavity, the urine does not become mature, and the fever will continue without any sweating or crisis for a long time. When the patient is close to his crises, do not give gruels if he is troubled, but only if the fever is remitting and he is changing for the better. You must also pay attention to the crises in other fevers, and discontinue gruels at those times. Fevers of this sort tend to become long and, if the lower regions are cold, to include abscesses about the ears and neck; if the lower regions are not cold, other resolutions occur: blood flows from the nose of such patients, and the cavity is set in motion.

18. Those who, during fevers, suffer from nausea and draw their hypochondrium tight, who on lying down cannot keep still, and whose extremities are all cold, require the greatest care and attention. Proceed with such patients by giving them nothing but dilute melicrat until the fever remits and the urine becomes mature. Put the person to bed in a dark room; let him lie on the softest bed-clothes, remain in the same position for a long time, and throw

---

36 MV: -ειν A.
37 A: Ὁπόσοισι δὲ πυρετοὶ MV.

καὶ ὡς ἥκιστα ῥιπτάζειν· μάλιστα γὰρ τοῦτο τοὺς
τοιούτους ὠφελέει. ἐπὶ δὲ τὸ ὑποχόνδριον λίνου σπέρμα ἐγχρίων ἐπιτίθει, φυλασσόμενος ὅπως μὴ φρίξῃ
προστιθέμενον· ἔστω δὲ ἀκροχλίαρον, ἐφθὸν ἐν ὕδατι
καὶ ἐλαίῳ.

19. Τεκμαίρεσθαι δὲ ἐκ τῶν οὔρων τὸ μέλλον ἔσε-
σθαι· ἢν μὲν γὰρ παχύτερα καὶ ὠχρότερα ᾖ, βελτίω·
ἢν δὲ λεπτότερα καὶ μελάντερα, πονηρότερα.[38] ἢν δὲ
μεταβολὰς ἔχῃ, χρόνον τε σημαίνει, καὶ ἀνάγκη τῷ
νοσέοντι[39] μεταβάλλειν καὶ ἐπὶ τὰ χείρω καὶ ἐπὶ τὰ
βελτίω τὴν ἀνωμαλίην.

20. Τοὺς δὲ ἀκαταστάτους τῶν πυρετῶν ἐᾶν, μέχρι
ἂν στῶσιν· ὁπόταν δὲ στῶσιν, ἀπαντῆσαι διαίτῃ καὶ
θεραπείῃ τῇ προσηκούσῃ, κατὰ φύσιν θεωρέων.

21. (9 L.) Εἰσὶ δὲ ὄψιες πολλαὶ τῶν καμνόντων· διὸ
436 προσεκτέον τῷ | ἰωμένῳ ὅπως μὴ διαλήσει τι[40] τῶν
προφασίων μήτε τῶν κατὰ λογισμὸν ὅσα τ'[41] ἐς
ἀριθμὸν ἄρτιον ἢ περισσὸν δεῖ φανῆναι. μάλιστα μὲν
οὖν δεῖ τὸν περισσὸν ἀριθμὸν εὐλαβεῖσθαι, ὡς αἵδε αἱ
ἡμέραι ἑτερορρεπέας ποιέουσιν τοὺς κάμνοντας.

22. Φυλάσσεσθαι οὖν δεῖ τὴν πρώτην ἡμέρην, ᾗ
ἦρκται ἀσθενεῖν ὁ κάμνων, ἰδόντα τὴν ἀρχὴν ἐξ ὅτου
καὶ δι' ὅτι· ἡγεῖται γὰρ τοῦτο πρῶτον εἰδῆσαι. ὁπόταν
438 δὲ ἔρῃ αὐτὸν καὶ διασκέψῃ | πάντα, πρῶτον μὲν
κεφαλὴν ὅπως ἔχει, εἰ ἀνάλγητος καὶ μὴ βάρος ἔχει
ἐν ἑωυτῇ· ἔπειτα ὑποχόνδρια καὶ πλευρά, εἰ ἀνάλγη-
τα·[42] ὑποχόνδριον μὲν γὰρ ἢν ἐπίπονον ᾖ ἢ ἐπηρμέ

246

himself about as little as possible, for these things, especially, help such patients. Anoint the hypochondrium with an application of linseed, taking care that the patient does not have a chill as a result of the application; let the linseed be luke-warm and boiled in water and oil.

19. Judge from urines what is about to happen, for if thicker and more pale yellow, they are more favourable, but if thinner and darker, more grievous; if they undergo changes, this indicates chronicity, and that the patient will experience changes both for the worse and for the better.

20. Leave unsettled fevers alone until they settle; when they do, counter them with regimen and suitable treatment, taking the patient's constitution into consideration.

21. Patients have many aspects. Therefore, the person treating must pay attention to see that none of the immediate signs escape his notice, nor any of those that you know by reckoning must come to light on an even- or an odd-numbered day. Actually, the odd-numbered, especially, must be observed, since on them patients tend to incline in one direction or the other.

22. Now, you must mark well the first day on which any patient began to be ill, observing whence and why the disease starts; for it is of the utmost importance to first learn this. When you question the patient and examine each thing carefully, do so first with regard to the state of his head, whether it is free of pain and has no heaviness in it; then the hypochondrium and the sides, whether they are free of pain; for if the hypochondrium is painful or swollen

---

38 MV: πονηρά A.   39 A: νοσήματι MV.
40 Reinhold: διαλύσεται A: διαλήσεταί τις MV.
41 A: μήτε ὁκόσα MV.   42 MV: -αλγῇ A.

νον[43] ἔχῃ τινὰ σκολιότητα ἢ κόρον, ἢ πλευροῦ ἀλγηδὼν ἐνῇ καὶ ἅμα τῷ ἀλγήματι βηχίον ἢ στρόφος ἢ
πόνος κοιλίης· ὅταν τι τούτων παρῇ, ὑποχονδρίων μὲν
μάλιστα, λύειν κοιλίην κλυσμοῖσιν· πινέτω δὲ μελίκρητον θερμὸν ἀφεψημένον. καταμανθάνειν δὲ καὶ ἐν
τῇσιν ἐξαναστάσεσιν, εἰ λειποθυμέει ἢ[44] εἰ τοῦ πνεύματος εὐφορίη αὐτὸν ἔχει. ἰδεῖν[45] δὲ τὴν διαχώρησιν,
μή τι μέλαν διεχώρησεν ἰσχυρῶς[46] χρῶμα ἢ εἰ[47] καθαρόν, ὁποῖα ὑγιαίνοντος ἂν εἴη διαχώρημα τα· καὶ ὁ
440   πυρετὸς ⟨εἰ⟩[48] ἐς τὴν τρίτην ἐπιπαροξυνόμενος· | κατ-
ιδὼν δὲ εὖ μάλα τοὺς τοιούσδε ἐν ταύτῃσι τῇσι νούσοισι τριταίους, πρὸς ταύτην ἤδη καὶ τἆλλα συνορᾶν·
καὶ ἢν ἡ τετάρτη τῇ τρίτῃ ἡμέρῃ ὁμοῖόν τι ἔχῃ τῶν
αὐτῶν τούτων, κινδυνώδης ὁ κάμνων γίνεται.

23. Τὰ δὲ σημεῖα· ἡ μὲν μέλαινα διαχώρησις θάνατον σημαίνει, ἡ δὲ ὁμοίη τῷ ὑγιαίνοντι, ὁπόταν ἁπά-
σας τὰς ἡμέρας φαίνηται, σωτήριον· ὁπόταν δὲ μὴ
ὑπακούῃ τῇ βαλάνῳ, ἐπῇ[49] δὲ καὶ[50] τοῦ πνεύματος
ἀπορίη,[51] ἢ διαναστὰς ἐπὶ θρόνον ἢ αὐτοῦ ἐν τῇ κλίνῃ,
ἢν ἀψυχίη ἐγγένηται, ταῦτα ὁπόταν προσῇ τῷ κάμνοντι ἢ τῇ καμνούσῃ,[52] παραφροσύνην ἐσομένην
442   προσδέχου. προσέχειν δὲ χρὴ καὶ τῇσι | χερσίν· ἢν
γὰρ τρομεραὶ ἔωσιν, προσδέχου τῷ τοιῷδε ἀπόσταξιν
αἵματος ἐκ ῥινῶν ἐσομένην. ὁρᾶν δὲ χρὴ τοὺς μυκτῆρας ἀμφοτέρους· ἢν γὰρ[53] ὁμοίως τὸ πνεῦμα δι’ ἀμφοῖν ἕλκηται καὶ πολὺ φέρηται διὰ τῶν μυκτήρων,

---

43 MV add ἤ.   44 Ermerins: καὶ AMV.

with some unevenness or over-fullness, or if pain of the
side is present, and with the pain a mild cough, colic or
pain in the cavity—if any of these things is present, especially in the hypochondrium, open the cavity with enemas;
also have the patient drink hot boiled-down melicrat; observe carefully when the patient rises, too, whether he
faints, or whether his breathing is adequate; observe the
stools, whether anything dark is passed, of a vivid colour, or
whether they are clean, like those of a healthy person; and
the fever, whether it has another access on the third day.
Observing such fevers very closely on the third day in these
diseases, from that day on pay attention to other things as
well, and if the fourth day brings some of these same things
the third one did, the patient is in danger.

23. Signs: dark stools indicate death, whereas those like
a healthy person's, when they have this appearance every
day, betoken recovery. But when a patient fails to have a
movement after he has received a suppository, and there is
also difficulty in breathing, or if either on getting up into a
chair or right in his bed he loses consciousness—when
these things happen to the patient, either male or female,
expect derangement. You must also pay attention to the
hands, for if they tremble, expect such a patient to have a
flow of blood from his nostrils. You must look at both nostrils, for if the air is drawn equally through them both, and

---

45 Later mss: ἰδὼν AMV.　　46 A adds πάνυ.

47 εἰ om. A.　　48 Ermerins.

49 Kuehlewein: ἐπὴν A: ἐνῇ MV.

50 καὶ om. MV.　　51 V: εὐφορίη AM.

52 MV add κατ᾽ ἀρχὰς.

53 γὰρ om. MV.

φιλέει γίνεσθαι σπασμός· ἢν δὲ σπασμὸς ἐγγένηται
τῷ τοιῷδε, θάνατος προσδόκιμος, καὶ καλῶς ἔχει προλέγειν.

24. (10 L.) Ἢν δὲ ἐν πυρετῷ χειμερινῷ ἡ γλῶσσα
τρηχέη γένηται καὶ ἀψυχίαι ἐνέωσιν, φιλέει τῷ τοιῷδε
καὶ ἐπάνεσις εἶναι τοῦ πυρετοῦ· ἀλλ' ὅμως τὸν τοιόνδε
παραφυλάσσειν τῇ λιμοκτονίῃ καὶ ὑδατοποσίῃ[54] καὶ
μελικρήτου πόσει· καὶ χυλοῖσι παραφύλασσε μηδὲν
πιστεύων τῇ ἀνέσει τῶν πυρετῶν, ὡς οἱ τοιάδε ἔχοντες
σημεῖα ἐπικίνδυνοί εἰσι θνήσκειν. ὁπόταν δὲ ταῦτα
444   συνίδῃς, οὕτω προλέγειν, ἤν | σοι ἀρκέσῃ θεωρήσας
εὖ μάλα.

25. Ὅταν δ' ἐν[55] πυρετοῖσι φοβερόν τι γένηται
πεμπταίοις ἐοῦσιν—⟨ἢν⟩[56] ἡ κοιλίη ἐξαίφνης ὑγρὰ
διαχωρήσῃ καὶ ἀψυχίη ἐγγένηται ἢ ἀφωνίη ὑπολάβῃ
ἢ σπασμώδης γένηται ἢ λυγμώδης—ἐπὶ τούτοισιν
ἀσώδεα φιλέει γίνεσθαι καὶ περὶ ὑπορρίνιον καὶ μέτω-
πον ἱδρῶτες καὶ αὐχένα ὄπισθεν τῆς κεφαλῆς· οἱ δὲ
ταῦτα πάσχοντες θνήσκουσιν πνευματωθέντες οὐκ ἐς
μακρόν.

26. Οἷσι δ' ἐν πυρετοῖσι τὰ σκέλεα γίνεται φυμα-
τώδεα καὶ ἐγχρονιζόμενα μὴ ἐκπεπαίνεται, ἐόντος
ἐν πυρετοῖσι, ἢν καὶ προσπέσῃ πνιγμὸς φάρυγγι
ἰσχνῶν ἐόντων τῶν περὶ τὴν φάρυγγα καὶ μὴ πεπαί-

54 Papyrus (see p. xif.), MV: ὕδατος πόσει ἢ A.
55 ἐν om. A.
56 Later mss.

250

much is carried through, a convulsion is likely; if a convulsion does occur in such a patient, death is to be expected, and it is good to predict this.[5]

24. If, in a winter fever, the tongue becomes rough and the patient loses consciousness, there is likely to be an abatement of the fever. But all the same, protect him by having him avoid food, and drink water and melicrat; also protect him by using juices; put no trust in the disappearance of the fever, for patients with signs like these are in mortal danger. When you see these signs, you should give your prognosis, if after thinking it over it seems a good idea.

25. When, in fevers, something terrible befalls patients on the fifth day—if, for example, the cavity suddenly passes watery stools and the patient loses consciousness, or speechlessness occurs, or he becomes convulsive, or develops hiccups—nausea is likely to come over such patients, and sweating under the nose, between the eyes, and in the neck at the back of the head. Patients that suffer these things become suffocated and before long die.

26. Fevers in which the legs are covered with tubercles that persist and do not come to maturity: while the patient is in such fevers, if, in addition, choking befalls his throat, even though the region about the throat is not swollen, and if it[6] does not come to maturity but instead subsides, it is

---

[5] For the special significance of prognosis in Hippocratic medicine see e.g. *Prognostic* ch. 1 (Loeb *Hippocrates* vol. II, 7).

[6] The syntax of this sentence is unclear. The traditional interpretation has been to understand "tubercles" as the subject of πεπαίνηται and σβεσθῇ. Adams, whom I follow, provides a vague "it" as subject. I also see no compelling reason why τὰ περὶ τὴν φάρυγγα could not be taken as subject.

446 νηται, ἀλλὰ[57] | σβεσθῇ, φιλέει τῷ τοιῷδε αἷμα ῥεῖν ἐκ
ῥινῶν· καὶ ἢν μὲν πολὺ ῥυῇ, λύσιν σημαίνει τῆς
νούσου· ἢν δὲ μή, μακρήν· ὁπόσῳ δ᾽ ἂν ἔλασσον ῥυῇ,
τοσῷδε χεῖρον κατὰ[58] μῆκος. ἢν δὲ τὰ ἄλλα ῥήϊστα
γένηται, προσδέχεσθαι τῷ τοιῷδε ἐς πόδας ἀλγή-
ματα· ἢν δὲ ἅψηται τοῦ ποδὸς καὶ ἐπώδυνος γενόμενος
παραμένῃ πυριφλεγὴς [γενόμενος][59] καὶ μὴ λυθῇ,
κατὰ σμικρὸν ἥξει καὶ ἐς αὐχένα ἀλγήματα καὶ ἐς
κληῖδα καὶ ἐς ὦμον καὶ ἐς στῆθος καὶ ἐς ἄρθρον, καὶ
τοῦτο δεήσει φυματῶδες γενέσθαι· σβεννυμένων δὲ
τούτων ἢν αἱ χεῖρες ἐφέλκωνται ἢ τρομεραὶ γένωνται,
σπασμὸς τὸν τοιόνδε ἐπιλαμβάνει καὶ παραφροσύνη·
καὶ φλυζάκια ἐπὶ τὴν ὀφρὺν καὶ ἐρυθήματα ἴσχει, καὶ
τὸ[60] βλέφαρον τὸ ἕτερον παρὰ τὸ ἕτερον παραβλα-
448 στάνει, καὶ σκληρὴ φλεγ|μονὴ κατέχει, καὶ οἰδέει
ἰσχυρῶς ὁ ὀφθαλμός, καὶ ἡ παραφροσύνη μέγα τι
ἐπιδιδοῖ· αἱ δὲ νύκτες μᾶλλον σημαίνουσιν ἢ αἱ ἡμέραι τὰ περὶ τὴν παραφροσύνην.[61] τὰ δὲ σημεῖα μάλιστα γίνεται ἐπὶ τὸν περισσὸν ἀριθμὸν ἢ ἐπὶ τὸν
ἄρτιον· ἐν ὁποτέρῳ δ᾽ ἂν τῶν ἀριθμῶν τούτων γίνηται,
ὄλεθροι ἐπιγίνονται.

27. Τοὺς τοιούσδε ἢν μὲν ἐξ ἀρχῆς φαρμακεύειν
προαιρῇ, πρὸ τῆς πέμπτης, ἢν βορβορύζῃ ἡ κοιλίη· εἰ
δὲ μή, ἐὰν ἀφαρμάκευτον εἶναι· ἢν δὲ διαβορβορύζῃ
καὶ τὰ ὑποχωρήματα χολώδεα ᾖ, σκαμμωνίῃ ὑποκάθαιρε μετρίως· ἐν δὲ τῇ ἄλλῃ θεραπείῃ ὡς ἐλάχιστα

57 MV: μηδὲ A.   58 Kuehlewein: καὶ AMV.

252

likely in such a patient for blood to flow from the nostrils. If much flows, this indicates resolution of the disease; if not much, that it will be long; the less that flows, the worse it is for the length of the condition. If the other symptoms diminish, expect pains in such a person's feet. If these attack the foot, and it becomes painful and violently inflamed, and remains that way without any resolution, then little by little pains will spread to the neck, collar-bone, shoulder, chest and joints, and this spread will inevitably be accompanied by tubercles. If, when these go down, the hands are dragged behind[7] or tremble, convulsions and derangement befall such a patient; blisters and red patches form on the eyebrow, the one eye-lid sends forth shoots over the other one,[8] a stubborn inflammation prevails, the eye swells massively, and the derangement increases greatly; nights give more evidence of the derangement than do days. These signs occur more on odd-numbered days than on even-numbered ones, but on whichever of these numbers they occur, death follows.

27. If you prefer to give medications to such patients right from the beginning, give them before the fifth day if the cavity rumbles; if it does not rumble, leave the patient without a medication. If the rumbling continues and the stools passed are bilious, clean the patient moderately downwards with scammony. As far as the rest of his treat-

---

[7] I.e. there is a palsy.   [8] No satisfactory interpretation of this passage has yet been given.

---

59 Del. Ermerins.

60 τὸ om. MV.

61 Papyrus, MV: ἀφρ- A.

προσφέρειν ποτὰ καὶ ῥυφήματα ἕως[62] βελτιόνως ἔχῃ,
ἢν μὴ ὑπερβῶσι τὴν τεσσαρεσκαιδεκάτην ἐπανέντες.

28. Ὁπόταν πυρέσσοντι τεσσαρεσκαιδεκαταίῳ ἐόντι ἀφωνίη παραγένηται, οὐ φιλέει λύσις ταχεία οὐδ᾽
ἀπαλλαγὴ τοῦ νοσήματος γίνεσθαι, ἀλλὰ χρόνον τῷ
τοιῷδε σημαίνει· ὁπόταν γὰρ φανῇ ἐπὶ τῇ ἡμέρῃ
ταύτῃ, μακρότερον συμπίπτει. ὅταν πυρέσσοντι τε-
450 ταρταίῳ γλῶσσα | ἐκτεταραγμένα διαλέγηται καὶ ἡ
κοιλίη χολώδεα διαχωρέῃ ὑγρά, φιλέει παραληρεῖν ὁ
τοιόσδε· ἀλλὰ χρὴ παραφυλάσσειν παρεπόμενον τοῖσιν ἀποβαίνουσιν.

29. Θερινῆς καὶ μετοπωρινῆς ὥρης ἐπὶ τῶν ὀξέων
αἵματος ἀπόσταξις ἐξαπίνης συντονίην καὶ πολλὴν
θερμασίην[63] κατὰ φλέβας δηλοῖ καὶ ἐς τὴν ὑστεραίην
λεπτῶν οὔρων ἐπιφάσιας. καὶ ἢν ἀκμάζῃ τῇ ἡλικίῃ
καὶ τὸ σῶμα ἐκ γυμνασίων[64] [ἢ][65] εὐσαρκώσιος ἔχῃ ἢ
μελαγχολικὸς ᾖ ἢ ἐκ πόσιος χεῖρες τρομεραί, καλῶς
ἔχει παραφροσύνην προειπεῖν ἢ σπασμόν. κἢν μὲν ἐν
ἀρτίῃσιν ἐπιγένηται, βέλτιον.[66] ἐν κρίσει δὲ ὀλέθριον,
ἢν μὴ πολὺ ἁλὲς ἀποχυθὲν αἷμα ἐξόδους ποιήσηται |
452 τῆς πλεονεξίης κατὰ ῥῖνας ἢ κατὰ ἕδρην ἐμπλησθεί-
σης[67] ἢ[68] ἀπόστασιν ἢ πόνους ἐν ὑποχονδρίῳ ἢ ἐς
454 ὄρχιν ἢ | ἐς σκέλεα· πεφθέντων δὲ τούτων ἔξοδοι
γίνονται πτυσμῶν παχέων, οὔρων λείων λευκῶν[69] ἔξ-
οδοι.

62 Potter: ὡς A: ἵνα MV.
63 Regenbogen: θεραπείην AMV.

ment is concerned, offer as few drinks and gruels as possible, until the condition improves, if the period of recovery does not extend beyond the fourteenth day.

28. When speechlessness comes over a patient that has had fever for fourteen days, a swift resolution or relief from the disease is not likely to occur; on the contrary, this indicates chronicity in such a patient; for, whenever speechlessness appears on that day, it is present for a longer time. In a patient with fever, when on the fourth day the tongue becomes confounded in speech, and the cavity passes watery bilious stools, such a patient is likely to rave; indeed, you must watch carefully what follows these events.

29. During the summer and fall a sudden nose-bleed during acute diseases indicates tension, great heat through the vessels, and that thin urines will appear towards the next day. If the patient is in the prime of life, and his body is in good condition from exercises, or if he is subject to dark bile, or if his hands tremble from drink, it is good to predict derangement or convulsions. If this happens on even days, it is better, but during a crisis it is a fatal sign, unless much blood, being shed in a mass, brings about an exit through the nostrils or anus of the excess that has built up through accumulation, or gives rise to an abscess, or to pains in the hypochondrium or radiating to a testicle or the legs. When these diseases come to maturity, there are discharges of thick sputa and of thin white urines.

---

64 A adds ᾖ.   65 Del. Joly.
66 Later mss: βελτίω AMV.
67 AMV: ἐμποιήσῃ Ermerins after Galen.
68 ᾖ om. A.   69 MV: λεπτῶν λείων A.

30. Πυρετῷ λυγγώδει· ὀπὸν σιλφίου, ὀξύμελι, δαῦ-
κον τρίψας πιεῖν δίδου καὶ χαλβάνην ἐν μέλιτι καὶ
456 κύμινον ἐκλεικτόν, καὶ χυλὸν | πτισάνης ἐπὶ τούτοισι
ῥυφεῖν. ἄφυκτος δὲ ὁ τοιοῦτος, ἢν μὴ ἱδρῶτες κριτικοὶ
καὶ ὕπνοι ὁμαλοὶ ἐγγίνωνται καὶ οὖρα παχέα καὶ
δριμέα καταδράμῃ ἢ ἐς ἀποστήματα καταστηρίξῃ.[70]
κόκκαλος καὶ σμύρνα ἐκλεικτόν· πίνειν δὲ τοῖσι τοιού-
τοις ὀξύμελι δίδου ὡς ἐλάχιστον· ἢν δὲ διψώδης ᾖ[71]
σφόδρα, τοῦ κριθίνου[72] ὕδατος.

31. (11 L.) Τὰ δ' ἐν πνεύμονι καὶ πλευρίτιδι[73] ὧδε
χρὴ σκέπτεσθαι· ἢν ὀξὺς ὁ πυρετὸς ᾖ καὶ τὰ ὀδυνή-
ματα τοῦ ἑτέρου πλευροῦ ἢ καὶ ἀμφοῖν, καὶ τοῦ
458 πνεύματος δὲ | ἀναφερομένου ἢν πονῇ καὶ βῆχες
ἐνέωσιν καὶ τὰ πτύαλα πτύῃ πυρρὰ ἢ πελιδνὰ ἢ καὶ
λεπτὰ ᾖ[74] καὶ ἀφρώδεα καὶ ἀνθηρά, καὶ εἴ τι ἄλλο
διαφέρον ἔχοι παρὰ τὰ μεμαθηκότα, τούτοισιν οὕτω
χρὴ διάγειν· ἢν μὲν ἡ ὀδύνη ἄνω περαίνῃ πρὸς κληῗδα
ἢ περὶ μαζὸν ἢ ἐν βραχίονι, τάμνειν χρὴ τὴν ἐν τῷ
βραχίονι φλέβα τὴν ἔσω, τὴν[75] ἐφ' ὁπότερον ἂν ᾖ τῶν
μερέων, κατὰ τόδε· ἀφαιρέειν δὲ κατὰ τὴν τοῦ σώματος ἕξιν καὶ ὥρην καὶ ἡλικίην καὶ χροιὴν πλέον καὶ
460 θρασέων, ἢν ὀξὺ τὸ ἄλγημα ἐνῇ, ἄγειν πρὸς | λειπο-
ψυχίην· ἔπειτα κλύζειν μετὰ τοῦτο. ἢν δὲ ὑποκάτω
τοῦ θώρηκος τὸ ἄλγημα ᾖ καὶ συντείνῃ πλείω[76] τῷ
πλευριτικῷ, τὴν κοιλίην ὑποκάθαιρε· μεσηγὺ δὲ τῆς
καθάρσιος μηδὲν δίδου· μετὰ κάθαρσιν δὲ ὀξύμελι.

[70] A: ἀπόστασιν στηρίξῃ MV.

256

30. For a fever with hiccups: knead together silphium juice, oxymel and dauke, and give this to drink; also all-heal juice in honey, a lozenge of cummin, and after that barley-water gruel. Such a patient does not escape unless critical sweats and regular sleep occur, and thick sharp urines come down, or unless the disease settles into abscesses. A lozenge of pine-cone and myrrh. Give such patients a very little oxymel to drink; if one happens to be very thirsty, give a little barley-water.

31. Conditions in the lung and pleurisy you must evaluate thus: if the fever is high and there are pains in one or both sides, if the patient suffers pain when he draws his breath, if he has a cough and it produces yellow-brown, livid, or thin frothy bright-coloured sputa, and if the patient is different in other ways from what is normal, you must proceed as follows with such patients: if the pain radiates upwards to the collar-bone, or is located about the breast or in the arm, you must incise the inner vessel of the arm on the affected side as follows: draw off, according to the condition of the body, the season, and the patient's age and colour, a considerable amount, and even venture, if intense pain is present, to continue until the loss of consciousness; afterwards administer an enema. If there is pain below the thorax in a patient with pleurisy and it becomes more intense, clean the cavity downwards; while cleaning is taking place, give nothing, but afterwards give

---

71 A: δυψώδεις ἔωσι MV.
72 Later mss: κριθίου AM: κριθείου V.
73 ἐν . . . π. A: περιπλευμονικὰ καὶ πλευριτικὰ MV.
74 ἢ om. MV.   75 τὴν om. MV.
76 A: λίην MV.

φαρμακεύειν δὲ τεταρταῖον· τὰς δὲ ἐξ ἀρχῆς τρεῖς
ὑποκλύζειν, κἢν μὴ κουφίζῃ, οὕτω δ' ὑποκάθαιρε.
φυλακὴ δὲ ἔστω ἕως ἀπυρέτου καὶ ἑβδόμης. εἶτα, ἢν
ἀσφαλὴς ἐὼν φαίνηται, οὕτω χυλῷ ὀλίγῳ καὶ λεπτῷ
τὸ πρῶτον καὶ μέλιτι μίσγων δίδου. ἢν δὲ ἀνάγῃ τε
ῥηϊδίως καὶ εὔπνους ᾖ καὶ ἀνώδυνος τὰ πλευρὰ καὶ
ἀπύρετος,[77] κατὰ σμικρὸν παχυτέρῳ τε καὶ πλείονι καὶ
δὶς τῆς ἡμέρης. ἢν δὲ μὴ ῥηϊδίως ἀπαλλάσσῃ, ἔλασ-
462 σόν τε τὸ πόμα καὶ τὸ ῥύφημα, ὀλίγον χυλὸν λεπτὸν
καὶ ἅπαξ, ἐν ὁποτέρῃ ἂν ὥρῃ βέλτιον διάγῃ· γνώσῃ δ'
ἐκ τῶν οὔρων.

32. Δεῖ δὲ τὸ ῥύφημα προσφέρειν τοῖσιν ἐκ τῶν
νοσημάτων μὴ πρότερον ἢ πέπονα τὰ οὖρα ἢ
πτύσματα ἴδῃς γεγενημένα. ἢν δὲ[78] φαρμακευθεὶς
συχνὰ καθαρθῇ, ἀναγκαῖον διδόναι, ἔλασσον δὲ καὶ
λεπτότερον· οὐ γὰρ δυνήσεται ὑπὸ κενεαγγίης ὑπνώσ-
σειν οὐδὲ πέσσειν ὁμοίως οὐδὲ τὰς κρίσιας ὑπομένειν·
ἀλλ' ἐπειδὰν συντήξιες ὤμων φαίνωνται καὶ τὰ ἀντέχοντα ἀποβάλῃ, ἀνθέξει οὐδέν.

Πέπονα δ' ἐστὶν τὰ μὲν πτύαλα, ὁπόταν γένηται
464 ὁμοῖα τῷ πύῳ, τὰ δὲ οὖρα | τὰς ὑποστάσιας ὑπερ-
ύθρους ἔχοντα, ὁποῖον ὀρόβων.

33. Οὐδὲν δὲ κωλύει καὶ πρὸς τὰ ἄλλα ἀλγήματα
τῶν πλευρέων χλιάσματα προστιθέναι καὶ κηρώματα·
ἀλείφειν δὲ σκέλεα καὶ ὀσφῦν θερμῷ καὶ λίπος ἐγ-
καταλείφειν· ἐπὶ δὲ ὑποχόνδρια λίνου σπέρμα καταπλάσσειν ἕως μαζῶν.

oxymel. Give a medication on the fourth day; on the first three days from the beginning use enemas, but if these do not lighten the patient, clean him downwards with a medication. Observe carefully until the disappearance of fever, and for seven days. Then, if the patient appears to be safe, first give a little thin juice with honey; if he expectorates easily and breathes freely, and if he has no pains in his sides and no fever, gradually give more juice and of a thicker consistency, twice daily. But if the patient is not recovering easily, the drink is to be less in amount and the gruel a little thin juice once a day, at whatever time the patient is doing better. You will recognize this from his urines.

32. You must not administer gruel to patients recovering from diseases before you see that their urines or sputa have become mature. However, if a patient has been cleaned out often by the use of medications, gruel must be given, though less in amount and thinner in consistency; for otherwise, on account of his emptiness, he will not be able to sleep, to bring the disease to maturity, or to withstand the crises, but when wasting appears in his shoulders and he loses his resistance, he will not hold out at all.

Sputa are mature when they become like pus; urines, when they have reddish sediments like vetch-meal.

33. Nothing prevents applying fomentations and wax-salves against other pains in the sides as well; anoint the legs and lumbar region with warm oil, and rub in animal fat; apply linseed plasters to the hypochondrium up as far as the breasts.

---

77 καὶ ἀπύρετος om. MV.
78 A adds μὴ.

Ἀκμαζούσης δὲ τῆς περιπνευμονίης ἀβοήθετον μὴ
ἀνακαθαιρομένου, καὶ πονηρόν, ἢν δύσπνους ᾖ καὶ
οὖρα λεπτὰ καὶ δριμέα καὶ ἱδρῶτες περὶ τράχηλον καὶ
κεφαλὴν γίνωνται. οἱ τοιοίδε ἱδρῶτες πονηροί, ὑπὸ
πνιγμοῦ καὶ ῥωχμῆς καὶ βίης ἐπικρατεόντων τῶν
νοσημάτων, ἢν μὴ οὖρα παχέα καὶ πολλὰ ὁρμήσῃ καὶ
πτύσματα πέπονα ἔλθῃ. ὅ τι δ’ ἂν τούτων αὐτοματίσῃ,
λύσει τὸ νόσημα.

466   34. Περιπνευμονίης ἐκλεικτόν· χαλβάνη | καὶ κόκ-
καλος ἐν μέλιτι Ἀττικῷ· καὶ ἀβρότονον ἐν ὀξυμέλιτι
πιεῖν <καὶ>⁷⁹ πέπερι. ἐλλέβορον μέλανα ἀποζέσας
πλευριτικῷ ἐν ἀρχῇσι περιωδύνῳ ἐόντι δίδου. ἀγαθὸν
δὲ καὶ τὸ πάνακες ἐν ὀξυμέλιτι ἀναζέσαντα καὶ διη-
θέοντα διδόναι πίνειν καὶ ἡπατικοῖσι καὶ τῇσιν ἀπὸ
τῶν φρενῶν περιωδυνίῃσι.

Καὶ ὅσα δὴ⁸⁰ ἐς κοιλίην καὶ ἐς οὔρησιν, ἐν οἴνῳ καὶ
468   ἐν μέλιτι, | τὰ δ’ ἐς κοιλίην ξὺν ὑδαρεῖ μελικρήτῳ
πίνειν πλεῖον δίδου.

35. (12 L.) Δυσεντερίη ἀπόστημα ἢ ἔπαρμά τι
παυσαμένη ποιήσει, ἢν μὴ ἐς πυρετοὺς ᾖ⁸¹ ἱδρῶτας ἢ⁸¹
οὖρα παχέα καὶ λευκὰ⁸² ἐπιφανῇ ἢ ἐς τριταίους ἢ ἐς
κιρσόν, ἢ ἐς ὄρχιν ἢ ἐς σκέλεα⁸³ ἢ ἐς ἰσχία στηρίξῃ ἡ
ὀδύνη.

36. (13 L.) Ἐν πυρετῷ χολώδει πρὸ τῆς ἑβδόμης
μετὰ ῥίγεος ἴκτερος ἐπιγενόμενος λύει τὸν πυρετόν·
ἄνευ δὲ ῥίγεος ἢν ἐπιγένηται ἔξω τῶν καιρῶν, ὀλέθριον.

When pneumonia is at its height, it is incurable if the patient is not well cleaned out, and bodes ill if he has difficulty breathing, if his urines are thin and sharp, and if there is sweating about his neck and head. Sweats of this kind indicate a bad outcome, with the diseases overcoming by suffocation, wheezing, and violence, unless copious thick urines begin and mature sputa come; if either of these occurs spontaneously, it will resolve the disease.

34. Lozenge for pneumonia: all-heal juice and pinecone in Attic honey; also have the patient drink southernwood in oxymel, and pepper. Boil off black hellebore, and give it to the patient with pleurisy, at the beginning, when he is in great pain. It is also good to boil up all-heal in oxymel, and, sieving it, to give it to drink both to patients with liver complaints and to those with severe pains arising from the diaphragm.

Medications that act on the cavity and promote urine give to drink in wine and honey; those for the cavity must be given in greater amounts and with dilute melicrat.

35. Dysentery, when it ends, will give rise to an abscess or some swelling, unless it appears in fevers, in sweats, in thick white urines, in tertians, or in a varix, or unless a pain settles in a testicle, in the legs, or in the hips.

36. In a bilious fever, a jaundice with chills occurring before the seventh day resolves the fever; but if it occurs without chills outside this favourable period, it is a mortal sign.

---

79 Kuehlewein.   80 Ermerins: δεῖ AMV.
81 Kuehlewein: καὶ AMV.
82 MV add καὶ λεῖα.
83 A adds ἀλγήματα.

37. (14 L.) Τετάνου δὲ ὀσφύος καὶ ἀπὸ μελαγχολι-
κῶν διὰ φλεβῶν πνευμάτων ἀπολήψιες ὅταν ἔωσι,
470 φλε|βοτομίη ῥύεται. ὅταν δ' ἀπὸ τῶν τενόντων σφο-
δρῶς ἔμπροσθεν ἀντισπῶνται καὶ ἱδρῶτες περὶ τράχηλον καὶ πρόσωπον, ὑπὸ τοῦ πόνου δακνομένων καὶ
ξηραινομένων τῶν τενόντων τῶν ὀρρωδέων—οἱ παχύ-
τατοι τὴν ῥάχιν συνέχουσιν, ᾗ οἱ μέγιστοι σύνδεσμοι
καταπεφυκότες [ἕως][84] ἐς πόδας ἀποτελευτῶσι—τῷ
τοιῷδε, ἢν μὴ πυρετὸς ἐπιγένηται καὶ ὕπνος καὶ τὰ
ἑπόμενα οὖρα πέψιν ἔχοντα ἔλθῃ καὶ ἱδρῶτες κριτικοί,
πίνειν <δίδου>[85] οἶνον κιρρὸν οἰνώδεα καὶ ἄλητον
ἑφθὸν ἐσθίειν· καὶ κηρωτῇ ἀλείφειν καὶ ἐγχρίειν, τά τε
σκέλεα περιελίσσειν ἕως τῶν ποδῶν, θερμῷ προβρέ-
χων ἐν σκάφῃ, καὶ βραχίονας[86] κατελίσσειν, καὶ
ὀσφῦν ἀπὸ τοῦ τραχήλου ἕως τῶν ἰσχίων, λάσιον[87] |
472 ἐγκηρώσας, ὅπως καὶ τὰ ἔμπροσθεν περιέξει· καὶ
διαλιπὼν πυρία τοῖσιν ἀσκίοισι, θερμὸν ὕδωρ ἐγχέων,
καὶ περιτείνων σινδόνιον ἐπανάκλινε αὐτόν. (38.) κοιλίην δὲ μὴ λύσῃς, ἢν μὴ βαλάνῳ, ἢν πολὺς χρόνος ᾖ
ἀδιαχωρήτῳ ἐούσῃ. καὶ ἢν μέν τί σοι ἐπιδιδῷ ἐπὶ τὸ |
474 βέλτιον· εἰ δὲ μή, τοῦ μάδου τῆς ῥίζης τρίβων ἐν οἴνῳ
εὐώδει καὶ τοῦ δαύκου πίνειν δίδου πρωῒ νῆστι πρὸ
τοῦ βρέχειν, καὶ τάχα ἐπὶ τούτοισιν τὸ ἄλευρον ἑφθὸν
χλιερὸν ἐσθιέτω ὡς πλεῖστον καὶ οἶνον, ὅταν βούληται, εὔκρητον ἐπιπινέτω. καὶ ἢν μέν σοι ἐπιδιδῷ ἐπὶ τὸ
βέλτιον· εἰ δὲ μή, προλέγειν.

[84] Del. Kuehlewein.   [85] Ermerins.   [86] MV add ἕως

262

37. Phlebotomy relieves lumbar tetanus, and also from melancholic airs in the vessels, when there are stoppages. But when the patients are drawn excessively forward by their tendons and they sweat about the neck and face, the tendons of the lower back being irritated and dried out by the stress—that is, the thickest ones, which hold the spine together where the very large ligaments terminate that go down to the feet—to such a patient, unless fever and sleep supervene, and the urines that follow arrive in a state of maturity, or unless there are critical sweats, give strong light-coloured wine to drink, and boiled flour to eat. Smear and anoint with wax-salves: bind the legs as far down as the feet, after first soaking them in a warm basin, and wrap the arms and the back from the neck to the hips with a rough piece of cloth rubbed in wax, in such a way that the bandage also encloses the anterior region. Apply fomentations now and then with small leather skins into which you have poured hot water; cover the patient with linen, and have him lie down. (38.) Do not open the cavity except with a suppository, if it has been closed for a long time. If the condition changes for the better, as the result of your attention, fine; if not, grate bryony root and dauke into sweet-smelling wine, and give this to the patient to drink in the fasting state, early in the morning before any water is applied externally; immediately after that let him eat as much warm boiled meal as he can, and then drink wine well mixed with water whenever he wishes. If the condition improves with your measures, fine; if not, predict accordingly.

δακτύλων.   [87] Littré, from Erotian (Nachmanson p. 58) and Galen (Kühn XIX. 117): ἐσθίονον A: σίαλον MV.

39. (15 L.) Τὰ δὲ νοσήματα πάντα λύεται ἢ κατὰ στόμα ἢ κατὰ κοιλίην ἢ κατὰ κύστιν·[88] ἡ δὲ τοῦ ἱδρῶτος ἰδέη κοινὸν ἁπάντων.

40. (16 L.) Ἑλλεβορίζειν δὲ χρὴ οἷς ἀπὸ κεφαλῆς φέρεται ῥεῦμα· ὅσοι δὲ ἐξ ἀποστημάτων ἢ φλεβορρα- γίης ἢ δι᾽ ἀκρησίην ἢ δι᾽ ἄλλην τινὰ ἰσχυρὴν αἰτίην ἔμπυοι γίνονται, μὴ δίδου ἐλλέβορον·[89] οὐδὲν γὰρ |
476 ὠφελήσει, καὶ ἤν τι πάθῃ, αἴτιος δόξει εἶναι ὁ ἐλλέβο- ρος. ἢν δὲ διαλύηται τὸ σῶμα ἢ πόνος ἐν κεφαλῇ ᾖ ἢ ἐμπεπλασμένα τὰ ὦτα ἢ ῥίς, ἢ πτυαλισμὸς ἢ γου- νάτων βάρος ἢ σώματος ὄγκος παρὰ τὸ ἔθος, ὅ τι ἂν συμβαίνῃ μήτε ὑπὸ πότων μήτε ὑπὸ ἀφροδισίων μήτε ὑπὸ λύπης μήτε ὑπὸ φροντίδων μήτε ἀγρυπνιῶν· ἢν μέν τι τούτων ἔχῃ αἴτιον, πρὸς τοῦτο ποιέεσθαι τὴν θεραπείην.

41. (17 L.) Τὰ δ᾽ ἐκ πορείης ἀλγήματα πλευρέων, νώτου, ὀσφύος, ἰσχίων, καὶ ὅσα ἀναπνέοντες ἀλγέ- ουσι πρόφασιν ἔχοντες· πολλάκις γὰρ[90] μεμάθηκε
478 φοιτᾶν ἐκ κραι|παλέων καὶ βρωμάτων φυσωδέων ἀλ- γήματα καὶ ἐς ὀσφῦν καὶ ἐς ἰσχία· οἷσι δ᾽ ἂν ᾖ αὐτῶν τοιάδε, δυσουρέεται. τούτων δὲ πορείη αἰτίη καὶ κορυ- ζέων καὶ βράγχων.

42. (18 L.) Ὅσα δὲ ἀπὸ διαιτημάτων, τὰ μὲν πολλά, ἕκαστος ὡς ἂν παρὰ τὸ ἔθος διαιτηθῇ μάλιστα, ἐπι-

---

[88] MV add ἢ τινος ἄλλου τοιοῦδε ἄρθρου.
[89] MV add τοῖσι τ(οι)ουτέοισι.
[90] πολλάκις γὰρ om. A.

39. All diseases are resolved through either the mouth, the cavity, or the bladder; sweating is a form of resolution common to them all.

40. You must give hellebore to those in whom a flux from the head has occurred; but to those that suppurate internally from abscesses or from the bursting of a vessel, or because of intemperance, or because of any other potent condition, do not give hellebore; for it will do no good, and if anything happens to the patient, the hellebore will seem to have been to blame. If the body is weakened, if there is pain in the head, if the ears or the nose are stopped, or if there is ptyalism, heaviness of the knees, or fullness of the body beyond what is normal, give hellebore,[9] provided that the condition is not the result of drink, venery, grief, anxiety or sleeplessness; if the condition has one of these as its cause, let the treatment be directed against that.

41. Pains of the sides, back, loins and hips from walking, and pains connected with inspiration that patients suffer for some obvious reason; for pains regularly invade the lower back and hips as the result of drunkenness and flatulent foods: in whichever patients such things occur, there is dysuria. Walking is to blame for these things, and also for coryzas and sore throats.

42. Many effects resulting from the conduct of life show themselves particularly when a person conducts his life otherwise than is his habit. Thus, for example, if those

[9] Galen, in his commentary (Kühn XV. 867 = CMG V 9, 1 p. 339), indicates that in this passage the author advises the administration of hellebore; thus I understand δίδου ἐλλέβορον.

σημαίνει. καὶ γὰρ ὅσοι ἂν μὴ μεμαθηκότες ἀριστᾶν[91]
ἀριστήσωσιν, ὄγκος πολὺς αὐτοῖσιν τῆς γαστρὸς καὶ
νυσταγμὸς καὶ πληθώρη· ἢν δὲ δειπνήσωσι,[92] κοιλίη
ἐκταράσσεται. ξυμφέροι[93] δ' ἂν τούτοισιν ἐκλουσαμέ-
νοισι καθεύδειν· κοιμηθέντας δὲ περιπατῆσαι βρα-
δέως συχνὴν περίοδον. καὶ ἢν μὲν λαπαχθῇ, δειπνῆ-
σαι καὶ πιεῖν οἶνον ἐλάσσονα ἀκρητέστερον· ἢν δὲ μὴ
λαπαχθῇ, ὑποχρίσασθαι τὸ σῶμα θερμῷ. καὶ ἢν
διψῇ, ὑδαρέα οἶνον γλυκὺν ἢ λευκὸν ἐπιπιόντα ἀνα-
παύεσθαι· ἢν δὲ μὴ ἐγκοιμηθῇ, πλείω ἀναπαύεσθαι.[94]
τὰ δ' ἄλλα ὁμοίως τοῖς ἐκ κραιπάλης διαιτάσθω.

480   43. Τὰ δὲ | ἀπὸ πομάτων· ὅσα μὲν ὑδαρέα, βραδύ-
πορά[95] ἐστι καὶ ἐγκυκλεῖται καὶ ἐπιπολάζει περὶ ὑποχόνδρια καὶ ἐς οὔρησιν οὐ κατατρέχει. τοιούτου δὲ
πόματος πληρωθεὶς μηδὲν ἔργον ὀξέως διαπρήξῃ,
ὁπόσα τῷ σώματι συνταθέντι ἢ βίῃ ἢ τάχει πονεῖν
συμβαίνει· ὡς μάλιστα δὲ ἡσυχαζέτω, μέχρι κατα-
πεφθῇ μετὰ τῶν σιτίων. ὁπόσα δὲ τῶν πομάτων ἀκρη-
τέστερά ἐστιν ἢ αὐστηρότερα, παλμὸν ἐν τῷ σώματι
καὶ σφυγμὸν ἐν τῇ κεφαλῇ ἐμποιέει. τούτοισι καλῶς
ἔχει[96] ἐπικοιμᾶσθαι καὶ θερμόν τι ἐπιρρυφεῖν, πρὸς ὅ
τι μάλιστα ἥδιστα ἔχουσιν. νηστείη δὲ πονηρὸν πρὸς
τὴν κεφαλαλγίην καὶ κραιπάλην.

482   44. Ὁπόσοι δὲ μονοσιτεῦσι, | κενοὶ καὶ ἀδύνατοί

---

[91] MV add ἤν.   [92] A: ἐπι- MV.
[93] Kuehlewein: ξυμφέρει AMV.   [94] ἢν δὲ . . . ἀ. om. A.
[95] A: -πορώτερά MV.   [96] MV: ἔχοι A.

not used to taking breakfast do so, they experience a great heaviness of the belly, drowsiness, and fullness, and if they take dinner, their cavity is set in motion. It would benefit such patients to have an enema and to go to bed, and after sleeping to go for a long slow walk. If the patient is emptied, let him dine, and drink a smallish amount of nearly neat wine; but if he is not emptied, anoint his body with warm oil; if he is thirsty, let him drink dilute sweet or white wine, and rest; if he does not sleep, let him rest even longer. The remainder of his treatment is to be conducted as for those suffering the after-effects of drunkenness.

43. The effects of beverages: those that are aqueous advance slowly, are enclosed, and remain high up about the hypochondrium; they do not run down to pass off as urine. When he is filled with such a drink, let the person not soon perform any physical task of the kind that cause the body to labour, strained by exertion or speed, but rest as much as possible until the drink is digested with his food. Beverages that are more concentrated or harsher provoke palpitation in the body and throbbing in the head. It is good for persons in this case to sleep, and afterwards to drink some kind of warm gruel, whichever kind they like best. Fasting is harmful for headache and the after-effects of drunkenness.

44. Those who eat only once a day[10] become exhausted

[10] Understand "contrary to their usual habit" (see beginning of ch. 42 above). This chapter and chapter 42 share much with *Regimen in Acute Diseases* 28–33 and *Ancient Medicine* 10.

ΠΕΡΙ ΔΙΑΙΤΗΣ ΟΞΕΩΝ (ΝΟΘΑ)

εἰσι καὶ οὐρέουσι θερμὸν παρὰ τὸ ἔθος κενεαγγέοντες.
γίνεται δὲ καὶ τὸ στόμα ἁλμυρὸν ἢ[97] καὶ πικρόν, καὶ
τρέμουσιν ἐν παντὶ ἔργῳ καὶ κροτάφους ἐπισυν-
τείνονται καὶ τὸ δεῖπνον οὐ δύνανται πέσσειν, ὅπως
περ ἦν[98] ἠριστηκότες ἔωσιν. τούτους δὲ χρὴ δειπνεῖν
ἔλασσον ἢ μεμαθήκασι καὶ ὑγροτέραν μᾶζαν ἀντὶ
ἄρτου καὶ λαχάνων λάπαθον ἢ μολόχην[99] ἢ πτισάνην
ἢ τεῦτλα. πίνειν δὲ κατὰ τὸ σιτίον οἶνον, ὅσον σύμ-
μετρον, καὶ ὑδαρέστερον καὶ ἀπὸ δείπνου περιπατῆσαι ὀλίγον, ἕως οὖρα καταδράμῃ καὶ οὐρήσῃ. χρήσθω δὲ καὶ ἰχθύσιν ἐφθοῖσιν.

45. Βρώματα δὲ μάλιστα ἐπισημαίνει· σκόροδον
φῦσαν καὶ θέρμην περὶ τὸν θώρηκα καὶ κεφαλῆς
βάρος καὶ ἄσην, καὶ εἴ τι ἄλλο ἄλγημα εἴη μεμαθη-
κὼς πρόσθεν, παροξύνειεν | ἄν· οὐρητικὸν δέ, καὶ τοῦτ'
ἔχει ἀγαθόν· ἄριστον δ' αὐτοῦ φαγεῖν μέλλοντι ἐς
πόσιν ἰέναι ἢ μεθύοντι.

46. Τυρὸς δὲ φῦσαν καὶ στεγνότητα <καὶ>[100] σιτίων
ἔξαψιν ποιέει, τό τ'[101] ὠμὸν καὶ ἄπεπτον, κάκιστον δὲ
ἐν ποτῷ φαγεῖν πεπληρωμένοισιν.

47. Ὄσπρια δὲ πάντα φυσώδεα, καὶ ὠμὰ καὶ ἐφθὰ
καὶ πεφρυγμένα, ἥκιστα δὲ βεβρεγμένα ἢ | χλωρά.
τούτοισι δὲ μὴ χρῆσθαι, εἰ μὴ μετὰ σιτίων. ἔχει δὲ καὶ
ἰδίας μοχθηρίας ἕκαστον αὐτῶν. ἐρέβινθος μὲν φῦσαν, <καὶ>[102] ὠμὸς καὶ πεφρυγμένος, καὶ πόνον ἐμποι-

484

486

---

97 ἢ om. MV. 98 ἦν om. A.

268

and weak, and pass warm urine on account of their abnormal emptiness. Their mouth becomes salty, or even bitter, they tremble in every activity, have a feeling of tightness in their temples, and are unable to digest their dinner as they would have if they had had a breakfast. These persons must eat less at dinner than they are used to, replace bread with quite moist barley-cake, and of vegetables have dock, mallow, peeled barley or beets. With their food, let them drink wine in a reasonable amount and quite dilute, and after dinner walk a little until urine runs down and is passed. Let the person also eat boiled fish.

45. Foods, in most cases, show their effects. Garlic, for example, produces flatus and heat about the thorax, heaviness of the head, and nausea, and, if any other pain was habitually present before, garlic will exacerbate it; it is diuretic, and this is of advantage; it is best of all to eat some if one is about to go drinking or is already drunk.

46. Cheese produces flatus, constipation, and heating of the other foods, and also gives rise to raw and undigested substances. The worst thing is for those already full to eat it together with their drink.

47. All pulses produce flatulence, whether they are raw, boiled or roasted, but least when steeped in water, or green. They are not to be employed except together with other foods. Also, each one of them has its own particular dangers. The chick-pea, both raw and when roasted, pro-

---

99 A: μαλάχην MV.
100 Later mss.
101 Later mss: τὸ δ' A om. MV.
102 Ermerins.

έει· φακὸς δὲ στύφει καὶ ἄραδον ἐμποιέει, ἢν μετὰ τοῦ
φλοιοῦ ᾖ. θέρμος δὲ τούτων ἥκιστα[103] κακὰ ἔχει.

48. Σίλφιον δὲ καὶ ὀπός· ἔστι μὲν οἷσι μάλιστα,
488 τοῖσι δὲ | ἀπείροις οὐ διέρχεται ἡ κοιλία,[104] ἀλλὰ
καλέεται ξηρὴ χολέρη. μάλιστα δὲ γίνεται, ἢν μετὰ
πολλοῦ τυροῦ μειχθῇ ἢ κρεηφαγίης κρεῶν βοείων· τὰ
μὲν γὰρ μελαγχολικὰ παροξυνθείη ἂν παθήματα ὑπὸ
βοείων κρεῶν· ἀνυπέρβλητος γὰρ ἡ φύσις αὐτῶν, καὶ
490 οὐ τῆς τυχούσης | κοιλίης καταπέψαι. βέλτιστα δ' ἂν
ἀπαλλάξαιεν, εἰ διέφθοισί τε χρέοιντο καὶ ὡς παλαι-
οτάτοισιν.[105]

49. Αἴγεια δὲ κρέα ὅσα τε ἐν βοείοις ἔνι κακὰ πάντ'
492 ἔχει τήν τε | ἀπεψίην, καὶ φυσωδέστερα καὶ ἐρευγμα-
τώδεα καὶ χολέρην ποιέει. ἔστι δὲ τὰ εὐωδέστατα[106]
καὶ ἥδιστα. ταῦτα ἄριστα δίεφθα καὶ ψυχρά· τὰ δ'
ἀηδέστερα δυσώδεα καὶ σκληρά· ταῦτα κάκιστα. καὶ
τὰ πρόσφατα βέλτιστα[107] δ' ἐστὶ τῇ θερινῇ, μετοπώρου δὲ κάκιστα.

50. Χοίρου δὲ πονηρά, ὅταν ᾖ ἐνωμότερα ἢ περι-
καῆ· χολερώδεα δ' ἂν εἴη καὶ ταρακτικά.[108] ὕεια δὲ
494 βέλτιστα τῶν κρεῶν | πάντων· κράτιστα δὲ τὰ μήτε
ἰσχυρῶς πίονα μήτε λεπτὰ μήτε ἡλικίην παλαιοῦ
ἱερείου. ἐσθίειν δὲ ἄνευ τῆς φορίνης καὶ ὑπόψυχρα.

---

[103] τ. ᾖ. MV: ᾖ. τ. ἐλάχιστα A.
[104] AMV: τῇ -ίῃ Galen.
[105] AMV: ἀπαλωτάτοισιν Coray in Littré.
[106] MV: -τερα στερεὰ A.
[107] MV: βέλτιον A.

duces flatulence and pain; the lentil contracts and is laxative, if it has its hull. The lupin is the least injurious of the pulses.

48. Silphium and its juice pass through the cavity very well in some persons, but not in others, who are unused to them, and produce what is called dry cholera. This occurs especially if the silphium is mixed with much cheese or eaten with beef, since melancholic affections are aggravated by beef, owing to its obdurate nature, and to the fact that it is not digested by just any cavity. The sufferer would be best relieved if he were to eat beef thoroughly boiled and well-aged.

49. Goat's meat possesses all the same disadvantages and the indigestibility of beef, but is more flatulent and produces belching and cholera. The most fragrant is also the most wholesome; this is best well-boiled and cold. The less pleasant kind is foul-smelling and tough; this is the worst kind.[11] Fresh goat's meat is best in summer, worst in autumn.

50. The meat of young pig is injurious when it is either too raw or scorched, since then it is likely to produce cholera and to set the cavity in motion. Pork is the best of all meats; the most nutritious is that which is neither very fat nor very lean, and which has not the age of an old slaughter-animal; eat it without the skin, and slightly cooled.

---

[11] The traditional punctuation connects ταῦτα κάκιστα with what follows rather than with what precedes, e.g. Joly: *la plus mauvaise est la plus fraîche*.

---

108 A: ἐκ- MV.

51. (19 L.) Χολέρης δὲ ξηρῆς ἡ γαστὴρ πεφύσηται, καὶ ψόφοι ἔνεισι καὶ ὀδύνη πλευρέων καὶ ὀσφύος, διαχωρέει δ' οὐδὲν κάτω, ἀλλ' ἀπεστέγνωται. τὸν τοιόνδε διαφύλαξον, ὅπως μὴ ἐμέεται, ἀλλὰ κοιλίη ὑπελεύσεται. κλύσον οὖν ὅτι τάχιστα[109] θερμῷ καὶ ὡς λιπαρωτάτῳ· καὶ ἐς ὕδωρ ἀλείφων ὡς πλείστῳ κάθιε θερμόν, ἐν σκάφῃ κατακλίνων, καὶ τοῦ θερμοῦ παρ-
496 άχει κατὰ σμικρόν·[110] καὶ ἢν | θερμαινομένῳ αὐτῷ ἡ κοιλίη ὑπίῃ λέλυται. ξυμφέρει δὲ καὶ ἐγκοιμᾶσθαι τῷ τοιῷδε καὶ πίνειν οἶνον λεπτὸν καὶ παλαιὸν καὶ ἀκρη-τέστερον· καὶ ἔλαιον δίδου, †ὥστε ἡσυχίη καὶ ἡ κοι-λίη ὑπίῃ, καὶ λέλυται·† σίτων δὲ καὶ τῶν ἄλλων ἀπεχέσθω. ἢν δὲ μὴ ἀνῇ[111] ὁ πόνος, ὄνου γάλα δίδου πίνειν ὅπως[112] καθαρθῇ. ἢν δὲ ὑγρὴ ᾖ ἡ κοιλίη καὶ χολὴ ὑποχωρῇ, καὶ στρόφοι καὶ ἔμετοι καὶ πνιγμοὶ καὶ δηγμοί, τούτοισι δὴ[113] κράτιστον ἀτρεμίζειν· πί-νειν δὲ μελίκρητον καὶ μὴ ἐξεμέειν.

52. (20 L.) Ὑδρώπων δύο μὲν φύσιες, ὧν ὁ μὲν ὁ ὑπὸ
498 τῇ σαρκὶ | ἐγχειρέων γίνεσθαι ἄφυκτος, ὁ δὲ μετ' ἐμφυσημάτων πολλῆς εὐτυχίης δεόμενος, μάλιστα μὲν ταλαιπωρίης καὶ πυρίης καὶ ἐγκρατείης.[114] ξηρὰ δὲ καὶ δριμέα ἐσθιέτω· οὕτω γὰρ ἂν οὐρητικώτατος εἴη καὶ ἰσχύοι μάλιστα. ἢν δὲ δύσπνους γένηται καὶ ἡ

109 A: τάχος MV.   110 MV: σμικρὸν παράχει A.
111 Later mss: ἀνείη AMV.
112 A: ἕως MV.   113 A: δὲ MV.
114 Galen: -ίη καὶ -ίη καὶ -ίη A: ὑπὸ ταλαιπωρίης καὶ ἐγκρατείης MV.

51. In dry cholera the belly is distended with air, rumbling sounds are heard, pain occupies the sides and the loins, and nothing passes off below, for there is a blockage. Take care that such a patient does not vomit, but that his cavity is evacuated downwards; clean him out as quickly as possible with an enema of warm very oily water. Lower him into hot water, anoint with plentiful oil, and, laying him down in the tub, pour in hot water a little at a time. If, when he is warmed, his cavity evacuates, resolution has taken place. It also benefits such a patient to sleep, and to drink thin old wine almost undiluted; also give him olive oil [so that there is peace and his cavity evacuates, and resolution has taken place];[12] let the patient refrain from cereals and other foods. If the pain does not remit, give ass's milk to drink, in order to clean the patient out. If the cavity is moist and bile passes down, and colic, vomiting, choking and a gnawing pain are present, it is best for these patients to remain still; let them drink melicrat, and not vomit it up.

52. Of dropsies there are two forms: the one, attacking beneath the tissues, is inescapable; the other, accompanied by tympanites, requires very good fortune, and above all exercise, vapour-baths, and self-control. Let the patient eat foods that are dry and sharp, for thus will he pass the most urine, and have the greatest strength. If he has dif-

[12] Many attempts have been made to solve this crux, none very convincing. Littré reads the text of the manuscripts and translates: *afin qu'il se calme et qu'il ait des évacuations: car alors il est guéri.* I suspect that "resolution has taken place" has mistakenly been repeated here from the previous sentence.

ΠΕΡΙ ΔΙΑΙΤΗΣ ΟΞΕΩΝ (ΝΟΘΑ)

ὥρα θερινὴ[115] ἐοῦσα τύχῃ καὶ ἡ ἡλικίη ἀκμάζῃ,[116] ἀπὸ
τοῦ βραχίονος αἷμα ἀφαιρεῖν· εἶτα θερμοὺς ἄρτους ἐξ
500 οἴνου μέλανος καὶ ἐλαίου | ἀποβάπτων ἐσθιέτω καὶ
ὡς ἐλάχιστα πίνων ὡς πλεῖστα πονείτω. καὶ κρέα ὕεια
σαρκώδεα ἐσθιέτω ἐξ ὄξους ἐφθά, ὅπως πρὸς τοὺς
ἀνάντεας περιπάτους ἀντέχῃ.

53. (21 L.) Ὁκόσοι κοιλίας τὰς κάτω θερμὰς ἔχουσι
καὶ δριμέα τὰ ὑποχωρήματα καὶ ἀνώμαλα διέρχεται
ὑπὸ συντήξιος αὐτοῖς, ἢν μὲν δυνατοὶ ἔωσιν, ἀντι-
σπάσαι ἐλλεβόρῳ·[117] ἢν δὲ μή, ὁ χυλὸς τῶν σητανίων
πυρῶν, παχύς, ψυχρὸς καὶ τὸ φάκινον ἔτνος καὶ ἄρτοι
ἐγκρυφίαι καὶ ἰχθύες πυρέσσοντι μὲν ἐφθοί, ἀπυρέτῳ
δὲ[118] ὀπτοί· καὶ οἶνος μέλας ἀπυρέτῳ, εἰ δὲ μή, ὕδωρ
ἀπὸ μεσπίλων ἢ μύρτων ἢ οὔων ἢ μήλων ἢ φοίνικος
βαλάνων[119] ἢ οἰνάνθης ἀμπέλου. ἢν δὲ πυρετὸς[120] ἔχῃ
502 καὶ στρόφοι ἔχωσι, γάλα ὄνειον[121] | ὀλίγον θερμὸν τὸ
πρῶτον, ἔπειτα δὲ ἐκ προσαγωγῆς πλεῖον καὶ λίνου
σπέρμα καὶ πύρινα ἄλφιτα καὶ τῶν Αἰγυπτίων κυάμων ἐξελὼν τὰ πικρά, καταλέσας, ἐπιπάσσων πινέτω
ἢ ὠὰ ἡμιπαγέα ἐσθιέτω ὀπτὰ καὶ σεμίδαλιν καὶ κέγ-
χρον καὶ χόνδρον ἐφθὸν ἐν γάλακτι· ἐφθὰ ψυχρὰ
ἐσθίειν. καὶ τὰ τούτοις ὅμοια καὶ ποτὰ καὶ ἐδέσματα
προσφερέσθω.

54. (22 L.) Τῆς διαιτητικῆς ἐστι μέγιστον παρα-
τηρεῖν καὶ παραφυλάσσειν[122] ἐν τοῖσι μακροῖσιν ἀρ-
504 ρωστήμασι καὶ τὰς | ἐπιτάσιας τῶν πυρετῶν καὶ τὰς

ficulty breathing, it happens to be the summer season, and he is in the prime of life, draw blood from his arm. Then let him eat warm bread dipped in dark wine and olive oil, drink as little as possible, and exercise himself as much as possible. Have him eat lean pork, boiled in vinegar, in order to be able to withstand up-hill walks.

53. Let those with hot lower cavities, and in whom sharp irregular evacuations occur because of colliquation, check this, if they can stand it, with a dose of hellebore. If not, administer thick cold water from spring-wheat, thick lentil soup, bread baked in ashes and, to those with fever, boiled fish, to those without, roasted fish; also give dark wine to those without fever, but otherwise water made from medlars, myrtle-berries, sorbs, apples, dates, or the flowers of grape-vines. If fever and colic are present, have the patient first drink a little warm ass's milk, and then more and more ass's milk over which have been sprinkled linseed, wheat meal, and ground Egyptian beans from which the bitter parts have been removed; or he can eat medium-fried eggs, and the finest wheat flour, millet, and spelt groats boiled in milk; let these be boiled and cooled. In addition, both drinks and foods similar to these are to be administered.

54. The most important part of the dietetic art is to observe and watch closely during long illnesses for the exacerbations of the fevers and for their remissions, so as

---

115 A: ἐαρινὴ MV.   116 MV add καὶ ῥώμη ᾖ.
117 A: ἐ. τῷ λευκῷ ἀ. MV.   118 MV add ἐόντι.
119 A: φοινικοβαλάνων MV.   120 MV add τε μὴ.
121 A: βόειον MV.
122 π. A: φυλάσσειν ὥσπερ ἐν τοῖς ὀξέσι καὶ MV.

ἀνέσιας, ὥστε τοὺς καιροὺς διαπεφυλάχθαι, ὁπότε μὴ
δεῖ τὰ σιτία προσενεγκεῖν, καὶ ἀσφαλέως ὁπότε δεῖ
προσενεγκεῖν εἰδέναι, ἔστι δ' ὅταν[123] πλεῖστον ἀπέχω-
σι τῆς ἐπιτάσιος.[124]

506   55. (23 L.) Εἰδέναι δὲ τοὺς κεφαλαλγικοὺς ἐκ γυμνασίων ἢ δρόμων ἢ πορειῶν ἢ κυνηγίων ἢ ἄλλου τευ
πόνου ἀκαίρου ἢ ἐξ ἀφροδισίων, τοὺς ἀχρόους, τοὺς
βραγχαλέους, τοὺς σπληνώδεας, τοὺς λειφαίμους,
τοὺς πνευματώδεας καὶ ξηρὰ βήσσοντας καὶ διψώ-
δεας, τοὺς φυσώδεας, φλεβῶν ἀπολήψιας, ἐντεταμέ-
νους ὑποχόνδρια καὶ πλευρὰ καὶ μετάφρενον, τοὺς
ἀπονεναρκωμένους καὶ ἀμαυρὰ βλέποντας καὶ οἷς
ἦχοι τῶν ὤτων ἐμπίπτουσιν καὶ τῆς οὐρήθρης ἀκρα-
τῶς διακειμένους, τοὺς ἰκτεριώδεας καὶ ὧν αἱ κοιλίαι
508   ὠμὰ ἐκβάλλουσιν | ἢ αἱμορραγέοντας ἐκ ῥινὸς ἢ καθ'
ἕδρην σφοδρῶς, ἢν ἐν ἐμφυσήμασιν ἔωσιν ἢ πόνος
αὐτοῖς ἐπιτρέχῃ σφοδρὸς καὶ μὴ ἐπικρατέωσιν. τῶν
τοιῶνδε μηδένα φαρμακεύειν. κίνδυνόν τε γὰρ ἕξει[125]
καὶ οὐδὲν ὠφελήσεις τάς τε ἀπὸ τοῦ αὐτομάτου κρίσιας ἀφαιρήσεις.

        56. (24 L.) Ἢν δὲ αἷμά τινι ξυμφέρῃ ἀφαιρέειν,
στερεὴν πρότερον ποιέειν τὴν κοιλίην καὶ οὕτως ἀφαιρέειν[126] καὶ λιμοκτονέειν καὶ οἶνον ἀφαιρέειν αὐτῷ.[127]
ἔπειτα τῇ διαίτῃ τὰ ἐπίλοιπα αὐτὸν καὶ πυρίησιν
510   ἐνίκμοισι | θεράπευε. ἢν δέ σοι κατάπυκνος ἡ κοιλίη
δοκέῃ εἶναι, μαλθακῷ κλύσματι ὑπόκλυζε.

        57. (25 L.) Ἢν δὲ φαρμακεῦσαι δόξῃ, ἐλλεβόρῳ
ἀσφαλέως ἄνω κάθαιρε· κάτω δὲ μηδενὶ τῶν τοιῶνδε.

276

carefully to avoid the particular times when foods must not be given, and to know when they may be given with safety, namely, when patients are furthest from the exacerbation.

55. Take note of patients with headaches that have arisen from physical exercises, running, walking, hunting, some other untimely exertion, or venery, and of those with poor colour, a sore throat, disease of the spleen, lack of blood, asthma, a dry cough, excessive thirst, flatulence, stoppage of the vessels, tension of the hypochondrium, sides and back, numbness, dullness of vision, ringing in the ears, loss of command over the urethra, jaundice, the passage of undigested stools, excessive bleeding from the nose or through the anus, tympanites, or an attack of severe pain they do not overcome: do not treat any of these with a medication, for that would be dangerous, and your effect would not be to help the patient, but only to deprive his crises of their spontaneity.

56. If it would benefit some patient to draw blood, first make his cavity firm, and then draw the blood, have the patient fast, and take away his wine. Then in the time that follows treat him by regimen and vapour-baths. If the cavity seems to be very costive, administer a gentle enema.

57. If it seems advisable to administer a medication, clean cautiously upwards with hellebore, but never clean

---

123 A adds τι.
124 M: ἀποστάσιος A: ἐπιστάσηος V.
125 Vassaeus from Galen: ὀξέες A: ἕξεις MV.
126 πρότερον . . . ἀφαιρέειν om. A.
127 Joly from the later ms. P of Galen: αὐτῶν AMV.

κράτιστον δὲ ἐς οὔρησιν καὶ ἐς ἱδρῶτας καὶ ἐς περιπάτους ἄγειν. καὶ τρίψει ἡσύχῳ χρέω, ἵνα μὴ πυκνώ-
σῃς τὴν ἕξιν· ἢν δὲ κλινοπετὴς ᾖ, ἄλλοι τριβόντων
αὐτόν. καὶ ἢν μὲν ἐν τῷ θώρηκι ὑπὲρ τῶν φρενῶν
λυπέῃ τὸ πάθος, ἀνακαθιζέτω[128] ὡς πλειστάκις, καὶ ὡς
ἥκιστα προσκλινέσθωσαν, ἐς ὅ τι δυνατοί εἰσιν, καὶ
καθίζοντα ἀνατριβόντων πολὺν χρόνον θερμῷ πολλῷ.
ἢν δ᾽ ἐν τῇ κάτω κοιλίῃ ὑπὸ φρένας ἴσχῃ τὰ ἀλγήματα, ἀνακεῖσθαι ξυμφέρει καὶ μηδεμίαν κίνησιν
512 κινεῖσθαι. τῷ τοιῷδε | σώματι μηδὲν προσφέρεσθαι
ἔξω τῆς ἀνατρίψιος. τὰ δ᾽ ἐκ τῆς κάτω κοιλίης λυόμε-
να δι᾽ οὔρων καὶ ἱδρώτων, ἢν ὀλίσθῃ μετρίως, ὑπὸ
αὐτοματισμοῦ λύεται, τὰ σμικρά· τὰ σφοδρὰ δὲ πονη-
ρόν· οἱ τοιοίδε γὰρ ἢ ἀπόλλυνται ἢ ἄνευ ἄλλων κακῶν
οὐ γίνονται ὑγιέες, ἀλλ᾽ ἀποστηρίζεται[129] τὰ τοιουτό-
τροπα.

58. (26 L.) Πόμα ὑδρωπιῶντι· κανθαρίδας τρεῖς,
ἀφελὼν τὴν κεφαλὴν καὶ πόδας ἑκάστης καὶ πτερά,
τρίψας ἐν τρισὶ κυάθοισιν ὕδατος τὰ σώματα· ὅταν δὲ
πονέῃ ὁ πίων, θερμῷ βρεχέσθω ὑπαλειψάμενος πρότε-
ρον· νῆστις δὲ πινέτω· ἐσθιέτω δὲ ἄρτους θερμοὺς ἐξ
ἀλείφατος.

514 59. (27 L.) Ἴσχαιμον·[130] ὀπὸν συκῆς ἐν εἰρίῳ προσ-
θεῖναι εἴσω πρὸς τὴν φλέβα ἢ πυτίην[131] συστρέψαντα
βῦσαι εἰς τὸν μυκτῆρα ἢ χαλκίτιδος τῷ δακτύλῳ
ἐπισπασάμενος πίεσον· καὶ τοὺς χόνδρους ἔξωθεν

---

[128] A: αὐτὸν ἀνακαθίζειν MV.

278

downwards in patients of this sort; it is best to get them to pass urine, sweat, and take walks. Also, employ gentle massage in order not to make the body too contracted; if the person is bed-ridden, have others massage him. If the affection produces pain in the thorax above the diaphragm, let the patient sit up as much as possible, and the attendants support him as little as they can; when he is sitting up, let them rub him for a long time with plentiful warm oil. If the pains are in the lower cavity below the diaphragm, it helps for the patient to recline and to lie perfectly still; to such a patient administer nothing except massage. Pains resolved from the lower cavity, through urine and sweat that flow in moderation, are resolved spontaneously as long as they are mild; but severe ones are bad, for such patients either die or fail to escape without other evils, since such conditions become chronic.

58. Beverage for a patient with dropsy: remove the head, legs and wings from three blister-beetles, and grind the bodies into three cyathoi of water. When the person that has drunk this beverage suffers pain, let him first be anointed and then bathed in hot water; let him drink the medication in the fasting state, and eat warm bread that has been dipped in oil.

59. Against nosebleed: apply fig juice, in a piece of wool, inside against the vessel; or roll together some rennet and stuff it into the nostril, or take up some copper ore with your fingers, insert it, and squeeze the cartilages on

---

129 Potter: ἀποστηρίζει καὶ AMV.
130 MV: περὶ πρὸς τὴν ἐκ ῥινῶν αἱμορραγίαν ἴσχ. A.
131 M: πιτύην AV.

προσπίεζε ἑκατέρωθεν. καὶ τὴν κοιλίην λῦσον ὄνου
γάλακτι ἑφθῷ καὶ τὴν κεφαλὴν ξυρῶν ψυκτικὰ πρόσφερε, ἢν ἐν ὥρῃ θερμῇ γίνηται.

60. (28 L.) Σησαμοειδὲς ἄνω καθαίρει· ἡ πόσις,
ἡμιόλιον δραγμῆς [σταθμὸς][132] ἐν ὀξυμέλιτι τετριμμέ-
νον. συμμίσγεται δὲ καὶ τοῖ|σιν ἑλλεβόροισιν, καὶ
ἧσσον πνίγει, τὸ τρίτον μέρος τῆς πόσιος.

61. (29 L.) Τριχώσιος· †ὑποθεὶς τὸ ῥάμμα τῇ βελόνῃ[133] τῇ τὸ κύαρ ἐχούσῃ κατὰ τὸ ὀξὺ τῆς ἄνω τάσιος
τοῦ βλεφάρου ἐς τὸ κάτω[134] διακεντήσας δίες, καὶ[135]
ἄλλο ὑποκάτω τούτου.† ἀνατείνας δὲ τὰ ῥάμματα
ῥάψον, καὶ κατάδησον,[136] ἕως ἂν ἀποπέσῃ. κἢν μὲν
ἱκανῶς ἔχῃ· εἰ δὲ μή, ἢν ἐλλείπῃ, ὀπίσω ποιέειν τὸ
αὐτό.[137]

62. Καὶ τὰς αἱμορροΐδας τὸν αὐτὸν τρόπον· τῇ
βελόνῃ διώσας ὡς[138] παχύτατον εἰρίου οἰσυπηροῦ
ῥάμμα καὶ ὡς μέγιστον ἀποδήσεις· ἀσφαλεστέρη[139]
γὰρ γίνεται ἡ θεραπείη. εἶτα ἀποπιέσας τῷ σήπτῳ
χρέω· καὶ μὴ βρέχε, πρὶν ἀποπέσῃ· καὶ αἰεὶ μίαν
καταλίμπανε. καὶ μετὰ ταῦτα ἀναλαβὼν ἑλλεβορίσαι.
εἶτα γυμναζέσθω καὶ ἀφιδρούτω· γυμνασίου δὲ τρί-
ψις, πάλη[140] ἀπὸ ὀρθοῦ· δρόμου δὲ ἀπεχέσθω καὶ

---

[132] Del. Joly after Helmreich.   [133] MV: ὑποθείστω ῥάμ-
ματι βελόνῃ A.   [134] MV: ἄνω A.   [135] MV: εἶναι δ’ A.
[136] A: -δει MV.   [137] A: τὰ αὐτά MV.
[138] Reinhold: τῇ βέλτιον ἡδίως εἴσως A: διώσεις τῇ βελόνῃ ὡς MV.   [139] MV: -τάτη A.
[140] δὲ . . . πάλη A: τε . . . πολλὴ MV.

both sides from the outside. Open the cavity with boiled ass's milk, and, if it happens in summer, shave the head and apply cooling agents to it.

60. Sesamoid cleans upwards: the potion consists of one-and-a-half drachmas ground into oxymel. It can also be mixed with hellebore, and this chokes less; in that case you employ one-third the amount of the potion.

61. Trichiasis: pass a thread through the eye of a needle; then, piercing through along the angle of the upper extension of the eye-lid in a downward direction, draw the needle through; do the same again below this.[13] Now pulling the threads tight, stitch them together, and keep them bound fast until they fall off. If this suffices, fine; if it fails, do the same again.

62. Treat haemorrhoids, too, in the same way: first thrust through a very long thick thread of greasy wool, and then tie them off, for this makes the treatment more effective. Then, after squeezing them out, use a putrefacient; do not wash until the haemorrhoids have fallen off, and always leave one behind. After that, restore the patient and administer hellebore. Then let him do exercises and be cleaned by sweating; use the massage of the gymnastic school and wrestling from the standing position; he must avoid running, drunkenness, and all sharp substances ex-

[13] In spite of J. Hirschberg's explanation (*Geschichte der Augenheilkunde im Alterthum*, Leipzig, 1899, p. 140), I agree with Francis Adams (*Paulus Aegineta*, London, 1844–7, II. 262) that "the description is so obscure that we must confess our inability to explain it."

ΠΕΡΙ ΔΙΑΙΤΗΣ ΟΞΕΩΝ (ΝΟΘΑ)

μέθης καὶ τῶν δριμέων ἔξω ὀριγάνου. ἐμείτω δὲ δι᾽
ἑπτὰ ἡμερέων ἢ τρὶς ἐν τῷ μηνί· οὕτω γὰρ ἂν ἔχοι
518 ἄριστα τὸ | σῶμα. οἶνον δὲ κιρρὸν αὐστηρὸν ὑδαρέα,
καὶ ὀλίγον τὸ ποτὸν πινέτω.

63. (30 L.) Τοῖσι δ᾽ ἐμπύοισι σκίλλης καταταμὼν
κυκλίσκους ἕψε ἐν ὕδατι, καὶ ἀποζέσας εὖ μάλα ἀπό-
χεον, καὶ ἐπιχέας ἄλλο ἕψε, ἕως ἂν ἁπτομένῳ δίεφθον
καὶ μαλθακὸν φανῇ· εἶτα τρίψας λεῖον σύμμισγε
κύμινον πεφρυγμένον καὶ λευκὰ σήσαμα καὶ ἀμυγδάλας λείας,¹⁴¹ τρίψας ἐν μέλιτι ἐκλεικτὸν δίδου καὶ ἐπὶ
τούτῳ γλυκύν. ῥυφήματα δέ, μήκωνος τῆς λευκῆς
ὑποτρίψας ὅσον λεκίσκιον, ὕδατι διείς, ἢ σητανίου
πλύματι ἀλεύρου ἑψήσας, μέλι ἐπιχέας, χλιερὸν ἐπιρ-
ρυφέων, οὕτω διαγέτω τὴν ἡμέρην. εἶτα ἐς τὰ ἀπο-
βαίνοντα λογιζόμενος τὸ δεῖπνον δίδου.

64. (31 L.) Δυσεντερίης· κυάμων καθαρῶν τεταρτη-
520 μόριον καὶ | ἐρυθροδάνου δώδεκα κάρφεα λεῖα συμ-
μείξαντα καὶ ἑψήσαντα λιπαρὸν διδόναι ἐκλείχειν.

65. (32 L.) Ὀφθαλμῶν· σποδὸς πεπλυμένη, λιπαρῷ
πεφυρημένη, ὡς σταῖς μὴ ὑγρόν, λεῖον τρίψας, ὀμφα-
κίῳ τῷ τῆς πικρῆς ὄμφακος ἀνυγρήνας, ἐν ἡλίῳ ξηρή-
νας, ὑγραίνειν ὡς ἐνάλειπτον· ὅταν δ᾽ αὖτις¹⁴² ξηρὸν
γένηται, λείῳ τετριμμένῳ ξηρῷ ὑπόχριε καὶ παρά-
πασσε τοὺς κανθούς.

66. (33 L.) Ὑγρῶν·¹⁴³ ἐβένου δραγμὴν μίαν¹⁴⁴ χαλκοῦ κεκαυμένου ἐννέα ὀβολοὺς ἐπ᾽ ἀκόνης τρίβων,

¹⁴¹ A: νέας MV.

282

cept marjoram. Have the patient vomit every seven days, or three times a month, for thus will his body be in the best condition. Have him drink dry light-coloured wine, diluted with water, and little of it.

63. For patients that are suppurating internally: cut up a squill and boil the discs in water; boil them well, pour off the water, replace it with new water, and boil again until they seem boiled through and soft to the touch. Then mash them smooth, and mix in roasted cummin, white sesame and ground almonds; knead this into honey, and administer as a lozenge, followed by a sweet wine. Gruels: grind up a small bowl of white poppy, either soak it in water or boil it in an infusion of this year's meal, add honey, and have the patient drink it warm, over the whole day. Then, taking the consequences into account, give him dinner.

64. For dysentery: mix together one quarter cotyle of cleaned beans and twelve twigs of madder until they are smooth, boil, and give in fat as a lozenge.

65. For the eyes: mix washed metallic ashes into a paste with fat in such a way that the fat is no longer liquid; knead it smooth, moisten with the oil of bitter unripe olives, dry in the sun, and moisten again to give it the consistency of an ointment; when it has become dry once more, anoint the eyes with this smoothly-ground dry mixture, and dust it into the corners.

66. For watery eyes: one drachma of ebony, nine obols of burnt copper ground in a mortar, three obols of saffron:

---

142 Kuehlewein: αὐτῆς A: om. MV.
143 Later mss: Ὑγρόν AMV.
144 μίαν om. MV.

κρόκου τριώβολον· ταῦτα τρίψας λεῖα, παράχει οἴνου
γλυκέος κοτύλην Ἀττικήν, κἄπειτα ἐς τὸν ἥλιον θείς,
κατακαλύψας, ὅταν συνεψηθῇ τούτῳ χρέω.

522   67. (34 L.) Πρὸς τὰς περιωδυνίας καὶ τὰ ῥεύματα·
ἔστω χαλκίτιδος δραγμή, σταφυλῆς, ὅταν δυσὶν ἡμέ-
ραις πεφθῇ,[145] ἐκπιέσας, σμύρναν καὶ κρόκον τρίψας
καὶ συμμείξας τὸ γλεῦκος, ἕψησον ἐν τῷ ἡλίῳ, καὶ
τούτῳ ὑπάλειφε τοὺς περιωδυνέοντας· ἔστω δὲ ἐν χαλ-
κῷ ἀγγείῳ.

68. (35 L.) Ὑπὸ ὑστερικῶν πνιγομένων γνῶσις·
πιέσαι τοῖσι τρισὶ δακτύλοισι· κἢν αἴσθηται, ταῦτα
ὑστερικά ἐστιν· ἢν δὲ μή, σπασμώδεα.

69. (36 L.) Τοῖσιν ὑπνωτικοῖσι[146] μηκωνίου[147] λεκίσκιον Ἀττικὸν στρογγύλον πόσις.

524   70. (37 L.) Λεπίδος μῆλαι τρεῖς τῷ πλάτει καὶ
526   ἁλήτου ση|τανίου κόλλης, πάντα ταῦτα λεῖα τρίψας,
καταπότια ποιήσας δίδου· κάτω ὕδωρ καθαίρει.

71. (38 L.) Κοιλίην ἐκκοπροῖ· ἐς ἰσχάδια[148] τοῦ
τιθυμάλλου ἀπόσταζε ὅσον ἑπτάκις ἐς ἕκαστον, καὶ
παιδίοισιν, εἶτα εἰς καινὸν ἄγγος συνθεὶς ταμιεύ-
εσθαι· δίδου πρὸ τῶν σιτίων.

72. (39 L.) Καὶ τὸ μηκώνιον τρίβων, ὕδωρ ἐπιχέων
528   καὶ διη|θέων, ἄλευρον φυρῶν, ἴτριον ὀπτῶν, μέλι

---

[145] Α: ὁκόταν δύο μέρεα λειφθῇ MV.
[146] Α: ὑδρωπιώδεσι(ν) MV.
[147] Α: -ιον MV.
[148] Kuehlewein: ἰσχάδα Α: ἰσχάδα ὀποῦ MV.

grind these smooth, pour in an Attic cotyle of sweet wine, and then cover and place in the sun; when it has combined, apply.

67. Against sharp pains and fluxes: let there be a drachma of copper ore, and grapes; when this has fermented for two days, press it out, knead into it myrrh and saffron, and add new wine; heat in the sun, and with this anoint persons suffering from sharp pains; let it be kept in a copper vessel.

68. Recognition of hysterical suffocations: pinch with three fingers: if the patient feels it, they are hysterical, if not, convulsive.

69. For sleepy patients: a spherical Attic lekiskion of spurge as a potion.[14]

70. Of copper scale three probes,[15] and a paste made from this year's flour; grind all this fine, form it into pills, and give; it is diuretic.

71. It empties the cavity of faeces: on to a dried fig distil spurge, about seven drops on each, also for children; place these together in a new vessel and store them; give before meals.

72. Also grind spurge, pour water over it, sieve, and into this mix meal; then bake as a cake, and pour honey over it;

---

[14] The meaning of this chapter is very doubtful. In place of A's "sleepy patients" MV read "patients with dropsy". μηκώνιον could be either the cathartic juice of spurge, or poppy juice (opium); cf. *Diseases of Women II* 201 where μηκώνιον is referred to as "sleep producing". A lekiskion is apparently a small pot, used here as an indication of volume.

[15] Presumably as much copper scale as can be picked up on a probe.

285

ἑφθὸν παραχέων, τοῖς ὑδρωπικοῖσιν τρώγειν δίδου
καὶ ἐπιπίνειν γλυκὺν ὑδαρέα ἢ μελίκρητον ὑδαρές [,
τὸ ἀπὸ τῶν κηρίων].[149] ἢ μηκώνιον συλλέγων ταμιεύ-
ου, καὶ θεράπευε.

[149] Del. Ermerins.

give to patients with dropsy, to eat; with it have them drink sweet wine diluted with water, or dilute melicrat. Or collect and store spurge, and treat with that.

TABLES AND INDEXES

WEIGHTS & MEASURES[1]
(with estimated equivalents)

i) Weights: 6 obols = 1 drachma; 100 drachmas = 1 mina

Equivalents	Attic	Aeginetan
obol	0.728 g.	1.0 g.
drachma	4.366 g.	6.03 g.
mina	436.6 g.	602.6 g

ii) Distance: Attic stade : 185 meters.

iii) Liquid Measures:

$$\left. \begin{array}{l} 12 \text{ cheramydes}[2] \\ 6 \text{ cyathoi} \\ 4 \text{ oxybapha} \end{array} \right\} = 1 \text{ cotyle}$$

12 cotylai = 1 chous (pl. choes)

[1] See F. Hultsch, *Griechische und Römische Metrologie* (2nd ed.), Berlin, 1882.

[2] I take χηραμύς to mean the same as κόγχη ἐλάττων. This latter unit is set equal to one half cyathos in Cleopatra, Περὶ σταθμῶν καὶ μέτρων 22 (*Scriptores Metrologici*, ed. F. Hultsch, Leipzig, 1866, I. 256).

Equivalents	Attic	Aeginetan
cheramys	22.8 ml.	
cyathos	45.6 ml.	
oxybaphon	68.4 ml.	
cotyle	0.2736 l.	0.379 l.
chous	3.283 l.	4.55 l.

iv) Dry Measures: 4 cotylai = 1 choenix

Equivalents	Attic	Aeginetan
cotyle	0.2736 l.	0.379 l.
choenix	1.094 l.	1.515 l.

NOTE

In the Indexes of Symptoms and Diseases and of Foods and Drugs, the references are to chapter numbers in the works included in volumes V and VI. The works are identified as follows:

A = *Affections*
B = *Diseases I*
C = *Diseases II*
D = *Diseases III*
E = *Internal Affections*
F = *Regimen in Acute Diseases (Appendix)*

A, B and C are in volume V.
D, E and F are in volume VI.

INDEX OF SYMPTOMS
AND DISEASES[1]

(The references in bold type indicate passages where the symptom or disease is described rather than merely mentioned by name.)

Alopecia, ἀλώπηξ, a disease in which the hair falls out; cf. Celsus 6.4.2: A35. Cf. C48

[1] This is an index of ancient diseases, i.e. diseases as they are described and defined in the texts. In most cases they correspond only distantly to their modern counterparts. Discussions of Hippocratic pathology and disease names are to be found in the following works: Daniel Le Clerc, *Histoire de la médecine*, La Haye, 1729, pp. 165–84; Ch. Daremberg, *Oeuvres choisies d'Hippocrate* (2nd ed.), Paris, 1855; Aug. Hirsch, *Handbuch der historisch-geographischen Pathologie* (2nd ed.), Stuttgart, 1881–1886; Reinhold Strömberg, *Griechische Wortstudien: Untersuchungen zur Benennung von Tieren, Pflanzen, Körperteilen und Krankheiten*, Göteborg, 1944, pp. 68–103; Eulalia Vintró, *Hipócrates y la nosología hipocrática*, Barcelona, 1972, pp. 147–74; M. D. Grmek, "La réalité nosologique au temps d'Hippocrate" in *La Collection Hippocratique et son rôle dans l'histoire de la médecine (Colloque de Strasbourg, 1972)*, Leiden, 1975, pp. 237–55; M. D. Grmek, *Les maladies à l'aube de la civilisation occidentale*, Paris, 1983.

---

[2] The suffix -ῖτις (-itis) changes a noun into an adjective.
Where it is joined with a noun that refers to a bodily organ, the resulting adjective often understands νοῦσος (disease), to acquire
the meaning "disease of that particular organ", e.g. ἀρθρῖτις (ar-
thritis) = "disease of the joints". In the Hippocratic writings -ῖτις
does not necessarily indicate the presence of inflammation.

INDEX OF SYMPTOMS AND DISEASES

E39, E41, E48 ter, E49, F1 bis, F2 bis, F8, F9, F17 bis, F29, F30, F32, F37, F55

Dark disease, μέλαινα: **C73, C74**
Darkening of vision, σκότωσις: F7
Diarrhoea, διαρροίη: **A25, A27**, B7, C48, E35, E42, E48
Dislocation, ἔκπτωμα: B6
Dizziness, σκοτοδινίη, a combination of darkening of vision (σκότος) and vertigo (δῖνος): A2, C4 bis, C15, C18
Dropsy, ὕδρωψ (= ὕδερος): A19, A20, **A22**, A36, B3 ter, **C61**, E12, E13, E19, **E22, E23, E24, E25, E26**, E42, E44, E49, **F52**, F58, F72
Drowsiness, νυσταγμός: F42
Dysentery, δυσεντερίη: A22, **A23**, A24, A25, A26, B3 ter, **F35**, F64
Dysuria, δυσουρίη, difficult micturition: C3, C14, E13, F41

Epileptic, ἐπίληπτος (ἐπιλαμβάνω, seize): F7
Epinyctis, ἐπινυκτίς; cf. Celsus 5.28.15c and Pliny *N.H.* 20.44: E31
Erysipelas, ἐρυσίπελας: B7; in the uterus B3; in the lung B18, C55, E6; E7 is a f.l.

Favus, κηρίον (honeycomb), a skin disease marked by the formation of round, cup-shaped yellow crusts resembling honeycomb. Cf. Celsus 5.28.13: A35
Fever, πυρετός (πῦρ, fire): passim, e.g. A10, B15, **B23, C40**, D11, E34, F25; bilious fever **F36**; summer fever **A14**; winter fever **A12, F24.** See also ardent, quartan, and tertian fevers.
Fistula, σῦριγξ: B3
Fracture, κάτηγμα: B6, B10, C24

Gargareon, see staphylitis.
Gout, ποδάγρη: **A31**, B3 bis

Haemorrhoids, αἱμορροίδες: B3, **F62**
Hepatitis, ἡπατῖτις: B3, **E27**; cf. F34 (ἡπατικός). Cf. **E28, E29**
Hiccups, disease with, λυγγώδης: **C64**; cf. F30
Hysteria, τὰ ὑστερικά: F68

295

INDEX OF SYMPTOMS AND DISEASES

Ileus, εἰλεός: **A21** and **D14** are intestinal obstruction. **E44, E45** and **E46** are quite different, but do share some symptoms with Aretaeus S.A. 2.6 (Ileus). Cf. Daremberg p. 293 n. 202.

Jaundice, ἴκτερος: **A32**, C2, C13, **C38, C39**, C41, **D11**, D17, **E35, E36, E37, E38, E45**, E49, F36, F55

Lepra, λέπρη (λεπίς, scale), an unidentifiable skin disease, but not leprosy; cf. Paulus 4.2: A35, B3
Lethargy, λήθαργος: **C65, D5**
Lichen, λειχήν, a papular skin disease; cf. Paulus 4.3: A35, B3
Lientery, λειεντερίη; cf. Strömberg pp. 90f.: A22, **A24**, A25 bis, B3 bis
Livid disease, πελίη: **C68**
Look awry, ἰλλαίνειν: D12, E52

Malignant disease, φονώδης: **C67**
Melancholy, μελαγχολίη, either an abnormal mental state, or an imbalance of dark bile, or both. Cf. Wittern p. 192 n.15: A36, B3, **B30**, D13, F16, F29, F37, F48

Nails curved, ὄνυχες ἕλκονται, a deformity in which the nails are unusually broad, and curved slightly over the tips of the fingers or toes; clubbing: C47, C48, C50, C61, E10, E23
Nephritis, νεφρῖτις: B3 bis, E18; nephritic consumption E15. Cf. **E14, E15, E16, E17**
Numbness, νάρκη: B20, C12, C16, C55, E6, F55

Ophthalmia, ὀφθαλμίη, an eye disease. Cf. Daremberg p. 584 n. 9: B3
Opisthotonus, ὀπισθότονος: **D13, E53**
Orthopnoea, ὀρθοπνοίη, the symptom of being able to breathe only in an upright posture: A7, C44, C57, C58, C59, D7, D16 bis, E3, E4, E7, E23, E40, F4, F10
Overfill with blood, ὑπεραιμεῖν: **C4, C17, C18.** Cf. **E18, E19**
Oxyrygmia, ὀξυρεγμίη, acid eructation: A26, A47 ter; cf. E6 (ἐρεύγεται ὀξύ)

296

INDEX OF SYMPTOMS AND DISEASES

Panting, ἆσθμα: C51, C58, E11; cf. D7 (ἀσθμαίνω). Cf. E5, E11
  (φῦσα)
Panus, φύγετρον, inflammatory swelling of the lymphatic glands.
  Cf. Celsus 5.28.10: A35
Paralysed, ἀπόπληκτος (= παραπλήξ); cf. Daremberg p. 258 n. 18:
  B3 bis, B4, C6, F7
Parangina, παρακυνάγχη: D10
Phlegmasia, φλεγυασίη; I take this term to mean a swelling or sup-
  posed collection of phlegm, not necessarily inflammation: D1,
  D7, F4; cf. C70 (φλεγματώδης); E20, E22, E49 (φλέγμα). See
  also white phlegm.
Phrenitis, φρενῖτις (φρήν, diaphragm, mind): A6, A10, A12, B3
  passim, B30, B34, C72, D9, D15, D16. Cf. E48
Pleurisy, πλευρῖτις (πλευρόν, side): A6, A7, A9 passim, A10, A12,
  A15, B3 passim, B7, B15 bis, B22, B26, B27, B28, B29, B31, B32,
  C44, C45, C46, D15 bis, D16, E7 bis, E27 ter, E39, F3, F31, F34
Pneumonia, περιπλευμονίη: A6, A9, A10, A11 bis, A12, B3 passim,
  B7, B12, B22, B26, B27, B28, B29, B31, B32, C27, C47, C63
  ter, D5, D6 bis, D9 bis, D15, D16 ter, F33, F34. Cf. E3, F31
Polyp, πώλυπος: A5, C33, C34, C35, C36, C37
Prurigo, κνισμός, a skin disease characterized by intense itching; cf.
  Paulus 4.4: A35
Psora, ψώρη (ψάω, rub away), an unidentifiable skin disease; cf.
  Paulus 4.2: A35

Quartan fever, τεταρταῖος πυρετός: A18, B3, C43, E40

Rage, μανίη / μαίνεσθαι: B30, C22, E29; cf. D13 (μανικός)

Sciatica, ἰσχιάς: A29, B3 ter, E51
Scrofula, χοιράδες, chronic enlargement and degeneration of the
  lymphatic glands; cf. Paulus 4.33: A35
Shadowy vision, σκιαυγεῖν. E48
Sleeplessness, ἀγρυπνίη: F1, F2, F40
Sore throat, βράγχος: C55, E6, E41, F55. Cf. A4
Speechlessness, ἀφωνίη / ἄφωνος: B4, C6, C21, C22, D4, D8,
  D13, E48, F6, F7, F25, F28. Cf. B3
Sphacelus, σφακελισμός, gangrenous necrosis. In the Hippocratic
  Collection this process is primarily confined to bones (e.g. *Frac-*

297

INDEX OF SYMPTOMS AND DISEASES

Tubercle, φῦμα (φύω, grow), a nodule containing purulent or caseous material; cf. Daremberg p. 282 n.144: **A34,** B6, B8, B11, B17, **B19, B20,** C30, C32, **C57, C60, E3, E9,** E23, F26 bis

Typhus, τῖφος, not the modern disease: **E39, E40, E41, E42, E43**

Unclean(ness), ἀκάθαρτος / ἀκαθαρσίη a state in which disease-producing impurities are present in the body: A18 bis, A19, A20, **A22,** C16, C41, C43, C63, **D2, D3,** D16

Varix, κιρσός: B14, B20, **E4/5,** F35
Vertigo, δῖνοι: F17

White phlegm, φλέγμα λευκόν: **A19,** A22, B3 bis, B7, **C71, E21, E50**
Withering disease, αὐαντή: **C66.** Cf. **E13**

GREEK NAMES OF SYMPTOMS AND
DISEASES INCLUDED IN INDEX

ἀγρυπνίη	Sleeplessness
αἱμορροΐδες	Haemorrhoids
ἀκάθαρτος	Unclean
ἀλφός	Alphos
ἀλώπηξ	Alopecia
ἄνθραξ	Anthrax
ἀντιάς	Tonsillitis
ἀπόπληκτος	Paralysed
ἀρθρῖτις	Arthritis
ἄσθμα	Panting
αὐαντή	Withering Disease
ἄφωνος	Speechless
βλητός	Stricken
βράγχος	Sore Throat
γαργαρεών	Staphylitis
διαρροίη	Diarrhoea
δῖνοι	Vertigo

δοθιήν	Boil
δυσεντερίη	Dysentery
δυσουρίη	Dysuria
εἰλεός	Ileus
ἔκπτωμα	Dislocation
ἔμπυος	Suppurate internally
ἐπίληπτος	Epileptic
ἐπινυκτίς	Epinyctis
ἐρυγματώδης	Belching Disease
ἐρυσίπελας	Erysipelas
ἡπατῖτις	Hepatitis
ἴκτερος	Jaundice
ἰλλαίνειν	Look awry
ἰσχιάς	Sciatica
κάτηγμα	Fracture
καῦσος	Ardent Fever
κέδματα	Swellings at the Joints
κηρίον	Favus
κιρσός	Varix
κνισμός	Prurigo
κόρυζα	Coryza
κρίσις	Crisis
κυνάγχη	Angina

πλευρῖτις	Pleurisy
ποδάγρη	Gout
πυρετός	Fever
πώλυπος	Polyp
σκιαυγεῖν	Shadowy vision
σκοτοδινίη	Dizziness
σκότωσις	Darkening of vision
σπληνῖτις	Splenitis
σταφύλη	Staphylitis
στραγγουρίη	Strangury
στρόφος	Colic
σύναγχος	Angina
σῦριγξ	Fistula
σφακελισμός	Sphacelus
τεινεσμός	Tenesmus
τερηδών	Teredo
τέτανος	Tetanus
τεταρταῖος πυρετός	Quartan Fever
τῖφος	Typhus
τριταῖος πυρετός	Tertian Fever
τρίχωσις	Trichiasis
ὕδρωψ	Dropsy
ὑπεραιμεῖν	Overfill with blood
ὑπογλωσσίς	Tongue, affection of
ὑστερικά	Hysteria

INDEX OF FOODS AND DRUGS[1]

Acorn, βάλανος, fruit of the oak, *Quercus Robur*: C13
All-heal, πάνακες, *Ferulago galbanifera*: F34
All-heal juice, χαλβάνη: F30, F34
Almond, ἀμυγδάλη (= κάρυον Θάσιον), fruit of the *Prunus Amygdalus*: C64, D11, D15, F63
Angel-fish, ῥίνη, *Squalus squatina*: E1, E12 bis, E49
Anise, ἄννισον, *Pimpinella Anisum*: D11
Apple, μῆλον: D17, F53
Aristolochia, ἀριστολοχία, *Aristolochia rotunda*: D16, E23
Arsenic, red, σανδαράκη, As₂S₂: C14
Asphalt, ἄσφαλτος: D10
Asphodel, ἀσφόδελος, *Asphodelus ramosus*: C38, E30
Ass, ὄνου κρέας: E6, E22

[1] Two comprehensive studies of Hippocratic materia medica are: J. H. Dierbach, *Die Arzneimittel des Hippokrates*, Heidelberg, 1824; R. von Grot, "Über die in der hippokratischen Schriftensammlung enthaltenen pharmakologischen Kenntnisse" in *Historische Studien aus dem Pharmakologischen Institute der Kaiserlichen Universität Dorpat*, I. 58–133, Halle, 1889.

Also important for Greek materia medica are: Dioskurides, *Arzneimittellehre . . . übersetzt und mit Erklärungen . . . von* J. Berendes, Stuttgart, 1902; D'Arcy W. Thompson, *A Glossary of Greek Birds*, Oxford, 1936, and *A Glossary of Greek Fishes*, London, 1947.

Dioscurides is quoted from the edition of M. Wellmann, Berlin, 1906–14.

INDEX OF FOODS AND DRUGS

Ass's milk, ὄνου γάλα: passim, e.g. A29, A30, C13, C40, C70, E10, E46, E51, F1, F8, F53

Barley, κριθή, *Hordeum sativum*: D17 quater
Barley, parched, κάχρυς (= κριθὴ ὀπτή): C67, D17, E51
Barley, peeled, πτισάνη: F44. See also barley-water.
Barley groats, κρίμνα κριθέων: D17 bis
Barley meal, ἄλφιτον: A52, E12
Barley-cake, μᾶζα: A41, A43, A50, A52, A61, C50, C55, E12 bis, E20, E27, E35, E41, E42, E51, F44
Barley-water, χυλὸς πτισάνης (= ὕδωρ κρίθινον): passim, e.g. A41, C40, D17, E35, F17, F30
Basil, ὤκιμον, *Ocymum basilicum*: A43, A54, E12
Bay oil, δάφνινον ἔλαιον: C13
Bayberry, δαφνίς, fruit of the bay-tree: C13
Bay-tree, δάφνη, *Laurus nobilis*: E52
Bean, Egyptian, κύαμος Αἰγύπτιος, *Nelumbium speciosum*: F53. Cf. F64
Beef, βόεια κρέα: A52, C47, C49, C52, E12, F48
Beet, σεῦτλον (= τεῦτλον), *Beta maritima*: A38, A41, A43, A55, C12, C26, C27, C44, C48, C55, C67, C69, C71, C74, E7, E12 bis, E21, E26, E30, E31, E41, E42, E46, E48, E51 bis, F44
Berry, Cnidian, κόκκος Κνίδιος, fruit of the θυμελαία, *Daphne Cnidium*: C48, E1, E7, E10, E13, E21, E22, E25, E26 bis, E30, E32, E38, E40, E42, E44, E47, E51 ter. Cf. spurge flax.
Black-tail, μελάνουρος, *Oblata melanurus*: E12
Blister-beetle, κανθαρίς, possibly *Cantharis vesicatoria* or *Meloë Cichorei*. See F. Adams on Paulus 7.3 κανθαρίδες (III.153 f.) and Berendes pp. 170f.: E36, F58
Blite, βλίτον, *Amaranthus Blitum*: A41, A43, A55
Bottle-gourd, σικύη, *Lagenaria vulgaris*: E51
Braize, φάγρος, *Pagrus vulgaris*; cf. Thompson *(Fish)* pp. 273ff.: E1
Bramble, βάτος, *Rubus ulmifolius*: A38
Bread, ἄρτος: A41, A43, A50, A51, A52, A61, C28, C48, C50, C71, E12 ter, E20 bis, E21, E22 bis, E25, E27, E30 ter, E35, E41, E42, E44, E49, E52, F53, F58. Cf. barley-cake.
Brine, ἅλμη: E32
Bryony, μάδος: F38

306

INDEX OF FOODS AND DRUGS

Buckthorn, ῥάμνος, various species of *Rhamnus*: A38

Bull's gall, χολή ταύρου: D14

Cabbage, κράμβη, *Brassica cretica*: A55, C19, E12 bis

Caper-plant, κάππαρις, *Capparis spinosa*: root bark D15; capers D16

Celery, σέλινον, *Apium graveolens*: A38 bis, A43, A54 bis, C19, C26, C28, C38 bis, C56, D1, D17 sexiens, E12, E22, E30, E35

Centaury, κενταύριον (= κενταυρίη), *Centaurea salonitana*: C54, C59, E1

Chaste-tree, ἄγνος, *Vitex Agnus-castus*: A38, E30

Cheese, τυρός: A47, A55, C50, E12, E35, E40, E41, F46, F48. See also goat's cheese.

Chicken, νεοσσὸς ἀλεκτρυόνος, young of *Gallus gallinaceus*: E27, E35, E41, E49. Cf. fowl.

Chick-pea, ἐρέβινθος, *Cicer arietinum*: A27, F47; white C38, D17, E14, E35, E45

Cinquefoil, πεντάφυλλον, *Potentilla reptans*: C42, D17

Clover, τρίφυλλον, *Psoralea bituminosa* or a member of the genus *Trifolium*: C42, C43 bis, D16, D17

Copper, burnt, χαλκὸς κεκαυμένος, cuprous oxide, possibly mixed with various other copper compounds: F66

Copper, flower of, ἄνθος χαλκοῦ, small grains of cuprous oxide made by quenching heated copper: C19, C25, C30, C33, C34, C36, C37, C47, D16 ter

Copper, scales of, λεπίς, flakes of cuprous oxide made by hammering: F70

Copper ore, χαλκῖτις, contradictory reports in antiquity make closer identification impossible: cf. e.g. Dioscurides 5.99, Pliny *H.A.* 34.117–120, Galen XII.226–229 (Kühn): F59, F67

Coriander, κορίαννον (= κόριον), *Coriandrum sativum*: A38, A54, C50, D17, E12, E35

Cow's milk, βόειον γάλα: C47, C51, E1, E2, E3, E6, E10 bis, E13, E28, E32 bis, E43, E46 bis, E49, E51

Cress, κάρδαμον, *Lepidium sativum*: D16; seed C26, E40

Cuckoo-pint, ἄρον τὸ μέγα, *Arum italicum*: C47, D15 bis, D16

Cucumber, σίκνος, *Cucumis sativus*: A57, C64, E35

Cummin, κύμινον, *Cuminum Cyminum*: A41, A43, D16, D17, E35, E51, F30, F63

INDEX OF FOODS AND DRUGS

Cyceon, κυκεών, a mixed drink containing meal, cheese, herbs and wine: C15, C43, C50, E4, E12 bis

Cyclamen, κυκλάμινος, *Cyclamen graecum*: C47

Cypress, κυπάρισσος, *Cupressus sempervirens*: C13

Date, βάλανος φοίνικος, fruit of the date-palm, *Phoenix dactylifera:* F53

Dauke, δαῦκος, perhaps *Athamanta Cretensis*: C54, C59, D15, F30, F38

Dill, ἄνηθον, *Anethum graveolens*: A43, C50, E12

Dock, λάπαθον, *Rumex conglomeratus*: A43, A55, F44

Dodder of thyme, ἐπίθυμον, *Cuscuta Epithymum*, a parasitic plant growing on thyme: E10

Dog, κυνὸς κρέας: A43, A52, E6, E22, E30. See also puppy.

Dogfish, γαλεός: E1, E12 bis, E24, E27

Dragon arum, δρακόντιον, *Dracunculus vulgaris*: E1 bis

Ebony, ἔβενος, wood of the *Diospyrus Ebenum*: F66

Eel, ἔγχελυς, *Anguilla vulgaris*: E6, E12, E30

Elder, ἀκτῆ, *Sambucus nigra*: A38, C19, E44

Fat, στέαρ: C47; of a sheep's kidneys E12. See also lard.

Fennel, μάραθον, *Foeniculum vulgare*: C56, E35 bis

Fig, σῦκον: C28, E20 bis

Fig, dried, ἰσχάς: E35, F71

Fig decoction, συκίον: C28, C31

Fig-tree, συκῆ, *Ficus carica*: juice F59; leaves A38

Fish, ἰχθύς: A43, A50, A52, A61, C48, E1, E6, E12, E21, E22, E24, E27, E30, E41, F11, F44, F53; of the rocks A43; of the shore A52, C74; of the marsh and river A52. See also sea-food.

Fishing-frog, βάτραχος, *Lophius piscatorius*: E12

Fleabane, κονύζα, some species of *Erigeron* or *Inula*; cf. Dioscurides 3.121 and Berendes pp. 343 f.: E44

Flour, ἄλητον: A14, A40, A41, C28, C29, C64, F37; this year's flour C20, F70

Fowl, ἀλεκτρυών (= ἀλέκτωρ = ὄρνις), *Gallus gallinaceus*: A41, A43 bis, A52, C44, C46, C48, C50, C56, C69, D12, E1, E9, E21, E22, E24, E41. See also chicken, hen's egg.

INDEX OF FOODS AND DRUGS

INDEX OF FOODS AND DRUGS

Kid, ἔριφος: A43. See also goat.

Lamb, κρέας οἰὸς ὡς νεωτάτης: A41, A43. See also mutton.
Lard, ὕειον ἄλειφα: C13
Lead, white, ψιμύθιον, Pb(OH)$_2$.2PbCO$_3$: C14
Lead monoxide, see silver, flower of.
Leek, πράσον, Allium Porrum: A38, A54, C38, E9, E12
Lentil, φακός, the Ervum lens and its fruit: A27, A41 bis, E12, E23, E42, E44, F47
Lentil decoction, φάκιον: A27, C15, C43, C48 ter, C49, C50 ter, C52, C55 bis, C70
Lentil-soup, φακῆ (= φάκινον ἔτνος): E7, E21 ter, E22 bis, E30 ter, E41, E42, E44, E47, F53
Linseed, λίνου καρπός (= λίνου σπέρμα), the seed of flax, Linum usitatissimum: D17, E1, F18, F33, F53
Litharge, see silver, flower of.
Lupin, θέρμος, Lupinus alba: F47

Madder, ἐρυθρόδανον, Rubia tinctorum: F64
Magnetic stone, λίθος Μαγνησίης; for the property of magnetic stone to purge phlegm cf. Dioscurides 5.130: E21
Maiden-hair, ἀδίαντον, Adiantum Capillus-Veneris: D17, E35
Mallow, μολόχη, Malva silvestris: F44
Mandrake, μανδραγόρας, Mandragora officinarum: C43
Mare's milk, ἵππειον γάλα: E3, E6, E28, E32
Marjoram, ὀρίγανον, various species of Origanum: A43, C19, C26 bis, C28, C47 bis, C48, C50, C52, C55, C64, C71, E1, E6, E7, E9, E10, E24, E27 bis, E41, F62
Meal, ἄλευρον (= ἄλφιτον): A38, A52, C42, C54, C55, C64, E1 passim, E6, E16, E21 bis, E30 bis, E40 ter, E41, E42 bis, E44, E51, F38, F72; this year's meal F63. See also barley-meal and wheat-meal.
Meal, bruised, of raw grain, ὠμήλυσις: C30, C31
Medlar, μέσπιλον, fruit of the medlar-tree, Mespilus germanica: F53
Melicrat, μελίκρητον, a mixed drink of honey and water: passim, e.g. A15, A40, C21, C67, D11, D17, E6, E21, F1, F72
Melon, σίκυος πέπων, Cucumis Melo: A57, D17 bis
Mercury (herb), λινόζωστις, Mercurialis annua: C12 bis, C69

310

Metallic ashes, σποδός; cf. Dioscurides 5.75: F65

Milk, γάλα: passim, e.g. A23, A25, C73, E8, E18. See also ass's milk, cow's milk, goat's milk, mare's milk, and woman's milk.

Millet, κέγχρος, *Panicum miliaceum*: A14, A40, A41 bis, C12, C19, C22, C40, C42, C43, C44 ter, C45, C46, C56, C64, C67 bis, C70, E1, E49, F53

Mint, μίνθη, *Mentha viridis*: C26, C28, D17, E12

Mulberry, μόρον, *Morus nigra*: E35

Mustard, νᾶπυ, *Brassica alba*: D15, D16

Mutton, κρέα μήλεια (= οἴεια): A52, C47, C48, C49, C50, C52, C69, C71, E12 bis, E22, E30, E41. See also lamb.

Myrrh, σμύρνα, the gum of an Arabian tree, *Balsamodendron Myrrha*: C13, C19, C25, F30, F67

Myrtle-berry, μύρτον, fruit of the myrtle, *Myrtus communis*: F53

Nightshade, στρύχνος: D1, E27

Oak-gall, κηκίς, an excrescence produced on some species of oak by punctures of the gall-fly (genus *Cynips*): C13

Olive, ἐλαία, *Olea Europaea*: A38

Olive oil, ἔλαιον: passim, e.g. A41, A55, C30, C60, D2, D16, E20, E52, F18, F51, F52; from unripe olives (ὀμφάκιον) F65

Onion, κρόμμυον, *Allium Cepa*: A54, C22, E35

Orchis, διδυμαῖον, perhaps some species of *Orchis*: E30

Oxymel, ὀξύμελι, a mixed drink of honey and vinegar: F11 bis, F18, F30 bis, F31, F34 bis, F60

Pennyroyal, γληχώ, *Mentha Pulegium*: A38, A41, D17 bis, E44

Peony, γλυκυσίδη, *Paeonia officinalis*: E40

Pepper, πέπερι, *Piper nigrum*: D12, D16, F34

Pigeon, πελειάς, species of the genus *Columba*: E27, E41

Pine-cone, κόκκαλος, fruit of the *Pinus pinea*: F30, F34

Piper, κόκκυξ, species of the genus *Trigla*: E21, E22, E30, E49

Plover, χαραδριός, *Charadrius oedicnemus;* it was popularly believed that a person with jaundice could be cured by looking at this bird. See Thompson *(Birds)* pp. 311–314: E37

Polyp, πωλύπους, octopus or cuttle-fish: E22, E40, E44

Pomegranate, ῥοιή (= σίδη), *Punica Granatum*: A38, A54, C46, C65, D16, E12

INDEX OF FOODS AND DRUGS

Pomegranate-peel, σίδιον: C47

Poppy, μήκων. Possible identifications in particular Hippocratic passages have been much discussed, generally with little agreement: cf. e.g. Dierbach pp. 235 ff., Littré II.523 f., Ermerins I.362ff., and von Grot pp. 108f.: D16, E12, E40, F63

Pork, κρέα ὕεια / χοίρεια: A52 bis, C47, C49, C52, E12, E21, E22, E30, E41, F50 bis, F52

Puppy, σκύλακος κρέας: A41, A43, C44, C56, E9, E24, E27. See also dog.

Quince, μῆλον τὸ Κυδώνιον, Cydonia vulgaris: D17

Radish, ῥαφανίς, Raphanus sativus: C47 bis, E6, E22, E30, E40

Rain-water, ὕδωρ ὄμβριον: D17; cf. E26

Raisin, ἀσταφίς (= σταφίς): A55, C32, D17 bis, E35 bis

Rennet, πυτίη: F59

Resin, ῥητίνη: C47

Rue, πήγανον, Ruta graveolens: A43, C26, C47, C50, C64, D15, E12, E30

Safflower, κνῆκος, Carthamus tinctorius: D17

Saffron, κρόκος, Crocus sativus: F66, F67

Salt, ἅλς: A41, A43, A52, C18, C47, E12 bis, E14, E26, E27, E35, E40, E41, E42, E44 bis, E46, E51. See also brine.

Salt-fish, τάριχον, generally tunny, Thynnus thynnus, or horse-mackerel, Scomber trachurus: C50, E12, E25, E30. Cf. saperdes.

Salvia, ἐλελίσφακος, a member of the genus Salvia, possibly S. officinalis, sage: A38, C47 bis, C54 bis, C64

Saperdes, σαπέρδης, some variety of salt-fish; cf. Thompson (Fish) p. 226: E25, E30

Savory, θύμβρη, Satureia Thymbra: C26, C47, C48, C52, C64, C71

Scammony, σκαμ(μ)ωνίη, Convolvulus Scammonia: E2, E14, E16, E18, E28, E43, E51, F27

Scorpion fish, σκορπίος, Scorpaena porcus: C48, C50, C71, E21, E22, E30, E49

Sea-food, θαλάσσια: A43, C47, C64, C69. See also fish.

Sea-spurge, μηκωνίς (= τιθυμαλλίς), Euphorbia Paralias; cf. Dioscurides 4.164.6: E1, E7, E10. Cf. spurge.

INDEX OF FOODS AND DRUGS

Selachian, σελάχιον: C48, C50, C71, C74, E12. See also angel-fish, dogfish, fishing-frog, skate, sting-ray, and torpedo.

Sesame, σήσαμον, *Sesamum indicum*: A47, A55, C50, C64, D15, E1 bis, E35, F63

Sesamoid, σησαμοειδές, possibly the seeds of some variety of hellebore: F60

Silphium, σίλφιον, *Ferula tingitana*: C42, C47, D15, D16 ter, E6, E23, E24, E27, E40, E42, E44, F30, F48

Silver, flower of, ἄνθος ἀργύρου, probably λιθάργυρος, lead monoxide, which Dioscurides (5.87) mistakenly believed to be a silver compound: C13, C14

Skate, βατίς, species of the genus *Raia*: E12 bis, E27

Soda, λίτρον, native sodium carbonate: C13, C26 ter, C28, C32, D16, E20, E26, E31, E48, E51

Sorb, οὖον, fruit of the service-tree, *Sorbus domestica*: F53

Southernwood, ἀβρότονον, *Artemisia arborescens*: F34

Spelt groats, χόνδρος, from *Triticum spelta*, a species of grain akin to wheat: A40, A41, A44, C51, F53

Spurge, τιθύμαλλος (= μηκώνιον), various species *of Euphorbia*: A38, F69, F71, F72 bis. See also sea-spurge.

Spurge flax, κνέωρον, according to Dioscurides (4.172) leaves of the θυμελαία, *Daphne Cnidium*: E21, E22, E25, E26 ter, E44, E47. Cf. berry, Cnidian.

Square-berry, τετράγωνον, fruit of the *Euonymus latifolius*: E45, E49

Squill, σκίλλη, *Urginea maritima*: F63

Squirting cucumber, σίκυος ἄγριος, *Ecballium Elaterium*: E26, E46

Squirting cucumber juice, ἐλατήριον: D10, D15, E18, E37

Star-gazer, καλλιώννμος, *Uranoscopus scaber*: E21, E22, E30, E49

Stinging-nettle, ἀκαλήφη (= κνίδη), *Urtica urens*: C47, D15

Sting-ray, τρυγών, *Trygon pastinaca*: E12, E27

Sulphur, θεῖον: D10

Sumach, ῥόος, *Rhus Coriaria*: C28 bis, E1

Thapsia, θαψίη, *Thapsia garganica*: D15, D16, E42; juice D8; root E18

Thyme, θύμον, *Thymbra capitata*: A43, C50

313

GREEK NAMES OF FOODS
AND DRUGS
INCLUDED IN INDEX

ἀβρότονον	Southernwood
ἄγνος	Chaste-tree
ἀδίαντον	Maiden-hair
αἴγειον κρέας	Goat
αἴγειος τυρός	Goat's cheese
αἰγὸς γάλα	Goat's milk
αἰγὸς κέρας	Goat's horn
αἰγὸς ὀρός	Goat's whey
ἀκαλήφη	Stinging-nettle
ἀκτῆ	Elder
ἀλεκτρυών	Fowl
ἄλευρον	Meal(wheat)
ἄλητον	Flour
ἄλμη	Brine
ἅλς	Salt
ἄλφιτον	Meal(barley)
ἄμπελος	Grape-vine
ἀμυγδάλη	Almond
ἄνηθον	Dill

διδυμαῖον	Orchis
δρακόντιον	Dragon Arum
δράκων	Weever
ἔβενος	Ebony
ἔγχελυς	Eel
ἐλαία	Olive
ἔλαιον	Olive oil
ἐλατήριον	Squirting-cucumber juice
ἐλελίσφακος	Salvia
ἐλλέβορος	Hellebore
ἐπίθυμον	Dodder of thyme
ἐρέβινθος	Chick-pea
ἔριφος	Kid
ἐρυθρόδανον	Madder
ἐρύσιμον	Hedge-mustard
θαλάσσια	Sea-food
Θάσιον κάρυον	Almond
θαψίη	Thapsia
θεῖον	Sulphur
θέρμος	Lupin
θύμβρη	Savory
θύμον	Thyme
ἵππειον γάλα	Mare's milk
ἱππόφεως	Hippopheos
ἰσάτις	Woad

μήκων	Poppy
μηκώνιον	Spurge
μηκωνίς	Sea-spurge
μήλειον κρέας	Mutton
μῆλον	Apple
μῆλον τὸ Κυδώνιον	Quince
μίνθη	Mint
μολόχη	Mallow
μόρον	Mulberry
μύρτον	Myrtle-berry
νᾶπυ	Mustard
νάρκη	Torpedo
νεοσσὸς ἀλεκτρυόνος	Chicken
οἴειον κρέας	Mutton
οἶνος	Wine
οἰὸς ὡς νεωτάτης κρέας	Lamb
ὄνου γάλα	Ass's milk
ὄνου κρέας	Ass
ὀξύμελι	Oxymel
ὀρίγανον	Marjoram
ὄρνις	Fowl
ὄροβος	Vetch
ὀρός	Whey
οὖον	Sorb
πάνακες	All-heal

πελειάς	Pigeon
πεντάφυλλον	Cinquefoil
πέπερι	Pepper
πέπλιον	Wild purslane
πήγανον	Rue
πράσον	Leek
πτισάνη	Barley, peeled
πυρός	Wheat
πυτίη	Rennet
πωλύπους	Polyp
ῥάμνος	Buckthorn
ῥαφανίς	Radish
ῥητίνη	Resin
ῥίνη	Angel-fish
ῥοιή	Pomegranate
ῥόος	Sumach
σανδαράκη	Arsenic, red
σαπέρδης	Saperdes
σελάχιον	Selachian
σέλινον	Celery
σεμίδαλις	Wheat flour
σεῦτλον	Beet
σησαμοειδές	Sesamoid
σήσαμον	Sesame
σίδη	Pomegranate
σίδιον	Pomegranate-peel

τρυγών^I	Turtle-dove
τρυγών^{II}	Sting-ray
τρὺξ οἰνηρὴ κεκαυμένη	Wine lees, burnt
τυρός	Cheese
ὕδωρ ὄμβριον	Rain-water
ὕειον ἄλειφα	Lard
ὕειον κρέας	Pork
ὑοσκύαμος	Henbane
ὑπερικόν	Hypericum
ὕσσωπος	Hyssop
φάγρος	Braize
φακῆ	Lentil-soup
φάκιον	Lentil decoction
φακός	Lentil
χαλβάνη	All-heal juice
χαλκῖτις	Copper ore
χαλκὸς κεκαυμένος	Copper, burnt
χαλκοῦ ἄνθος	Copper, flower of
χαραδριός	Plover
χοίρειον κρέας	Pork
χόνδρος	Spelt groats
ψιμύθιον	Lead, white
ᾠὸν ἀλεκτορίδος	Hen's egg